Lecture Notes in Computer Sci

T0238605

Commenced Publication in 1973
Founding and Former Series Editors:
Gerhard Goos, Juris Hartmanis, and Jan van Leeuwen

Shinji Kikuchi   Shelly Sachdeva
Subhash Bhalla (Eds.)

Databases in Networked Information Systems

6th International Workshop, DNIS 2010
Aizu-Wakamatsu, Japan, March 29-31, 2010
Proceedings

 Springer

Volume Editors

Shinji Kikuchi
Shelly Sachdeva
Subhash Bhalla
University of Aizu
Graduate Department of Computer and Information Systems
Ikki Machi, Aizu-Wakamatsu, Fukushima 965-8580, Japan
E-mail: {d8111106, d8111107, bhalla}@u-aizu.ac.jp

Library of Congress Control Number: Applied for

CR Subject Classification (1998): H.2, H.3, H.4, H.5, C.2

LNCS Sublibrary: SL 3 – Information Systems and Application, incl. Internet/Web and HCI

ISSN   0302-9743
ISBN-10   3-642-12037-7 Springer Berlin Heidelberg New York
ISBN-13   978-3-642-12037-4 Springer Berlin Heidelberg New York

springer.com

© Springer-Verlag Berlin Heidelberg 2010
Printed in Germany

Typesetting: Camera-ready by author, data conversion by Scientific Publishing Services, Chennai, India
Printed on acid-free paper   06/3180

Preface

Large-scale information systems in public utility services depend on computing infrastructure. Many research efforts are being made in related areas, such as mobile computing, cloud computing, sensor networks, high-level user interfaces and information accesses by Web users. Government agencies in many countries plan to launch facilities in education, health-care and information support as a part of e-government initiative. In this context, information interchange management has become an active research field. A number of new opportunities have evolved in design and modeling based on the new computing needs of the users. Database systems play a central role in supporting networked information systems for access and storage management aspects.

The 6th international workshop on Databases in Networked Information Systems (DNIS) 2010 was held during March 29–31, 2010 at University of Aizu in Japan. The workshop program included research contributions and invited contributions. A view of research activity in information interchange management and related research issues was provided by the sessions on related topics. The keynote address was contributed by Divyakant Agrawal. The workshop session on "Networked Information Systems: Infrastructure" had invited papers by Harumi Kuno and Malu Castellanos. The session on "Accesses to Information Resources" had an invited contribution from Joachim Biskup. The following section on "Information and Knowledge Management Systems" included invited contributions from Toyoaki Nishida and Tetsuo Kinoshita. The session on "Information Extraction from Data Resources" included the invited contribution of Polepalli Krishna Reddy. The section on "Geospatial Decision Making" comprised contributions by Cyrus Shahabi and Yoshiharu Ishikawa. We would like to thank the members of the Program Committee for their support and all authors who contributed to DNIS 2010.

The sponsoring organizations and the Steering Committee deserve praise for the support they provided. A number of individuals contributed to the success of the workshop. I thank Umeshwar Dayal, Joachim Biskup, Divyakant Agrawal, Cyrus Shahabi, and Mark Sifer for providing continuous support and encouragement.

The workshop received invaluable support from the University of Aizu. In this context, I thank Shigeaki Tsunoyama, President of University of Aizu. Many thanks also go to the faculty members at the university for their cooperation and support.

March 2010

Shinji Kikuchi
Shelly Sachdeva
Subhash Bhalla

Organization

DNIS 2010 was organized by the Graduate Department of Information Technology and Project Management, University of Aizu, Aizu-Wakamatsu, Fukushima, PO 965-8580, (JAPAN).

Steering Committee

Divyakant Agrawal	University of California, USA
Umeshwar Dayal	Hewlett-Packard Laboratories, USA
Toyoaki Nishida	Graduate School of Informatics, Kyoto University, Japan
Krithi Ramamritham	Indian Institute of Technology, Bombay, India
Cyrus Shahabi	University of Southern California, USA
Executive Chair	Nadia Bianchi-Berthouze, University College London, UK
Program Chair	Subhash Bhalla, University of Aizu, Japan
Publicity Committee Chair	Shinji Kikuchi, University of Aizu, Japan
Publications Committee Chair	Shelly Sachdeva, University of Aizu, Japan

Program Committee

D. Agrawal	University of California, USA
S. Bhalla	University of Aizu, Japan
V. Bhatnagar	University of Delhi, India
P.C.P. Bhatt	Indian Institute of Information Technology, Banglore, India
P. Bottoni	University La Sapienza of Rome, Italy
L. Capretz	University of Western Ontario, Canada
M. Capretz	University of Western Ontario, Canada
G. Cong	Aalborg University, Denmark
U. Dayal	Hewlett-Packard Laboratories, USA
V. Goyal	Indraprastha Institute of Information Technology (IIIT D), Delhi, India
W.I. Grosky	University of Michigan-Dearborn, USA
J. Herder	University of Applied Sciences, Fachhochschule Düsseldorf, Germany
Y. Ishikawa	Nagoya University, Japan
Q. Jin	University of Aizu, Japan
A. Kumar	Pennsylvania State University, USA
H. Kuno	Hewlett-Packard Laboratories, USA

A. Mondal	Indraprastha Institute of Information Technology (IIIT D), Delhi, India
T. Nishida	Kyoto University, Japan
L. Pichl	International Christian University, Tokyo, Japan
P.K. Reddy	International Institute of Information Technology (IIIT), Hyderabad, India
C. Shahabi	University of Southern California, USA
M. Sifer	Sydney University, Australia

Sponsoring Institution

Center for Strategy of International Programs, University of Aizu, Aizu-Wakamatsu City, Fukushima P.O. 965-8580, Japan.

Table of Contents

Information Extraction from Data Resources

Geo-spatial Decision Making

Data Management Challenges in Cloud Computing Infrastructures*

Divyakant Agrawal, Amr El Abbadi, Shyam Antony, and Sudipto Das

University of California, Santa Barbara
{agrawal,amr,shyam,sudipto}@cs.ucsb.edu

Abstract. The challenge of building consistent, available, and scalable data management systems capable of serving petabytes of data for millions of users has confronted the data management research community as well as large internet enterprises. Current proposed solutions to scalable data management, driven primarily by prevalent application requirements, limit consistent access to only the granularity of *single objects, rows, or keys*, thereby trading off consistency for high scalability and availability. But the growing popularity of "cloud computing", the resulting shift of a large number of internet applications to the cloud, and the quest towards providing data management services in the cloud, has opened up the challenge for designing data management systems that provide consistency guarantees at a granularity larger than *single rows and keys*. In this paper, we analyze the design choices that allowed modern scalable data management systems to achieve orders of magnitude higher levels of scalability compared to traditional databases. With this understanding, we highlight some design principles for systems providing scalable and consistent data management as a service in the cloud.

1 Introduction

Scalable and consistent data management is a challenge that has confronted the database research community for more than two decades. Historically, distributed database systems [16,15] were the first generic solution that dealt with data not bounded to the confines of a single machine while ensuring global serializability [2,19]. This design was not sustainable beyond a few machines due to the crippling effect on performance caused by partial failures and synchronization overhead. As a result, most of these systems were never extensively used in industry. Recent years have therefore seen the emergence of a different class of scalable data management systems such Google's Bigtable [5], PNUTS [6] from Yahoo!, Amazon's Dynamo [7] and other similar but undocumented systems. All of these systems deal with petabytes of data, serve on-line requests with stringent latency and availability requirements, accommodate erratic workloads, and run on cluster computing architectures; staking claims to the territories used to be occupied by database systems.

One of the major contributing factors towards the scalability of these modern systems is the data model supported by these systems, which is a collection of *key-value*

---

* This work is partially funded by NSF grant NSF IIS-0847925.

S. Kikuchi, S. Sachdeva, and S. Bhalla (Eds.): DNIS 2010, LNCS 5999, pp. 1–10, 2010.

pairs with consistent and atomic read and write operations only on *single keys*. Even though a huge fraction of the present class of web-applications satisfy the constraints of *single key* access [18,7], a large class of modern Web 2.0 applications such as collaborative authoring, online multi-player games, social networking sites, etc, require consistent access beyond *single key* semantics. As a result, modern *key-value* stores cannot cater to these applications and have to rely on traditional database technologies for storing their content, while scalable *key-value* stores drive the in-house applications of the corporations that have designed these stores.

With the growing popularity of the "cloud computing" paradigm, many applications are moving to the cloud. The *elastic* nature of resources and the *pay as you go* model have broken the infrastructure barrier for new applications which can be easily tested out without the need for huge upfront investments. The sporadic load characteristics of these applications, coupled with increasing demand for data storage while guaranteeing round the clock availability, and varying degrees of consistency requirements pose new challenges for data management in the cloud. These modern application demands call for systems capable of providing scalable and consistent data management as a service in the cloud. Amazon's SimpleDB (http://aws.amazon.com/simpledb/) is a first step in this direction, but is designed along the lines of the *key-value* stores like Bigtable and hence does not provide consistent access to multiple objects. On the other hand, relying on traditional databases available on commodity machine instances in the cloud result in a scalability bottleneck for these applications, thereby defeating the scalability and elasticity benefits of the cloud. As a result, there is a huge demand for data management systems that can bridge the gap between scalable *key-value* stores and traditional database systems.

At a very generic level, the goal of a scalable data management system is to sustain performance and availability over a large data set without significant over-provisioning. Resource utilization requirements demand that the system be highly dynamic. In Section 2, we discuss the salient features of three major systems from Google, Yahoo!, and Amazon. The design of these systems is interesting not only from the point of view of what concepts they use but also what concepts they eschew. Careful analysis of these systems is necessary to facilitate future work. The goal of this paper is to carefully analyze these systems to identify the main design choices that have lent high scalability to these systems, and to lay the foundations for designing the next generation of data management systems serving the next generation of applications in the cloud.

2   Analyzing Present Scalable Systems

Abstractly, a distributed system can be modeled as a combination of two different components. The *system state*, which is the distributed meta data critical for the proper operation and the health of the system. This state requires stringent consistency guarantees and fault-tolerance to ensure the proper functioning of the system in the presence of different types of failures. But scalability is not a primary requirement for system state. On the other hand is the *application state*, which is the application specific information or data which these systems store. The consistency, scalability and availability of the *application state* is dependent purely on the requirements of the type of application that the system aims to support, and different systems provide varying trade-offs between different attributes. In most cases, high scalability and high availability is given a higher priority. Early attempts to design distributed databases in the late eighties and early nineties made a design decision to treat both the *system state* and *applications state* as a cohesive whole in a distributed environment. We contend that the decoupling of the two states is the root cause for the high scalability of modern systems.

2.1 System State

We refer to the meta data and information required to correctly manage the distributed system as the *system state*. In a distributed data management system, data is partitioned to achieve scalability and replicated to achieve fault-tolerance. The system must have a *correct* and *consistent* view of the mappings of partitions to nodes, and that of a partition to its replicas. If there is a notion of the master amongst the replicas, the system must also be aware of the location of the master at all times. Note that this information is in no way linked to the data hosted by the system, rather it is required for the proper operation of the entire system. Since this state is critical for operating the system, a distributed system cannot afford any inconsistency or loss. In a more traditional context, this corresponds to the system state in the sense of an operating systems which has a global view about the state of the machine it is controlling.

Bigtable's design [5] segregates the different parts of the system and provides abstractions that simplify the design. There is no data replication at the Bigtable layer, so there is no notion of replica master. The rest of Bigtable's *system state* is maintained in a separate component called Chubby [3]. The *system state* needs to be stored in a consistent and fault-tolerant store, and Chubby [3] provides that abstraction. Chubby guarantees fault-tolerance through log-based replication and consistency amongst the replicas is guaranteed through a Paxos protocol [4]. The Paxos protocol [14] guarantees safety in the presence of different types of failures and ensures that the replicas are all consistent even when some replicas fail. But the high consistency comes at a cost: the limited scalability of Chubby. Thus if a system makes too many calls to Chubby, performance might suffer. But since the critical system meta data is considerably small and usually cached, even Chubby being at the heart of a huge system does not hurt system performance.

In PNUTS [6], there is no clear demarcation of the *system state*. Partition (or *tablet*) mapping is maintained persistently by an entity called the *tablet controller*, which is a single pair of active/standby servers. This entity also manages tablet relocation between different servers. Note that since there is only one *tablet controller*, it might become a bottleneck. Again, as with Chubby, an engineering solution to move the *tablet controller* away from the data path, and caching of mappings is used. On the other hand, the mapping of tablets to its replicas is maintained by the Yahoo! Message Broker (YMB) which acts as a fault-tolerant guaranteed delivery publish-subscribe system. Fault-tolerance in YMB is achieved through replication – at a couple of nodes, to commit the change, and more replicas are created gradually [6]. Again, better scalability is ensured through limiting the number of nodes (say two in this case) requiring synchronization. The per-record master information is stored as meta data for the record. Thus, the *system state* in PNUTS is split between the *tablet controller* and the *message broker*.

On the other hand, Amazon's Dynamo [7] uses an approach similar to peer-to-peer systems [17]. Partitioning of data is at a per-record granularity through consistent hashing [13]. The key of a record is hashed to a space that forms a ring and is statically partitioned. Thus the location of a data item can be computed without storing any explicit mapping of data to partitions. Replication is done at nodes that are neighbors of the node to which a key hashes to, a node which also acts as a master (although Dynamo is *multi-master*, as we will see later). Thus, Dynamo does not maintain a dynamic *system state* with consistency guarantees, a design different compared to PNUTS or Bigtable.

Even though not in the same vein as scalable data management systems, Sinfonia [1] is designed to provide an efficient platform for building distributed systems. Sinfonia [1] can be used to efficiently design and implement systems such as distributed file systems. The *system state* of the file system (e.g. the inodes) need to be maintained as well as manipulated in a distributed setting, and Sinfonia provides efficient means for guaranteeing consistency of these critical operations. Sinfonia provides the *minitransaction* abstraction, a light weight version of distributed transactions, supporting only a small set of operations. The idea is to use a protocol similar to Two Phase Commit (2PC) [10] for committing a transaction, and the actions of the transaction are piggy backed on the messages sent out during the first phase. The light weight nature of *minitransactions* allow the system to scale to hundreds of nodes, but the cost paid is a reduced set of operations.

Thus, when it comes to critical *system state*, the designers of these scalable data management systems rely on traditional mechanisms for ensuring consistency and fault-tolerance, and are willing to compromise scalability. But this choice does not hurt the system performance since this state is a very small fraction of the actual state (*application state* comprises the majority of the state). In addition, another important distinction of these systems is the number of nodes communicating to ensure the consistency of the *system state*. In the case of Chubby and YMB, a commit for a general set of operations is performed on a small set of participants (five and two respectively [3,6]). On the other hand, Sinfonia supports limited transactional semantics and hence can scale to a larger number of nodes. This is in contrast to traditional distributed database systems, which tried to make both ends meet, i.e., providing strong consistency guarantees for both *system state* and *application state* over any number of nodes.

2.2 Application State

Distributed data management systems are designed to host large amounts of data for the applications which these systems aim to support. We refer to this application specific data as the *application state*. The *application state* is typically at least two to three orders of magnitude larger than the *system state*, and the consistency, scalability, and availability requirements vary based on the applications.

Data Model and its Implications. The distinguishing feature of the three main systems we consider in this paper is their simple data model. The primary abstraction is a table of items where each item is a *key-value* pair. The value can either be an uninterpreted string (as in Dynamo), or can have structure (as in PNUTS and Bigtable). Atomicity is supported at the granularity of a single item – i.e., *atomic read/write* and *atomic* *read-modify-write* are possible to only individual items and no guarantee is provided across objects. It is a common observation that many operations are restricted to a single entity, identifiable with a primary key. However, the disk centric nature of database systems forces relatively small row lengths. Consequently, in a traditional relational design, logical single entities have to be split into multiple rows in different tables. The novelty of these systems lie in doing away with these assumptions, thus allowing very large rows, and hence allowing the logical entity to be represented as a single physical entity. Therefore, *single-object* atomic access is sufficient for many applications, and transactional properties and the generality of traditional databases are considered an overkill. These systems exploit this simplicity to achieve high scalability.

Restricting data accesses to a *single-object* results in a considerably simpler design. It provides designers the flexibility of operating at a much finer granularity. In the presence of such restrictions, application level data manipulation is restricted to a single compute node boundary and thus obviates the need for multi-node coordination and synchronization using 2PC or Paxos, a design principle observed in [11]. As a result, modern systems can scale to billions of data tuples using horizontal partitioning. The logic behind such a design is that even though there can be potentially millions of requests, the requests are generally distributed throughout the data set, and all requests are limited to accesses to a single object or record. Essentially, these systems leverage *inter-request* parallelism in their workloads. Once data has been distributed on multiple hosts, the challenge becomes how to provide fault-tolerance and load distribution. Different systems achieve this using different techniques such as replication, dynamic partitioning, partition relocation and so on. In addition, the *single key* semantics of modern applications have allowed data to be less correlated, thereby allowing modern systems to tolerate the non-availability of certain portions of data. This is different from traditional distributed databases that considered data as a cohesive whole.

Single Object Operations and Consistency. Once operations are limited to a single key, providing single object consistency while ensuring scalability is tractable. If there is no object level replication, all requests for an object arrive at a single node that hosts the object. Even if the entire data set is partitioned across multiple hosts, the *single key* nature of requests makes them limited to a single node. The system can now provide operations such as *atomic reads*, *atomic writes*, and *atomic read-modify-write*.

Replication and Consistency. Most modern systems need to support *per-object* replication for high availability, and in some cases to improve the performance by distributing the load amongst the replicas. This complicates providing consistency guarantees, as updates to an object need to be propagated to the replicas as well. Different systems use different mechanisms to synchronize the replicas thereby providing different levels of consistency such as *eventual consistency* [7], *timeline consistency* [6] and so on.

Availability. Traditional distributed databases considered the entire data as a cohesive whole, and hence, non availability of a part of the data was deemed as non-availability of the entire systems. But the *single-object* semantics of the modern applications have allowed data to be less correlated. As a result, modern systems can tolerate non-availability of certain portions of data, while still providing reasonable service to the rest of the data. It must be noted that in traditional systems, the components were cohesively bound, and non-availability of a single component of the system resulted in the entire system becoming unavailable. On the other hand, modern systems are loosely coupled, and the non-availability of certain portions of the system might not affect other parts of the system. For example, if a partition is not available, then that does not affect the availability of the rest of the systems, since all operations are *single-object*. Thus, even though the system availability might be high, record level availability might be lower in the presence of failures.

2.3   The Systems

In Bigtable [5], a single node (referred to as *tablet server*) is assigned the responsibility for part of the table (known as a *tablet*) and performs all accesses to the records assigned to it. The *application state* is stored in the Google File System (GFS) [9] which provides the abstraction of a scalable, consistent, fault-tolerant storage for user data. There is no replication of user data inside Bigtable (all replication is handled at the GFS level), hence it is by default *single master*. Bigtable also supports *atomic read-modify-write* on *single keys*. Even though scans on a table are supported, they are best-effort without providing any consistency guarantees.

PNUTS [6] was developed with the goal of providing efficient read access to geographically distributed clients while providing serial *single-key* writes. PNUTS performs explicit replication to ensure fault-tolerance. The replicas are often geographically distributed, helping improve the performance of web applications attracting users from different parts of the world. As noted earlier in Section 2.1, Yahoo! Message Broker (YMB), in addition to maintaining the *system state*, also aids in providing application level guarantees by serializing all requests to the same key. PNUTS uses a *single master* per record and the master can only process updates by publishing to a single broker, as a result providing *single-object time line consistency* where updates on a record are applied in the same order to all the replicas [6]. Even though the system supports *multi-object* operations such as range queries, no consistency guarantees are provided. PNUTS allows the clients to specify their consistency requirements for reads: a read that does not need the guaranteed latest version can be satisfied from a local copy and hence has low latency, while reads with the desired level of freshness (including read from latest version) are also supported but might result in higher latency.

Dynamo [7] was designed to be a highly scalable key-value store that is highly available to reads but particularly for writes. This system is designed to make progress even in the presence of network partitions. The high write availability is achieved through an asynchronous replication mechanism which acknowledges the write as soon as a small number of replicas have written it. The write is eventually propagated to other replicas. To further increase availability, there is no statically assigned coordinator (thereby making this a *multi master* system), and thus, the *single-object* writes also do not have a serial history. In the presence of failures, high availability is achieved at the cost of lower consistency. Stated formally, Dynamo only guarantees *eventual consistency*, i.e. all updates will be eventually delivered to all replicas, but with no guaranteed order. In addition, Dynamo allows multiple divergent versions of the same record, and relies on application level reconciliation based on vector clocks.

2.4 Design Choices

So far in this section, our discussion focussed on the current design of major internet-scale systems. We anticipate more such key-value based systems will be built in the near future, perhaps as commodity platforms. In such cases, there are a few issues that need to be carefully considered and considerable deviation from the current solutions may be appropriate.

Structure of Value. Once the design decision to allow large values in key-value pairs is made, the structure imposed on these values becomes critical. At one extreme, one could treat the value as an opaque blob-like object, and applications are responsible for semantic interpretation for read/writes. This is in fact the approach taken in Dynamo. Presumably this suits the needs of Amazon's workload but is too limited for a generic data serving system. On the other hand, PNUTS provides a more traditional flat row like structure. Again, the row can be pretty large and frequent schema changes are allowed without compromising availability or performance. Also, rows may have many empty columns as is typical for web workloads. In Bigtable, the schema consists of column families and applications may use thousands of columns per family without altering the main schema, effectively turning the value into a 2D structure. Other choices that should be considered include restricting the number of columns, but allowing each column to contain lists or more complex structures. This issue needs to be studied further since the row design based on page size in no longer applicable, and hence more flexibility for novel structures is available.

System Consistency Mechanism. As discussed earlier, maintaining consistency of the *system state* is important for these systems. One obvious problem is to how to keep track of each partition assignment and consensus based solutions seem to be a good solution. But to add more features to the system, there is a need for reliable communication between partitions, e.g. supporting batched blind writes. PNUTS resorts to a reliable message delivery system for this purpose and hence is able to support some features such as key-remastering. This issue also needs further study since it might bring unnecessary complexity and performance problems unless carefully designed.

Storage Decoupling. Given that data is partitioned with a separate server responsible for operations on data within each partition, it is possible to store the data and run the server on the same machine. Clearly this avoids a level of indirection. However we think such close coupling between storage and servers is quite limiting since it makes features such as secondary indexes very hard to implement and involves much more data movement during partition splitting/merging. It would be better to follow a design where data is replicated at the physical level with a level of indirection from the server responsible for that partition. This is applicable even if there are geographically separated logical replicas since each such replica can maintain local physical replicas which would facilitate faster recovery by reducing the amount of data transfer across data centers. This design will need some mechanism to ensure that servers are located as close as possible to the actual data for efficiency while not being dependent on such tight coupling.

Exposing Replicas. For systems which aim for availability or at least limited availability in the face of network partitions, it makes sense to allow applications to be cognizant of the underlying replication mechanism. For systems, with limited availability, allowing the application to specify freshness requirements allows easy load spreading as well as limited availability. This is the case in both PNUTS and Dynamo. But in these setting we think designers should strongly consider adding support for multi-versioning, similar to that supported in Bigtable. These versions are created anyway as part of the process and the design decision is to store them or not. Note that old versions are immutable anyway and when storage servers are decoupled as discussed above, this allows analysis applications to efficiently pull data without interfering with the online system and also allowing time-travel analysis.

3   The Next Generation of Scalable Systems

In this section, we summarize the main design principles that allow *key value* stores to have good scalability and elasticity properties. We then discuss the shortcomings of such *key value* stores for modern and future applications, and lay the foundation for discussion of design principles of the next generation of scalable data management systems supporting complex applications while providing scalable and consistent access to data at a granularity large than *single keys*. The design of such stores is paramount for the success of data rich applications hosted in the cloud.

3.1   Scalable Design Principles

In this section, we highlight some of the design choices that have lent scalability to the *key value* stores:

- **Segregate System and Application State.** This is an important design decision that allows dealing differently with different components of the system, rather than viewing it as one cohesive whole. The *system state* is critical and needs stringent consistency guarantees, but is orders of magnitude smaller than the *application state*. On the other hand, the *application state* requires varying degrees of *consistency* and operational flexibility, and hence can use different means for ensuring these requirements.
- **Limit interactions to a Single physical machine.** Limiting operations to the confines of a single physical machines lends the system the ability to horizontally partition and balance the load as well as data. In addition, failure of certain components of the system does not affect the operation of the remaining components, and allows for graceful degradation in the presence of failure. Additionally, this also obviates distributed synchronization and the associated cost. This design principle has also been articulated in [11] and forms the basis for scalable design.
- **Limited distributed synchronization is practical.** Systems such as Sinfonia [1] and Chubby [3] (being used at the core of scalable systems such as Bigtable [5] and GFS [9]) that rely on distributed synchronization protocols for providing consistent data manipulation in a distributed system have demonstrated that distributed synchronization, if used in a prudent manner, can be used in a scalable data management system. The system designs should limit distributed synchronization to the minimum, but eliminating them altogether is not necessary for a scalable design.

The above mentioned design principles will form the basis for the next generation of scalable data stores.

3.2 Moving beyond Single Key Semantics

A large class of current web-applications exhibit *single key* access patterns [18,7], and this is an important reason for the design of scalable data management systems that guarantee *single key* atomic access. But a large number of present and future applications require scalable and consistent access to more than a single key. For example, let us consider the example of an online casino game. Multiple players can join a game instance, and the profiles of the participants in a game are linked together. Every profile has an associated balance, and the balance of all players must be updated consistently and atomically as the game proceeds. There can be possibly millions of similar independent game instances which need to be supported by the system. Additionally, the load characteristics of these applications can be hard to predict. Some of these applications might not be popular and hence have low load characteristics, while sudden popularity of these applications can result in a sudden huge increase in the load on the system [8,12]. The cloud computing paradigm provides efficient means for providing computation for these systems, and for dealing with erratic load patterns. But since these applications cannot be supported by *key value* stores like Bigtable or Simple DB, they have to rely on traditional databases, and traditional database servers running on commodity machine instances in the cloud often become a scalability bottleneck. A similar scalability challenge is confronted by the movement of more and more web applications to the cloud. Since a majority of the web applications are designed to be driven by traditional database software, their migration to the cloud results in running the database servers on commodity hardware instead of premium enterprise database servers. Additionally, porting these applications to utilize *key value* stores is often not feasible due to various technical as well as logistic reasons. Therefore, modern applications in the cloud require a next generation data storage solution that can run efficiently on low cost commodity hardware, while being able to support high data access workload and provide consistency granularity and functionality at a higher granularity compared to *single key* access.

3.3 Concluding Remarks

Among the primary reasons for the success of the cloud computing paradigm for utility computing are *elasticity*, *pay as you go* model of payment, and use of commodity hardware in a large scale to exploit the economies of scale. Therefore, the continued success of the paradigm necessitates the design of a scalable and elastic system that can provide data management as a service. This system should efficiently and effectively run on commodity hardware, while using the *elasticity* of the cloud to deal with the erratic workloads of modern applications in the cloud, and provide varying degrees of consistency and availability guarantees as per the application requirements. The spectrum of data management systems has the scalable *key value* stores on one end, and flexible, transactional, but not so scalable database systems on the other end. Providing efficient data management to the wide variety of applications in the cloud requires bridging this gap with systems that can provide varying degrees of consistency and scalability. In this paper, our goal was to lay the foundations of the design of such a system for managing "clouded data".

References

1. Aguilera, M.K., Merchant, A., Shah, M., Veitch, A., Karamanolis, C.: Sinfonia: a new paradigm for building scalable distributed systems. In: SOSP, pp. 159–174 (2007)
2. Bernstein, P.A., Hadzilacos, V., Goodman, N.: Concurrency Control and Recovery in Database Systems. Addison Wesley, Reading (1987)
3. Burrows, M.: The Chubby Lock Service for Loosely-Coupled Distributed Systems. In: OSDI, pp. 335–350 (2006)
4. Chandra, T.D., Griesemer, R., Redstone, J.: Paxos made live: an engineering perspective. In: PODC, pp. 398–407 (2007)
5. Chang, F., Dean, J., Ghemawat, S., Hsieh, W.C., Wallach, D.A., Burrows, M., Chandra, T., Fikes, A., Gruber, R.E.: Bigtable: A Distributed Storage System for Structured Data. In: OSDI, pp. 205–218 (2006)
6. Cooper, B.F., Ramakrishnan, R., Srivastava, U., Silberstein, A., Bohannon, P., Jacobsen, H.A., Puz, N., Weaver, D., Yerneni, R.: PNUTS: Yahoo!'s hosted data serving platform. In: Proc. VLDB Endow., vol. 1(2), pp. 1277–1288 (2008)
7. DeCandia, G., Hastorun, D., Jampani, M., Kakulapati, G., Lakshman, A., Pilchin, A., Sivasubramanian, S., Vosshall, P., Vogels, W.: Dynamo: amazon's highly available key-value store. In: SOSP, pp. 205–220 (2007)
8. von Eicken, T.: Righscale Blog: Animoto's Facebook Scale-up (April 2008), http://blog.rightscale.com/2008/04/23/animoto-facebook-scale-up/
9. Ghemawat, S., Gobioff, H., Leung, S.T.: The Google file system. In: SOSP, pp. 29–43 (2003)
10. Gray, J.: Notes on data base operating systems. In: Flynn, M.J., Jones, A.K., Opderbeck, H., Randell, B., Wiehle, H.R., Gray, J.N., Lagally, K., Popek, G.J., Saltzer, J.H. (eds.) Operating Systems. LNCS, vol. 60, pp. 393–481. Springer, Heidelberg (1978)
11. Helland, P.: Life beyond distributed transactions: an apostate's opinion. In: CIDR, pp. 132–141 (2007)
12. Hirsch, A.: Cool Facebook Application Game – Scrabulous – Facebook's Scrabble (2007), http://www.makeuseof.com/tag/best-facebook-application-game-scrabulous-facebooks-scrabble/
13. Karger, D., Lehman, E., Leighton, T., Panigrahy, R., Levine, M., Lewin, D.: Consistent hashing and random trees: distributed caching protocols for relieving hot spots on the world wide web. In: STOC, pp. 654–663 (1997)
14. Lamport, L.: The part-time parliament. ACM Trans. Comput. Syst. 16(2), 133–169 (1998)
15. Lindsay, B.G., Haas, L.M., Mohan, C., Wilms, P.F., Yost, R.A.: Computation and communication in R*: a distributed database manager. ACM Trans. Comput. Syst. 2(1), 24–38 (1984)
16. Rothnie Jr., J.B., Bernstein, P.A., Fox, S., Goodman, N., Hammer, M., Landers, T.A., Reeve, C.L., Shipman, D.W., Wong, E.: Introduction to a System for Distributed Databases (SDD-1). ACM Trans. Database Syst. 5(1), 1–17 (1980)
17. Stoica, I., Morris, R., Karger, D., Kaashoek, M.F., Balakrishnan, H.: Chord: A scalable peer-to-peer lookup service for internet applications. In: SIGCOMM, pp. 149–160 (2001)
18. Vogels, W.: Data access patterns in the amazon.com technology platform. In: VLDB, p. 1. VLDB Endowment (2007)
19. Weikum, G., Vossen, G.: Transactional information systems: theory, algorithms, and the practice of concurrency control and recovery. Morgan Kaufmann Publishers Inc., San Francisco (2001)

Managing Dynamic Mixed Workloads for Operational Business Intelligence

Harumi Kuno[1], Umeshwar Dayal[1], Janet L. Wiener[1], Kevin Wilkinson[1],
Archana Ganapathi[2], and Stefan Krompass[3]

[1] Hewlett-Packard Laboratories
Palo Alto, CA
[2] UC Berkeley
Berkeley, CA
[3] Technische Universität München
Munich, Germany

Abstract. As data warehousing technology gains a ubiquitous presence in business today, companies are becoming increasingly reliant upon the information contained in their data warehouses to inform their operational decisions. This information, known as business intelligence (BI), traditionally has taken the form of nightly or monthly reports and batched analytical queries that are run at specific times of day. However, as the time needed for data to migrate into data warehouses has decreased, and as the amount of data stored has increased, business intelligence has come to include metrics, streaming analysis, and reports with expected delivery times that are measured in hours, minutes, or seconds. The challenge is that in order to meet the necessary response times for these operational business intelligence queries, a given warehouse must be able to support at any given time multiple types of queries, possibly with different sets of performance objectives for each type. In this paper, we discuss why these dynamic mixed workloads make workload management for operational business intelligence (BI) databases so challenging, review current and proposed attempts to address these challenges, and describe our own approach. We have carried out an extensive set of experiments, and report on a few of our results.

1 Introduction

Traditionally, business decisions can be divided into strategic, tactical, or operational levels. These can be thought of as long-term, medium-term, or short-term decisions. Until recently, enterprises have used BI almost exclusively for only offline, long-term, strategic decision-making. For example, in order to decide whether or not a company should expand its business into a particular new market, a small number of expert users might analyze historical data using well-understood queries that are run at assigned times of the day and/or night, for the benefit of a decision-making cycle lasting weeks or months.

However, as enterprises have become more automated, real-time, and data-driven, the industry is evolving toward adaptive, operational BI systems that support online, operational decision making at all levels in the enterprise [12,28]. For example, an on-line retailer would like to analyze a user's real-time click stream data and up-to-the-minute

S. Kikuchi, S. Sachdeva, and S. Bhalla (Eds.): DNIS 2010, LNCS 5999, pp. 11–26, 2010.
© Springer-Verlag Berlin Heidelberg 2010

inventory to offer dynamically priced product bundles, or a bank would like to detect and react in real-time to fraudulent transactions. To support such workloads, an operational BI system may need to maintain consistent throughput for OLTP-style queries that continuously load data into the data warehouse from operational systems (for instance, those which run cash register transactions), while also simultaneously providing fast response times for queries submitted by financial analysts seeking an immediate answer and reliable completion times for queries kicked off by monthly status report generation. That is to say, in general, operational BI systems must support dynamic mixed workloads, composed of different types of queries with different objectives and dynamic arrival rates.

Ideally, a workload management system should control resource contention and maintain an ideal execution environment for each workload. However, traditional workload management assumes that the resource requirements and performance characteristics of a workload are known, and counts on being able to use such information in order to make critical workload management decisions such as: Should we run this query? If so, when? How long do we wait for it to complete before deciding that something went wrong (so we should kill it)? Given an expected change to a workload, should we upgrade (or downgrade) the existing system?

As data warehouses grow bigger and queries become more complex, predicting how queries will behave, particularly under resource contention — for example, how long they will run, how much CPU time they will need, how many disk I/Os they will incur, — becomes increasingly difficult. Workload management decisions must then be made based on incomplete and possibly incorrect information, as sketched in Figure 1. Bad decisions can result in inappropriate levels of resource contention, requiring highly skilled human administrators to intervene.

Our research has focused on how to characterize and break this vicious cycle. We have carried out a systematic study of the effectiveness of various workload

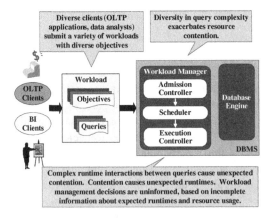

Fig. 1. Workload management decisions must be made in the absence of critical information when the wide diversity of operational BI queries and workload objectives prohibit accurate prediction of resource requirements and interactions

management policies and thresholds for diverse workloads including unexpectedly long-running queries. We have also studied how to improve the database system's ability to predict the resource requirements and performance of a potentially long-running query, so that performance is not entirely unpredictable. Our ultimate goal is to dramatically reduce the cost of ownership of BI database systems by simplifying or automating workload management decisions.

In the remainder of this paper, we first review prior work in Section 2 and describe our general approach to building a workload management system that can handle scenarios involving unexpectedly long-running queries in Section 3. We discuss our investigation of how the diverse resource requirements of BI queries impact the effectiveness of workload management policies in Section 4. We then describe our efforts to use machine learning techniques to predict resource usage *before* queries are actually run, in Section 5. For more detail, readers are referred to our earlier papers: [8,11,16,17]. We conclude with a description of our ongoing efforts in Section 6.

2   Related Work

The efforts described in this paper are informed by three primary areas of research: workload management, query progress indicators, and machine learning. In this section, we discuss each of these in turn.

2.1   Workload Management

Commercial database systems and professional database system administrators have been forced to develop their own strategies for dealing with long-running queries. Commercial tools tend to use simple rules based upon absolute thresholds. Our studies show that such rules are ineffective in the presence of unexpectedly long-running queries.

To our knowledge, few other researchers explicitly consider long-running queries in workload management. Benoit [3] presents a goal-oriented framework that models DBMS resource usage and resource tuning parameters for the purposes of diagnosing which resources are causing long-running queries and determining how to adjust parameters to increase performance. He does not address the evaluation of workload management mechanisms, nor does he model or manage the state of an individual query's execution. Weikum *et al.* [27] discuss what metrics are appropriate for identifying the root causes of performance problems (e. g., overload caused by excessive lock conflicts). They focus on tuning decisions at different stages such as system configuration, database configuration, and application tuning. Also, they address the OLTP context, not BI.

Regarding work on workload management techniques for resource allocation, we share a focus with researchers such as [4,7,15,23,25], who consider how to govern resource allocation for queries with widely varying resource requirements in multi-workload environments. Davison and Graefe [7] present a framework for query scheduling and resource allocation that uses concepts from microeconomics to manage resource allocation. Krompass *et al.* [15] present a framework for adaptive QoS management based on an economic model that adaptively penalizes individual requests. A major difference between such work and ours is that we consider the case where some problem queries are not entitled to resources, and therefore in addition to admission control and scheduling, we also consider actions such as kill, kill+requeue, and suspend+resume. Also, their focus is OLTP, not BI, and thus they make assumptions such as transaction-specific Service Level Agreements (SLAs).

2.2 Query Progress Indicators

Query progress indicators attempt to estimate a running query's degree of completion. We believe that work in query progress indicators is complementary to our goals and offers a means to identify our various types of long-running queries at early stages — potentially before the workload has been negatively impacted.

Prior attempts to predict database performance are all subject to one or more of the following limitations:

- They do not attempt to predict any actual performance metrics: they instead estimate the percentage of work done or produce an abstract number intended to represent relative "cost" (like the query optimizer's cost estimate) [1,10,14,18,24,29,30].
- They attempt to predict only a single performance metric, such as the elapsed time or actual cardinality of the underlying data [5,6,19,20,22,26,31].
- They assume that the progress indicator has complete visibility into the number of tuples already processed by each query operator [5,6,19,20]. Such operator-level information can be prohibitively expensive to obtain, especially when multiple queries are executing simultaneously.

We are sometimes asked why the predictions made by the query optimizer are insufficient. The primary goal of the database query optimizer is to choose a good query plan. To compare different plans, the optimizer uses cost models to produce rough cost estimates for each plan. However, the units used by most optimizers do not map easily onto time units, nor does the cost reflect the use of individual resources. Unlike the optimizer, our model bases its predictions on the relative similarity of the cardinalities for different queries, rather than their absolute values. As a result, our model is not as sensitive to cardinality errors.

On the opposite end of the spectrum, query progress indicators use elaborate models of operator behavior and detailed runtime information to estimate a running query's degree of completion. They do not attempt to predict performance before the query runs. Query progress indicators require access to runtime performance statistics, most often the count of tuples processed. This requirement potentially introduces the significant overhead of needing to instrument the core engine to produce the required statistics. Such operator-level information can be prohibitively expensive to obtain, especially when multiple queries are executing simultaneously.

With regard to workload management, Luo *et al.* [21] leverage an existing progress indicator to estimate the remaining execution time for a running query (based on how long it has taken so far) in the presence of concurrent queries. They then use the progress indicator's estimates to implement workload management actions. For example, they propose a method to identify a query to block in order to speed up another query. We believe their work is complementary to our own.

2.3 Machine Learning Predictions

A few papers use machine learning to predict a relative cost estimate for use by the query optimizer. In their work on the COMET statistical learning approach to cost estimation, Zhang *et al.* [31] use transform regression (a specific kind of regression) to produce a self-tuning cost model for XML queries. Because they can efficiently incorporate new training data into an existing model, their system can adapt to changing workloads, a very useful feature that we plan to address in the future. COMET, however, focuses on producing a single cost value intended to be used to compare query plans to each other as opposed to a metric that could be used to predict resource usage or runtime. Similarly, IBM's LEO learning optimizer compares the query optimizer's estimates with actuals at each step in a query execution plan, and uses these comparisons from previously executed queries to repair incorrect cardinality estimates and statistics [22,26]. Like COMET, LEO focuses on producing a better cost estimate for use by the query optimizer, as opposed to attempting to predict actual resource usage or runtime. Although a query optimizer that has been enhanced with LEO can be used to produce relative cost estimates prior to executing a query, it does require instrumentation of the underlying database system to monitor actual cost values. Also, LEO itself does not produce any estimates; its value comes from repairing errors in the statistics underlying the query optimizer's estimates.

The PQR [13] approach predicts ranges of query execution time using the optimizer's query plan and estimates. They construct a decision tree to classify new queries into time-range buckets. They do not estimate any performance metrics other than runtime, nor have they trained or tested with extremely long-running queries as do we. Even so, their work informed ours, because they found that there was correlation between cost estimates and query runtimes.

Many efforts have been made to characterize workloads from web page accesses [1,29], data center machine performance and temperature [24], and energy and power consumption of the Java Virtual Machine [9], to name a few. In databases, Elnaffar [10] observes performance measurements from a running database system and uses a classifier (developed using machine learning) to identify OLTP vs. Decision Support workloads, but does not attempt to predict specific performance characteristics. A number of papers [14,18,30] discuss how to characterize database workloads with an eye towards validating system configuration design decisions such as the number and speed of disks and the amount of memory. These papers analyze features such as how often indexes are used, or the structure and complexity of SQL statements, but they do not make actual performance predictions.

3 Our Approach

We attack the problem of how to manage dynamic mixed workloads from two directions, as sketched in Figure 2. First, we address the problem of how best to make workload management decisions in the absence of complete information about workloads characterized by great diversity in query resource requirements and great variation in resource contention. To this end, we have built an experimental framework and designed and carried out experiments based on actual workloads that contain a wide variety of queries.

Our investigation has produced models of workload management policies and an understanding of how different policies impact various workload management scenarios.

Second, we address the problem of how to obtain informative predictions (at compilation time) about query resource requirements. We have prototyped our approach, and have validated our methodology using actual workloads run on an HP Neoview system. With regard to this problem, we have developed a method that exploits machine learning to produce compile-time predictions of multiple characteristics of resource usage. Given a standard database system configuration, we enable the vendor to use training workload runs to develop models and tools that are then distributed to customer instances of the configuration and used to manage their workload.

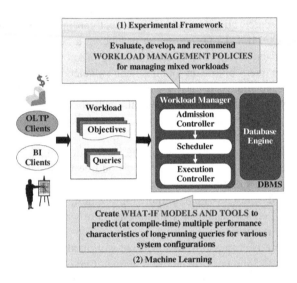

Fig. 2. We use (1) an experimental framework to evaluate policies for managing mixed workloads, and (2) machine learning to develop models and tools for "what-if" modeling and performance prediction

4 Evaluating Workload Management Policies

Workload management plays a critical role in the total cost of ownership of an Operational BI database system. Regularly-run, well-tuned queries may be well-behaved and predictable, but factors such as data skew, poorly-written SQL, poorly-optimized plans, and resource contention can lead to poorly-behaved, unpredictable queries. All of these conditions can contribute to unexpected contention for resources and undesirable impact on performance, which in turn require expensive human administrators.

We make three contributions towards reducing this cost of ownership. First, we provide a means to distinguish between different types of long-running queries. Second, we identify critical workload management decisions that administrators must make when faced with workloads that contain these long-running queries. Third, we carry out experiments to evaluate how effective known and proposed workload management policies are at making such decisions.

4.1   Workload Management Components

Database workload management is generally discussed in terms of the application of policies to workloads. The policies initiate control actions when specific conditions are reached. Thus far, commercial database vendors have led the state of the art in workload management, adding policies to respond to customer needs.

The goal of workload management is to satisfy workload objective(s). For example, a simple objective might be to complete all queries in the shortest time, and a more complex objective would be to provide fast response for short queries and to complete as many long queries as possible.

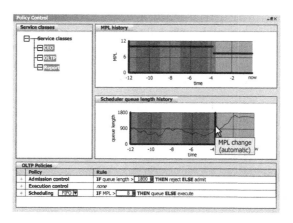

Fig. 3. Policy control window

For example, Figure 3 shows the policy control window from our workload management dashboard. In the figure, the "CEO" service class has high-priority, ad hoc queries written on behalf of a company executive. These queries must be completed promptly — as long as its cost estimates are accurate. The "OLTP" service class has queries that are short, well-understood, and have a fixed arrival rate. The objectives for these queries require the throughput to be above a certain transactions per second threshold and the average response time to be lower than another threshold, expressed in terms of milliseconds. Finally, the "Report" service class comprises medium-sized, roll-up report queries with an objective to complete all of the queries before a deadline. These queries are also well-understood.

Workload management systems use admission control, scheduling, and execution policies to meet performance objectives. These three control points, shown in Figure 1, enact policies that control which queries are admitted into the database management system, the order and number of admitted queries that are queued for the core database engine to run, and when to invoke execution management control actions at runtime.

Admission Control. Admission control decides whether a newly arriving query should be admitted into the system, i. e., passed to the scheduling component, or rejected. The primary goal of admission control is to avoid accepting more queries than can be executed effectively with available resources.

Admission control policies can place different kinds of limits on the system, e. g., the number of queries running concurrently, the number of concurrent users, or the expected costs of the submitted queries. Typical admission control actions are:

- Warn: accept the query but signal a warning
- Hold: hold a query until the DBA releases it
- Reject: reject the query

If a query passes all of the admission control policies then the query is admitted for scheduling. Some systems support policies that allow a high priority query to bypass admission control and scheduling and start executing immediately.

Scheduling. The main goal of the scheduling component is to avoid a state of system overload. The scheduler determines when to start the execution of a query. It maintains queues of pending queries and policies determine how the queues are managed. The most commonly used queue types used by schedulers include:

- separate queues for different query priorities
- separate queues for different expected runtimes
- one FIFO queue for all queries
- all queries start immediately

Some policies include parameters and thresholds, e. g., to map expected runtime to an appropriately sized queue. If the metric is below threshold, then another query may start.

Execution Control. Admission control and scheduling policies apply to queries *before* execution. Their decisions are based on compile-time information, such as the optimizer's cost estimates for queries, plus information about the current operating environment, such as system load. However, at runtime, a query may behave significantly differently from its cost estimates, and the operating environment might have changed dramatically. The task of execution control is to limit the impact of these deviations from expectations. Execution control uses both cost estimates and runtime information to make its decisions.

Different execution conditions may be evaluated by an execution policy, such as CPU time above a threshold or elapsed exceeding an estimate by an absolute or relative amount. The administrator must choose the thresholds in the conditions to achieve workload objectives. Some typical execution control actions include:

- None: Let the query run to completion
- Warn: Print a message to a log; query continues
- Reprioritize: Change the priority of the query
- Stop: Stop processing query; return results so far

- Kill: Abort the query and return an error
- Kill & Requeue: Abort the query, then put it in a scheduling queue to start over
- Suspend and Resume : Stop processing query, and put saved state in scheduling queue.

4.2   Workload Management for Operational BI

As illustrated by the call-out boxes in the figure, an operational BI system must handle mixed workloads ranging from OLTP-like queries that run for milliseconds to BI queries and maintenance tasks that last for hours. The diversity and complexity of the queries make it incredibly difficult to estimate resource requirements, and thus contention, prior to runtime. The subsequent inaccuracy of resource and contention estimates in turn make it difficult to predict runtime or gauge query progress, and force workload management decisions to be made in the absence of critical information.

In order to understand a variety of situations that involve resource contention in data warehouse environments, we interviewed numerous practioners with experience with multiple commercial database products. For example, we find that an effective workload management system should be able to distinguish between the following problem scenarios involving long-running queries:

- A query has heavy initial cost estimates. It should not be admitted to the database when system load is expected to be high.
- A particular query had reasonable initial estimates and was admitted to the database, but is much more costly than expected and is impeding the performance of other running queries. It should be stopped.
- The system is overloaded and one or more queries are making poor progress. However, all queries have roughly equal costs and are getting equal shares of resources. Stopping any particular query will not improve the situation if a new similar query would be admitted instead. The number of concurrent queries should be reduced.

4.3   Experimental Framework

We developed and built an experimental framework, described in a previous paper [16], to evaluate the effectiveness of existing and newly-developed workload management techniques in a controlled and repeatable manner. The architecture of this framework follows the generic architecture in Figure 1.

Our framework permits us to select from a variety of workload management policies and algorithms and insert them into key workload management modules — i. e., the admission controller, the scheduler, and the execution controller. We synthesized these policies and algorithms from the policies of current commercial systems, but the policies and algorithms are not limited to techniques already implemented by database systems and tools.

We implemented a simulator for the database engine that mimics the execution of database queries in a highly parallel, shared-nothing architecture. The simulator does not include components like the query compiler and the optimizer: we provide the query plans and the costs as input.

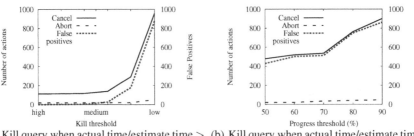

(a) Kill query when actual time/estimate time >   (b) Kill query when actual time/estimate time >
kill threshold   kill threshold and progress < progress threshold

Fig. 4. Impact of aggressive workload management thresholds

Using a simulated database engine was necessary. There are a number of reasons why we could not use an actual database engine for our experiments. First, we investigate workloads that run for hours. Our simulated database engine "runs" these workloads in seconds, which lets us repeat the workloads with many different workload management policies. Second, each workload management component in today's databases implements only a subset of the possible workload management features described in industrial and academic literature. Using a real database would limit us to the policies that a particular product provides, contradicting our goal to experiment with an exhaustive set of techniques *and* to model features that are currently not available.

We model a workload as composed of one or more jobs. Each job consists of an ordered set of typed requests and is associated with a performance objective. Each request type maps to a tree of operators, and each operator in a tree maps in turn to its resource costs. Our current implementation associates the cost of each operator with the dominant resource associated with that particular operator type (e. g., disk or memory).

4.4   Initial Results

Our experiments demonstrate that although it is most efficient to recognize the queries that will not complete and either not admit them or stop their execution, in certain situations recognizing such queries is difficult. For example, in [16], we explored the impact of handling problem queries aggressively.

In the first experiment, we killed a query when actual time exceeded expected time by a relative threshold. Figure 4(a) shows the results. The x-axis is the kill threshold: on the left side, the threshold is high and so few queries are killed. On the right, the threshold is low; a query is killed soon after it exceeds its time estimate. The left-hand y-axis denotes the number of actions taken and the right-hand y-axis denotes the number of false positives, queries that are not actually problem queries. Even with a high kill threshold, some false positives occur because the actual processing time for some normal queries is up to 1.3 times higher than predicted.

Since the number of false positives increased sharply as the kill threshold got more aggressive, in the second experiment we added a second threshold, designed not to kill queries that would complete soon. In this experiment, a query is killed only if it exceeds the kill threshold (held constant at "low") *and* has made low progress. The x-axis of Figure 4(b) shows different values for low progress. In this figure (particularly for the lower-progress thresholds), there are many fewer false positives than were incurred using a low kill threshold by itself. Together, Figures 4(a) and 4(b) demonstrate that rectifying workload management problems aggressively means executing more actions, but that combining metrics can reduce the number of queries cancelled or aborted by mistake. Informed by the results of our experiments, we are currently developing mechanisms to determine effective thresholds and policies for a variety of conditions.

5 Prediction of Resource Usage

While improving workload management algorithms to handle inaccurate predictions is half of our approach, the other half is to improve the quality and accuracy of predictions. In this section, we discuss our efforts to predict multiple aspects of query performance and resource usage for a wide variety of queries, using only information available at compile-time.

Our ultimate goal is to enable database vendors to build models and tools that can accurately predict resource requirements. Customers can then use these to tools to do what-if modeling, schedule and manage workloads, and select appropriate system configurations. Figure 5 overviews the uses of accurate prediction tools. Further motivation for our prediction work can be found in [11].

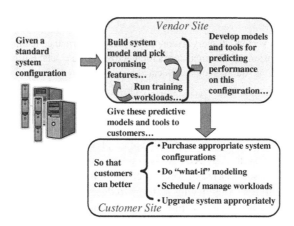

Fig. 5. Uses of accurate prediction tools

Because performance depends upon resource contention and resource availability, in order to control resource contention in a dynamic mixed workload, we need information about resource usage and resource contention just as much as we need information about query runtimes. Our work is unique in that one of our objectives is to predict multiple resource usage characteristics, such as CPU usage, memory usage, disk usage, and message bytes sent.

We considered several techniques for making such predictions, from linear regression to clustering techniques. The key feature of our chosen technique, called Kernel Canonical Correlation Analysis (KCCA) [2], is that it finds multivariate correlations among the query properties and query performance metrics on a training set of queries and then uses these statistical relationships to predict the performance of new queries. That is to say, given two data sets, one representing multiple compile-time query characteristics and one representing multiple runtime performance characteristics, we can train KCCA to find relationships between these two datasets so that later, given a new query we've never seen before, we can predict runtime performance characteristics at compile-time.

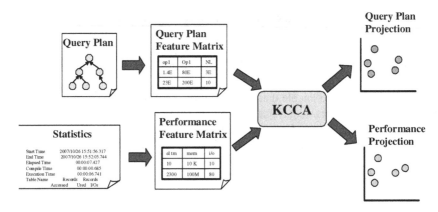

Fig. 6. Training: From vectors of query features and query performance features, KCCA projects the vectors onto dimensions of maximal correlation across the data sets. Furthermore, its clustering effect causes "similar" queries to be collocated.

Figure 6 illustrates the training process by which KCCA can build a predictive model of query performance from training data. Since KCCA is a generic algorithm, we needed to formalize the problem of performance prediction and map it onto the data structures and functions used by KCCA. This task involved two key design decisions.

First, we needed to summarize the pre-execution information about each query into a vector of "query features." Identifying features to describe each query was not a trivial task. Many machine learning algorithms require feature vectors, but there is no simple rule for defining them. Typically, features are chosen using a combination of domain knowledge and intuition. [11] describes our evaluation of potential feature vectors. In the end, we chose the simplest set of features that gave good results for our test cases — the operators and cardinality estimates described in the optimizer's query plan. Should it become necessary (e.g., if we should be faced with a new workload for which this simple predictive model does not work), it would be straightforward (though time-consuming) to substitute a different set of feature vectors and use KCCA to train a new predictive model.

We also needed to summarize the performance statistics from executing the query into a vector of "performance features." Selecting features to represent each query's performance metrics was a fairly straightforward task; we gave KCCA all of the performance metrics that we could get from the HP Neoview database system when running the query. The metrics we use for the experiments in this paper are elapsed time, disk I/Os, message count, message bytes, records accessed (the input cardinality of the file_scan operator) and records used (the output cardinality of the file_scan operator). The performance feature vector thus has six elements. (Other metrics, such as memory used, could be added easily when available.)

We then combine these two sets of vectors into a query feature matrix with one row per query vector and a performance feature matrix with one row per performance vector. It is important that the corresponding rows in each matrix describe the same query.

Second, we needed to define and compute a similarity measure between each pair of query feature vectors and between each pair of performance feature vectors. KCCA uses a kernel function to compute this similarity measure and we use the common "Gaussian" kernel, as described in [11].

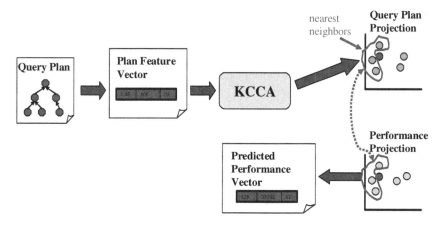

Fig. 7. Prediction: KCCA projects a new query's feature vector, then looks up its neighbors from the performance projection and uses their performance vectors to derive the new query's predicted performance vector

Given these features matrices and the similarity measures, KCCA then projects the feature vectors into new high-dimensional subspaces — a query projection and a performance projection — where nearness indicates similarity. These projections are correlated in the sense that queries that are neighbors in one prediction are neighbors in the other projection.

Figure 7 shows how we predict the performance of a new query from the query projection and performance projection in the KCCA model. Prediction is done in three steps. First, we create a query feature vector and use the model to find its coordinates on the query projection. We then infer its coordinates on the performance projection: we use the k nearest neighbors in the query projection to do so (we evaluate choices for k in [11]). Finally, we must map from the performance projection back to the metrics we want to predict. Finding a reverse-mapping from the feature space back to the

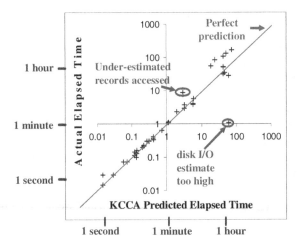

Fig. 8. Predicted vs. actual elapsed times for 61 test queries. We use a log-log scale to accommodate the wide range of query execution times from milliseconds to hours. The predictive risk value (a measure of how "good" a prediction is, where a predictive risk value close to 1 implies near-perfect prediction) for our prediction was 0.55 due to the presence of a few outliers (as marked in the graph). Removing the furthest outlier increased the predictive risk value to 0.61.

input space is a known hard problem, both because of the complexity of the mapping algorithm and also because the dimensionality of the feature space can be much higher or lower than the input space (based on the goal of the transformation function). We evaluated several heuristics for this mapping, discussed in [11].

We show sample results for an experiment with 1027 queries in the training set and a non-overlapping test set of 61 queries. Figure 8 illustrates that we are able to predict elapsed time within 20% of the actual time for at least 85% of the test queries in one experiment. Predictions for resource usage such as records used, disk I/O, message counts, etc., were similarly accurate and are very useful for explaining the elapsed time predictions. For example, for one prediction where elapsed time was much too high, we had greatly overpredicted the disk I/Os. This error is likely due to our parallel database's methods of cardinality estimation. When we underpredicted elapsed time by a factor of two, it was due to under-predicting the number of records accessed by a factor of three.

6   Conclusions

The extreme diversity of resource requirements and the potential for show-stopping resource contention raise the stakes for managing an Operational BI workload. The difficulty of predicting expected behavior means that workload management decisions must be made with a minimum of information, and poorly informed decisions can lead to dire consequences for system performance.

We address this challenge in two ways. We are carrying out a systematic study of workload management policies intended to limit the impact of inaccurate information.

We are also developing techniques for predicting multiple-query resource requirements of both short- and long-running queries more accurately. We hope that our approach will improve the effectiveness of critical tasks that rely on accurate predictions, including system sizing, capacity planning and workload management.

Acknowledgements

We would like to thank Stefan Krompass's advisor, Alfons Kemper, as well as Archana Ganapathi's advisors, who are Armando Fox, Michael Jordan, and Dave Patterson. We also thank Stefan Kinauer for his help implementing the workload management user interface.

References

1. Arlitt, M.F.: Characterizing Web user sessions. SIGMETRICS Performance Evaluation Review 28(2), 50–63 (2000)
2. Bach, F.R., Jordan, M.I.: Kernel Independent Component Analysis. Journal of Machine Learning Research 3, 1–48 (2003)
3. Benoit, D.G.: Automated Diagnosis and Control of DBMS Resources. In: EDBT PhD. Workshop (2000)
4. Carey, M.J., Livny, M., Lu, H.: Dynamic Task Allocation In A Distributed Database System. In: ICDCS, pp. 282–291 (1985)
5. Chaudhuri, S., Kaushik, R., Ramamurthy, R.: When Can We Trust Progress Estimators for SQL Queries? In: Proc. of the ACM SIGMOD Intl. Conf. on Management of Data, pp. 575–586 (2005)
6. Chaudhuri, S., Narasayya, V., Ramamurthy, R.: Estimating Progress of Execution for SQL Queries. In: Proc. of the ACM SIGMOD Intl. Conf. on Management of Data, pp. 803–814 (2004)
7. Davison, D.L., Graefe, G.: Dynamic Resource Brokering for Multi-User Query Execution. In: Proc. of the ACM SIGMOD Intl. Conf. on Management of Data, pp. 281–292 (1995)
8. Dayal, U., Kuno, H., Wiener, J.L., Wilkinson, K., Ganapathi, A., Krompass, S.: Managing operational business intelligence workloads. SIGOPS Oper. Syst. Rev. 43(1), 92–98 (2009)
9. Eeckhout, L., Vandierendonck, H., Bosschere, K.D.: How Input Data Sets Change Program Behaviour. In: 5th Workshop on Computer Architecture Evaluation Using Commercial Workloads (2002)
10. Elnaffar, S., Martin, P., Horman, R.: Automatically Classifying Database Workloads. In: Proc. of ACM Conference on Information and Knowledge Management (CIKM), pp. 622–624 (2002)
11. Ganapathi, A., Kuno, H., Dayal, U., Wiener, J., Fox, A., Jordan, M., Patterson, D.: Predicting Multiple Metrics for Queries: Better Decisions Enabled by Machine Learning. In: Proc. of the 21st Intl. Conf. on Data Engineering, ICDE (2009)
12. Gillin, P.: BI @ the Speed of Business. Computer World Technology (December 2007)
13. Gupta, C., Mehta, A.: PQR: Predicting Query Execution Times for Autonomous Workload Management. In: Proc. Intl Conf on Autonomic Computing, ICAC (2008)
14. Keeton, K., Patterson, D.A., He, Y.Q., Raphael, R.C., Baker, W.E.: Performance Characterization of a Quad Pentium Pro SMP using OLTP Workloads. In: The 25th Intl. Symposium on Computer Architecture (ISCA), pp. 15–26 (1998)

15. Krompass, S., Gmach, D., Scholz, A., Seltzsam, S., Kemper, A.: Quality of Service Enabled Database Applications. In: Dan, A., Lamersdorf, W. (eds.) ICSOC 2006. LNCS, vol. 4294, pp. 215–226. Springer, Heidelberg (2006)
16. Krompass, S., Kuno, H., Dayal, U., Kemper, A.: Dynamic Workload Management for Very Large Data Warehouses: Juggling Feathers and Bowling Balls. In: Proc. of the 33rd Intl. Conf. on Very Large Data Bases, VLDB (2007)
17. Krompass, S., Kuno, H., Wiener, J.L., Wilkinson, K., Dayal, U., Kemper, A.: Managing long-running queries. In: EDBT 2009, pp. 132–143. ACM, New York (2009)
18. Lo, J.L., Barroso, L.A., Eggers, S.J., Gharachorloo, K., Levy, H.M., Parekh, S.S.: An Analysis of Database Workload Performance on Simultaneous Multithreaded Processors. In: The 25th Intl. Symposium on Computer Architecture (ISCA), pp. 39–50 (1998)
19. Luo, G., Naughton, J.F., Ellmann, C.J., Watzke, M.W.: Toward a Progress Indicator for Database Queries. In: Proc. of the ACM SIGMOD Intl. Conf. on Management of Data, pp. 791–802 (2004)
20. Luo, G., Naughton, J.F., Ellmann, C.J., Watzke, M.W.: Increasing the Accuracy and Coverage of SQL Progress Indicators. In: Proc. of the 21st Intl. Conf. on Data Engineering (ICDE), pp. 853–864 (2005)
21. Luo, G., Naughton, J.F., Yu, P.S.: Multi-query SQL Progress Indicators. In: Ioannidis, Y., Scholl, M.H., Schmidt, J.W., Matthes, F., Hatzopoulos, M., Böhm, K., Kemper, A., Grust, T., Böhm, C. (eds.) EDBT 2006. LNCS, vol. 3896, pp. 921–941. Springer, Heidelberg (2006)
22. Markl, V., Lohman, G.: Learning Table Access Cardinalities with LEO. In: Proc. of the ACM SIGMOD Intl. Conf. on Management of Data, p. 613 (2002)
23. Mehta, M., DeWitt, D.J.: Dynamic Memory Allocation for Multiple-Query Workload. In: Proc. of the 19th Intl. Conf. on Very Large Data Bases (VLDB) (August 1993)
24. Moore, J., Chase, J., Farkas, K., Ranganathan, P.: Data Center Workload Monitoring, Analysis, and Emulation (2005)
25. Schroeder, B., Harchol-Balter, M., Iyengar, A., Nahum, E.M.: Achieving Class-Based QoS for Transactional Workloads. In: Proc. of the 22nd Intl. Conf. on Data Engineering (ICDE), p. 153 (2006)
26. Stillger, M., Lohman, G.M., Markl, V., Kandil, M.: LEO - DB2's LEarning Optimizer. In: Proc. of the 27th Intl. Conf. on Very Large Data Bases (VLDB), pp. 19–28 (2001)
27. Weikum, G., Hasse, C., Mönkeberg, A., Zabback, P.: The COMFORT Automatic Tuning Project. Information Systems 19(5), 381–432 (1994)
28. White, C.: The Next Generation of Business Intelligence: Operational BI. DM Review Magazine (May 2005)
29. Yoo, R.M., Lee, H., Chow, K., Lee, H.-H.S.: Constructing a Non-Linear Model with Neural Networks for Workload Characterization. In: IISWC, pp. 150–159 (2006)
30. Yu, P.S., Chen, M.-S., Heiss, H.-U., Lee, S.: On Workload Characterization of Relational Database Environments. Software Engineering 18(4), 347–355 (1992)
31. Zhang, N., Haas, P.J., Josifovski, V., Lohman, G.M., Zhang, C.: Statistical Learning Techniques for Costing XML Queries. In: Proc. of the 31st Intl. Conf. on Very Large Data Bases (VLDB), pp. 289–300 (2005)

A Study on Workload Imbalance Issues in Data Intensive Distributed Computing

Sven Groot, Kazuo Goda, and Masaru Kitsuregawa

University of Tokyo, 4-6-1 Komaba, Meguro-ku, Tokyo 153-8505, Japan

Abstract. In recent years, several frameworks have been developed for processing very large quantities of data on large clusters of commodity PCs. These frameworks have focused on fault-tolerance and scalability. However, when using heterogeneous environments these systems do not offer optimal workload balancing. In this paper we present Jumbo, a distributed computation platform designed to explore possible solutions to this issue.

1 Introduction

Over the past decade, the volume of data processed by companies and research institutions has grown explosively; it is not uncommon for data processing jobs to process terabytes or petabytes at a time. There has also been a growing tendency to use large clusters of commodity PCs, rather than large dedicated servers. Traditional parallel database solutions do not offer the scalability and fault-tolerance required to run on such a large system.

As a result, several frameworks have been developed for the creation of customized distributed data processing solutions, the most widely known of which is Google's MapReduce [1], which provides a programming model based on the map and reduce operations used in functional programming, as well as an execution environment using Google File System [2] for storage. Hadoop [3] is a well-known open-source implementation of GFS and MapReduce.

Microsoft Dryad [4] is an alternative solution which offers a much more flexible programming model, representing jobs as a directed acyclic graph of vertex programs. This extra flexibility can however make it more difficult to efficiently parallelize complex job graphs.

Workload balancing is an important aspect of distributed computing. However, MapReduce does not provide adequate load balancing features in many scenarios, and the nature of the MapReduce model makes it unsuited to do so in some cases. Microsoft has to our knowledge not published any data on how Dryad behaves in a heterogeneous environment, thus it is unfortunately not possible for us to provide a comparison to it.

In the following sections, we will outline the issues with workload balancing in the MapReduce model and introduce Jumbo, our distributed computation platform which is designed to further investigate and ultimately solve these issues.

S. Kikuchi, S. Sachdeva, and S. Bhalla (Eds.): DNIS 2010, LNCS 5999, pp. 27–32, 2010.

2 Workload Imbalance Issues in MapReduce

A MapReduce job consists of a map phase and a reduce phase. For the map phase, the input data is split and each piece is processed by a map task. Parallelism is achieved by running multiple map tasks at once in the cluster. Typically, there are far more map tasks in the job than there are nodes in the cluster, which means the map phase is well suited for load balancing. Faster nodes in the cluster take less time on the individual tasks, and therefore run more of them.

Data from the map phase is partitioned and distributed over several reduce tasks. In contrast to the map phase, the number of reduce tasks typically equals the capacity of the cluster. This means that if some nodes finish early, there are no additional reduce tasks for them to process. While it is possible to use more reduce tasks, this means that some tasks will not be started until the others complete. These tasks cannot do any of their processing in the background while the map tasks are still running, so doing this will typically reduce performance and is therefore not desirable.

Hadoop provides a mechanism for load balancing called speculative execution. Long-running map or reduce tasks will be started more than once on additional nodes, in the hope that those nodes can complete the task faster. This strategy works in some cases, but it is not optimal. Speculative execution will discard the work done by one of the two task instances, and the extra load caused by the additional instance - particularly in the case of a reduce task, which will need to retrieve all relevant intermediate data, causing additional disk and network overhead - can in some cases delay the job even further.

A further problem occurs with MapReduce's inflexible programming model. Many more complicated data processing jobs will need more than one MapReduce phase, done in sequence. For example, the frequent item set mining algorithm proposed in [5] consists of three consecutive MapReduce jobs. In this case, it is not possible to start consecutive jobs until the preceding job has finished completely. Even in cases where the next job could already have done some work with partial data, this is not possible. Any load balancing mechanisms available to MapReduce can only consider the tasks of one of the consecutive jobs at a time, rather than the whole algorithm.

3 Jumbo

In order to evaluate workload balancing and other issues in data intensive distributed computing, we have developed Jumbo, a data processing environment that allows us to investigate these issues. We have decided to develop our own solution, rather than building on Hadoop's existing open-source foundation, because some of the issues with the current MapReduce-based solutions are fundamental to the underlying model.

Jumbo consists of two primary components. The first of these is the Jumbo Distributed File System, which provides data storage. Jumbo DFS is very similar to GFS and Hadoop's HDFS in design.

Jumbo DFS uses a single name server to store the file system name space, which is kept in memory and persisted by using a log file and periodic checkpoints. Files are divided into large blocks, typically 64 or 128MB, which are stored on data servers. Each block is replicated to multiple servers, typically three, and the replicas are placed in a rack-aware manner for improved fault-tolerance.

The second component is Jumbo Jet, the data processing environment for Jumbo, providing a programming model as well as an execution environment.

Jumbo Jet represents jobs as a sequence of stages. The first stage reads data from the DFS, while each consecutive stage reads data from one or more preceding stages. Intermediate data from each of the stages is stored on disk to improve fault-tolerance. The final stage writes its output to the DFS. This sequence of stages forms a directed acyclic graph.

The stages are divided up into one or more tasks, each performing the same operation but on a different part of the data. The tasks in a stage can be executed in parallel.

In order to divide DFS input data across multiple tasks, the input data is simply split into pieces, typically using DFS blocks as a unit. Since each task reads data from a single block, the task scheduler can attempt to schedule that task to run on a node that has a local replica of that block, reducing network load.

When a stage reads input from another stage, the data from that input stage is partitioned by using a partitioning function. Every task in the input stage creates the same partitions, and each task in the stage reading that data will read all the pieces of just one partition from all the tasks in the input stage. Unlike in MapReduce, it is not required for the intermediate data to be sorted. Jumbo allows full flexibility in specifying how the data from each input task is processed. While you can perform a merge-sort like MapReduce does, you can also just process each piece in sequence, or process records in a round-robin fashion, or write a custom input processor that uses whatever method is required.

This design has many obvious similarities to MapReduce. Indeed, it is trivial to emulate MapReduce using this framework by creating a job with two stages, the first performing a map operation, the second a reduce operation, and sorting the intermediate data. However, Jumbo is not limited to this, and can more easily represent a larger variety of jobs.

Job scheduling is handled by a single job server, which performs a role similar to the JobTracker in Hadoop. Each server in the cluster will run a task server which receives tasks to execute from the job server. The job server also keeps track of failures, and reschedules failed tasks.

Currently, Jumbo does not yet contain any load balancing features beyond what Hadoop provides. However, Jumbo's more flexible design means that it will be much better suited for future experiments with load balancing than what we would be able to do with Hadoop. Jumbo's design allows us to implement complex algorithms such as the PFP algorithm from [5] as a single job so task scheduling decisions for load balancing can consider the entire job structure rather than just a part. It will also be possible to restructure jobs in different ways, besides the MapReduce structure, if this proves beneficial to distributing the workload.

Although we expect our solutions will also be applicable to other systems, using our own ensures we can fully control the design and implementation.

4 Example

In order to demonstrate the issue of workload balancing, we have run a simple experiment using both Hadoop and Jumbo. For this experiment we have used the GraySort (previously known as TeraSort) benchmark included with Hadoop, and created an equivalent job in Jumbo. Jumbo uses a sorting strategy that is very close to that of MapReduce. The job consists of two stages. The first stage partitions the input into N pieces (where N is the number of tasks in the second stage), and sorts each partition. The second stage performs a merge operation on all the input files for each partition from the first stage, and writes the result to the DFS.

Unfortunately we were not able to evaluate the behaviour of Microsoft Dryad in this experiment, as Dryad cannot run on our cluster.

The sort operation was executed on an increasing number of nodes, each time increasing the total amount of data so that the amount of data per node stays the same, 4GB per node. The number of reduce tasks (or in Jumbo, the number of second stage tasks) is also increased with the number of nodes. This means that ideally, the execution time should stay identical on a homogeneous cluster.

Two sets of nodes were used for this experiment: 40 older nodes, with 2 CPUs, 4GB RAM and one disk, and 16 newer nodes with 8 CPUs, 32GB RAM and two disks. At first we ran the job on only the older nodes. Once we were using all 40 older nodes, we added the 16 newer nodes.

Figure 1 shows the results of this. Neither Jumbo nor Hadoop quite achieve linear scalability for the first 40 nodes, as the execution time drops slightly. This is mainly because of the increasing number of map tasks or first stage tasks, which increases the number of network transfers and also affects the merge strategy to use. We are continuously reducing this overhead and improving the scalability of Jumbo.

It can also be seen that Jumbo is considerably faster than Hadoop. This difference is caused by some inefficient implementation choices in Hadoop, causing Hadoop to waste I/O operations which is very expensive, especially on the older nodes with just one disk.

However, the interesting part happens when going from 40 to 56 nodes. The final 16 nodes are much faster than the rest, which should lead to an overall improvement in performance. Although the execution time does drop, it doesn't drop as far as expected. Using Jumbo, the 40 old nodes alone take 616 seconds to sort 160GB, a total throughput of 273MB/s. We also executed the sort using only the 16 new nodes, which sorted 64GB in 223 seconds at 293MB/s. This means the total throughput for the 56 node cluster should be 566MB/s, which

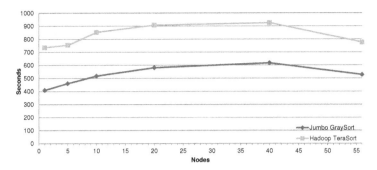

Fig. 1. Sorting performance of Hadoop and Jumbo. Up to 40 nodes, all nodes used are identical; only in the 56 nodes case was a heterogeneous environment used.

means that sorting 224GB on those 56 nodes should take 405 seconds, rather than the 527 seconds observed. In practice, 405 seconds is probably not realistic due to some additional overhead, but there is clearly room for improvement.

As indicated in Sect. 2, the issue in this particular scenario lies with the reduce phase, or in the case of Jumbo the second stage, of the job. For both Hadoop and Jumbo, the first stage finishes considerably faster with 56 nodes, because that stage consists of a very large number of tasks, 1792 in total, and the faster nodes are able to process more of them.

However, the reduce phase, or second stage in Jumbo, is where the bulk of the work is done. The I/O intensive merge operation takes up most of the job's total execution time, and because the number of tasks here equals the number of nodes in the cluster they cannot be balanced in any way.

This phenomenon can be clearly seen in Fig. 2, which shows the execution times for each node in the cluster. The 16 faster nodes finish their work considerably earlier than the 40 slower nodes. Because there are no additional tasks in the second stage, there is no way for Jumbo to assign additional work to those nodes after they finish.

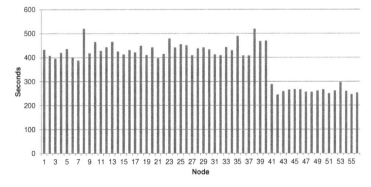

Fig. 2. Execution times of individual nodes in the cluster for Jumbo GraySort. Nodes 1-40 are the older, slower nodes, while 41-56 are the faster ones.

One apparently obvious solution is to simply assign a larger amount of records to the partitions on the faster nodes, but this requires a-priori knowledge of how to divide the records. It also requires knowing which partition will be processed by which node, and since failures may cause any task to be reassigned to a different node this is also not a desirable scenario.

It should be noted that even amongst the identical nodes, various external circumstances cause these nodes to also have varied execution times. Ideally, a load balancing solution would be able to reduce this effect as well.

5   Conclusion

We have given an overview of the workload balancing issues in data intensive distributed computing, particularly when using the MapReduce model. We have also introduced Jumbo, our own data processing system.

Improper workload balancing leads to a considerable waste of resources, with some nodes sitting idle while others are still working. Flexible methods to dynamically redistribute the workload will be required to solve this. However, naive methods of doing this such as Hadoop's speculative execution cause too much overhead, negating their potential benefits.

For our future work we intend to use Jumbo as a platform to develop and evaluate different methods for workload balancing so we can more fully utilize the resources available in a heterogeneous cluster.

References

1. Dean, J., Ghemawat, S.: Mapreduce: Simplified data processing on large clusters. In: OSDI 2004: Proceedings of the 6th conference on Symposium on Opearting Systems Design & Implementation, Berkeley, CA, USA, p. 10. USENIX Association (2004)
2. Ghemawat, S., Gobioff, H., Leung, S.T.: The google file system. In: SOSP 2003: Proceedings of the nineteenth ACM symposium on Operating systems principles, pp. 29–43. ACM Press, New York (2003)
3. Apache: Hadoop core, http://hadoop.apache.org/core
4. Isard, M., Budiu, M., Yu, Y., Birrell, A., Fetterly, D.: Dryad: distributed data-parallel programs from sequential building blocks. SIGOPS Oper. Syst. Rev. 41(3), 59–72 (2007)
5. Li, H., Wang, Y., Zhang, D., Zhang, M., Chang, E.Y.: Pfp: parallel fp-growth for query recommendation. In: RecSys 2008: Proceedings of the 2008 ACM conference on Recommender systems, pp. 107–114. ACM, New York (2008)

Information Extraction, Real-Time Processing and DW2.0 in Operational Business Intelligence

Malu Castellanos, Umeshwar Dayal, Song Wang, and Gupta Chetan

Hewlett-Packard Laboratories
Palo Alto, CA, USA
firsname.lastname@hp.com

Abstract. In today's enterprise, business processes and business intelligence applications need to access and use structured and unstructured information to extend business transactions and analytics with as much adjacent data as possible. Unfortunately, all this information is scattered in many places, in many forms; managed by different database systems, document management systems, and file systems. Companies end up having to build one-of-a-kind solutions to integrate these disparate systems and make the right information available at the right time and in the right form for their business transactions and analytical applications. Our goal is to create an operational business intelligence platform that manages all the information required by business transactions and combines facts extracted from unstructured sources with data coming from structured sources along the DW2.0 pipeline to enable actionable insights. In this paper, we give an overview of the platform functionality and architecture focusing in particular in the information extraction and analytics layers and their application to situational awareness for epidemics medical response.

1 Introduction

Today, organizations use relational DBMSs to manage (capture, store, index, search, process) structured data needed for on-line transaction processing (OLTP)[1] applications. However, typical business transactions (billing, order processing, accounts payable, claims processing, loan processing, etc.) need both structured and unstructured information (contracts, invoices, purchase orders, insurance claim forms, loan documents, etc.). This information is scattered in many places and managed by different database, document management, and file systems [1]. This makes it difficult to find all the information relevant to a business transaction, leading to increased transaction cost or loss of revenue. A good example is hospital billing (Fig. 1), where massive amounts of information (i.e., physician notes, file records, forms) are generated by different processes (admission, lab tests, surgery, etc.) during a patient's hospital stay. This immense amount of unstructured data spread across different entities needs to be captured, sorted, reconciled, codified and integrated with other structured data to process billing transactions. Unfortunately, the lack of a platform to extract structured data from these unstructured documents and integrate it with other structured data

---

[1] Also for on-line analytics processing (OLAP) but it is not the focus of this paper.

S. Kikuchi, S. Sachdeva, and S. Bhalla (Eds.): DNIS 2010, LNCS 5999, pp. 33–45, 2010.

causes errors in the billing process. These errors result in a 30% revenue loss according to an internal reliable source working with a major hospital[2]. Another example is
loan processing, where the inability to make all of the relevant information available
to business transactions has led to mismanagement of the whole process. In these
cases, companies have to build one-of-a-kind solutions to integrate disparate systems
and make the right information available at the right time in the right form for business users.

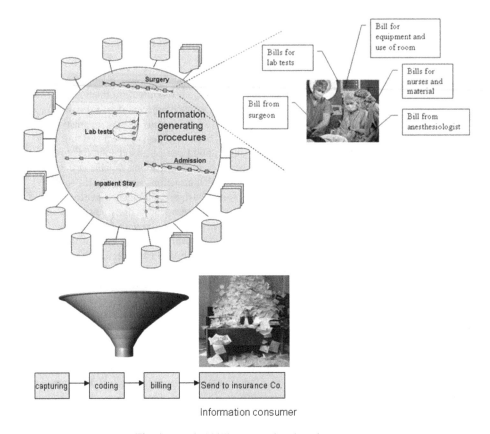

Fig. 1. Hospital billing operational environment

The problem described before is exacerbated by today's competitive and highly
dynamic environment where analyzing data to understand how the business is performing, to predict outcomes and trends, to improve the effectiveness of business
processes and to detect and understand the impact of external events on the internal
business operations has become essential. In fact, the next wave of competitive
differentiation will be based on the ability to harness unprecedented amounts of data
into actionable information. Analytics will give data-driven companies a powerful

---

[2] Due to confidentiality we cannot give the name of the hospital.

advantage in customer service, new product development, risk profiling, real-time pricing, pattern recognition, fraud detection and many other examples. The value of analytics is in enabling actions based on insights ("actionable insights") derived from information extracted from all sorts of data sources, that is, structured or unstructured and internal or external. Moreover, the trend towards operational BI where the speed to insight is critical to improve the effectiveness of business operations with minimal latency, is forcing companies to analyze structured and unstructured information as soon as it is generated, that is, as data streams. A good example is the case of epidemics awareness for medical response where hospitals analyze incident reports of incoming patients and news feeds about the onset of epidemics and compare them to determine the effect that the epidemics event is having on the hospital. This allows the hospital to be prepared to deal with the epidemics by ordering supplies, scheduling resources, etc. Once more, the lack of a platform that supports the required functionality, (analytics in this case) on the combination of extracted information forces companies to adopt ad-hoc solutions that are rather difficult to build, very costly, limited and often not even satisfactory.

The combination of streaming data with stored data on one dimension and structured and unstructured data in another dimension gives an enormous competitive advantage to companies that have the technology to exploit this information in a timely manner.

Our goal is to create an operational business intelligence platform for all data required by business transactions and BI applications. This is done in response to the need that organizations have to adopt technology that enables them to a) incorporate unstructured data into their business transactions, b) to gain "real-time" actionable insight derived from information that is extracted from structured and unstructured data sources, and c) doing all this in an integrated, scalable and robust way.

The rest of the paper is structured as follows: Section 2 presents the layered architecture of our operational BI platform. Section 3 describes the role of information extraction in DW 2.0, the information extraction pipeline and its fit into textual-ETL that is inherent to the DW 2.0 architecture. Section 4 illustrates the operation of the IE and analytics layers and their interaction with a concrete application to contractual situational awareness where real-time processing is of essence. Section 5 gives a brief overview of related work and Section 6 presents our conclusions.

2 Platform Architecture

Our operational business intelligence platform has two main functions:
  (a) It integrates structured and unstructured data to extend the scope of business transactions with adjacent data, for a better support of business processes' underlying business operations.
  (b) It integrates structured and unstructured data to extend the scope of BI applications and analytics with adjacent data to give the users a 360° data visibility.
The components of the platform are shown in Figure 2.

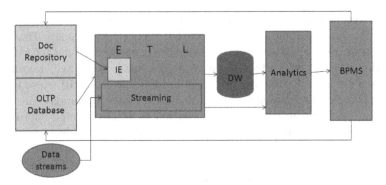

Fig. 2. Architectural components

OLTP DBMS - A massively parallel DBMS is often deployed in OLTP environments that require scalability and very high availability where its fault tolerant nature delivers the stability to guarantee that transactions are completed. The idea is to extend these benefits to all operational information –structured and unstructured - required by business transactions.

Document Repository - Unstructured data comes in many forms: scanned or electronic documents, email, audio, video, Web pages and many others. Our platform initially focuses on text-based documents, such as Word or PDF files. Text documents are sent to the platform repository along with any accompanying metadata. The metadata is stored in the search database and the document is parsed to populate the inverse index database that supports search functionality. All of the metadata about a document is stored in the DBMS tables and the actual document is treated as a Binary Large Object (BLOB), stored in the Large Object Storage subsystem.

Extract-Transform-Load (ETL) – The back-end of the architecture is a data integration pipeline for populating the data warehouse by extracting data from distributed and usually heterogeneous operational sources (OLTP database); cleansing, integrating and transforming the data; and loading it into the data warehouse. The integration pipeline supports batch and streaming processing as well as structured and unstructured data. The latter is known as textual ETL and is an essential component of the DW 2.0 architecture (see Section 3).

Information Extraction (IE) – It handles the "E" part of (textual) ETL of unstructured data sources. It provides the ability to automatically read through textual documents, retrieve business-relevant information (e.g., doctor's name, surgical procedure, symptoms), and present it in a structured fashion to populate the data warehouse so that searching and accessing this information is substantially simplified. Before extraction ever happens, a document model that reflects the structure of the documents and a domain model that specifies the items to be extracted need to be defined [2]. Also a subset of documents is manually tagged with the target information according to the domain model (the use of a GUI greatly simplifies this process). Once tagged, the documents go through a normalization process that may involve stemming, correcting misspellings, and eliminating stop words. Then, the documents are transformed into different representations that suit the input requirements of specialized algorithms that use these documents to learn extraction models (see Section 3). Once models are learned they are applied to production documents to mine target data from text. Notice that this is the data used to extend the scope of business transactions and BI applications with structured data extracted from unstructured sources.

Datawarehouse (DW) – It is a repository designed to facilitate reporting and analysis through a variety of front-end querying, reporting and analytic sources. It typically consolidates data from operational databases but in our platform it also consolidates structured data extracted from unstructured sources. It is also implemented on a parallel DBMS (in fact, it might be the same DBMS that runs the OLTP Database and the Document Repository.

Analytics – They consume data from structured sources along with data that the IE component extracted from unstructured ones. Moreover, since data may come from internal and external sources and may take the form of stored or streaming data, this layer needs to incorporate analytic algorithms that are capable of simultaneously dealing with streams of varying speeds and static data. This enables users to get better insight in a timely manner. In section 4 we introduce one of our streaming analytics techniques.

BPMS Layer - The business process management system (BPMS) controls transactions, requests to the Document Repository and analytics. At different steps of a business process OLTP transactions are carried out and it is these transactions whose scope is augmented by the data extracted by the information extraction layer in such a way that unstructured data is processed as part of the normal business processes.

3 Information Extraction and DW 2.0

Data warehousing began in the 1980s as a way to reduce users' frustration with their inability to get integrated, reliable, accessible data. On-line applications had no appreciable amount of historical data because they jettisoned their historical data as quickly as possible in the name of high performance. Thus, corporations had lots of data and very little information. For 15 years, people built different manifestations of data warehouses where the focus was on structured data extracted from OLTP databases. Increasingly, enterprises have come to realize that for business intelligence applications supporting decision making, information derived from unstructured and semi-structured data sources is also needed. By some accounts, over 80% of an enterprise's information assets are unstructured, mainly in the form of text documents (contracts, warranties, forms, medical reports, insurance claims, policies, reports, customer support cases, etc.) and other data types such as images, video, audio, and other different forms of spatial and temporal data. This led to the evolution of the underlying architecture of the data warehouse into what Bill Inmon has called DW 2.0 [3]. The core idea of DW 2.0 is the integration of structured and unstructured data. Thus, one of the main differences between the first generation of data warehouses and DW 2.0 is that in the latter unstructured data is a valid part of the data warehouse. Unstructured data exists in various forms in DW 2.0: actual snippets, edited words and phrases and extracted structured data. The latter being the interesting one here as described below.

The information extraction layer of our platform architecture serves for the purpose of extracting structured data from unstructured sources. The goal is to make this data available to BI applications and analytics and to OLTP applications that carry out business operations. For illustration, imagine a company that has a BI application that creates user profiles and makes purchase recommendations based on these profiles. To improve the accuracy of a profile, it may be important to consider not just the transactional data in the various operational databases involved in the ordering process, but also to look at the content of web pages that the user has browsed, his reviews or blog postings, and any contracts he may have with the company. This example makes evident that unstructured data provides valuable contextual information for more informed operational decision making.

To incorporate unstructured data into the DW architecture (i.e., DW 2.0), it is not sufficient to merely store the data as is (i.e., text) into the warehouse. Rather, it is important to extract useful information from the unstructured data and turn it into structured data that can be stored, accessed and analyzed efficiently along with other structured data. However, this is not an easy task. Take, for example, the text in customer reviews, which are written in an ad-hoc manner. The lack of structure makes it hard to find the product and the features referred to in the review, and especially which features the customer likes and which he doesn't like.

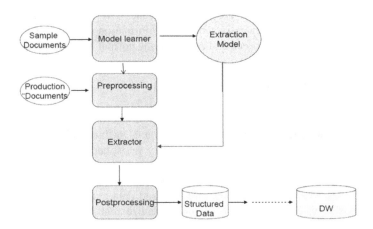

Fig. 3. Information extraction pipeline for unstructured data

Numerous *information extraction* (IE) techniques have been developed that try to learn models for the retrieval of relevant entities, relationships, facts, events, etc. from text data [4]. Some of the most popular techniques are based on rule learning [5], Hidden Markov Models [6] and more recently Conditional Random Fields [7]. We experimented with these techniques but did not obtain good results so we developed our own technique based on genetic algorithms (GA) [8] to learn the most relevant combinations of prefixes and suffixes of tagged instances of the role-based entity types of interest. To this end, a bag of terms is built from all the prefixes in the context of the tagged entities in the training set. Another bag is built from their suf-fixes. For example, given the tagged sentence *symptoms appeared on* <expiration-Date> *May 31, 2009,* </expirationDate> *after returning from a trip toMexico....* The terms *"symptoms"*, *"appeared"*, *"on"* are added to a bag of prefixes of the role-based entity type "SymptomsOnsetDate" whereas the terms *"after"*, *"returning"*, *"from"*, *"trip"*, *"Mexico"* are added to its bag of suffixes. The bags are then used to build indi-viduals with N random prefixes and M random suffixes in the first generation and for injecting randomness in the off-springs in later generations. Since only the best indi-viduals of each generation survive, the fitness of an individual is computed from the number of its terms (i.e., prefixes and suffixes) that match the context terms of the tagged instances. The best individual in a pre-determined number of iterations repre-sents a context pattern given by its terms and is used to derive an extraction rule that recognizes entities of the corresponding type. The GA is run iteratively to obtain more extraction rules corresponding to other context patterns. The process ends after a given number of iterations or when the fitness of the new best individual is lower than a given threshold. The rules are validated against an unseen testing set and those with the highest accuracy (above a given threshold) constitute the final rule set for the given role-based entity type.

Figure 3 shows the pipeline of information extraction from text data sources that is built into our platform's IE layer with the goal of extracting relevant data from a col-lection of documents; e.g., contract number, customer, and expiration date from con-tracts. Whatever the information extraction algorithm used, the source data always needs to be pre-processed to get rid of noise, transform it to the representation required by the extraction method, etc before the actual extraction takes place. In addition, the output also needs to be post-processed to gather the structured data in the form of at-tribute-value pairs, which can then be transformed and loaded into the data warehouse. The pipeline of tasks to extract, transform and load data from unstructured sources into a data warehouse so that it can be integrated into the structured world corresponds to what Inmon has named *textual-ETL* [9]. The pipeline involves numerous operations from which those belonging to IE correspond to the 'E' (extract) part of ETL. The extracted data from the unstructured sources often relates to data in structured sources so it is staged into a landing area and is then loaded into the data warehouse, where the information from both structured and unstructured sources is consolidated (e.g., con-tract data is related to customer data) for use by BI applications.

The challenge in textual-ETL consists in identifying how to abstract all the above tasks into operators that can be used to design, optimize and execute these flows in the same way as for structured data.

Using a uniform approach for ETL of structured and unstructured data makes it possible to use the same data warehouse infrastructure. The question is whether there should be separate pipes for the structured and unstructured data in the ETL flows or only one. Having separate pipes (as shown in Figure 4a) is conceptually simpler to design. However, having a single pipe with two separate extraction stages but a single integrated transformation-load stage (as shown in Figure 4b) creates better opportuni-ties for end-to-end optimization.

In the next section we present an application of our operational BI platform. Spe-cifically, we describe how information extraction and analytics help to get actionable insight in a timely manner in the context of epidemics awareness for hospital response.

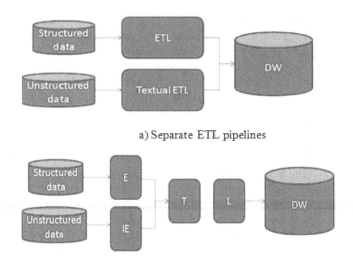

a) Separate ETL pipelines

b) Integrated pipeline

Fig. 4. ETL Pipeline for Structured and Unstructured Data

4 An Application: Situational Awareness

An important function of our operational BI platform is extracting and correlating different sources of unstructured data. Here we describe how correlating unstructured stored data with unstructured streaming data from the web enables situational awareness. We apply it to the medical domain where hospital needs to be alert about epidemics onsets and how they are affecting the patients arriving at the hospitals. This allows the hospitals to be prepared to deal with the epidemic by ordering medical supplies, scheduling resources, etc.

Data buried in incident reports of patients arriving at a hospital and in articles of news feeds provide valuable information that when extracted and correlated can provide actionable insight that can enable the hospital to be prepared for an epidemic situation. It is easy to imagine the complexity and unfeasibility of manually correlating news feeds with patient incident reports. Not only is the amount of news articles immense (and the amount of reports might be quite large as well) but the streams arrival rate might be too fast to cope with. Furthermore, it is practically impossible to keep track of the many details of the incidents in the reports to correlate them to the news articles about the epidemics onset which would require analyzing every report and every news article.

Our platform facilitates this by using novel techniques to extract relevant information from disparate sources of unstructured data (in this case the patient incident reports and the news feeds) and determine which elements (i.e., documents) in one stream are correlated to which elements in the other stream. It also has the capability to do inner correlation to compute reliability scores of the information extracted from fast streams. The need for this feature can easily be seen for news streams: the more news articles report on a given event, the more reliable it is that the event indeed occurred. Consequently, as the streams are being processed, correlations are updated by factoring in a reliability score.

The process follows several steps:

i.   First, the IE layer extracts relevant data from the patient incident reports. This includes symptoms and history facts like places where the patient has been, people who have been in contact with the patient, date when the symptoms appeared, etc. Notice that this is not a simple recognition of words or entities like a simple date, instead it is entity recognition at a higher semantic level where it is necessary to make distinctions like start date of symptoms versus other dates in the report (e.g., date of visit to the hospital). Given that there is ample variability in the way reports are written, it is necessary to learn models that recognize these kind of semantic entities which we call *role-based-entities*.

ii. Second, a categorizer in the Analytics layer classifies the articles from news feeds (e.g., New York Times RSS feeds) into interesting (e.g., "epidemics") and non-interesting (the others) categories.

iii. From those articles in the interesting categories the IE layer extracts relevant data about the location of the onset, the symptoms, the transmission channel, etc. Again, this is a high semantic level of entity recognition.

iv. Inner correlations of the descriptor formed from the extracted data with descriptors of previous articles within a time window are computed in the Analytics layer to derive reliability scores.

v.   Finally the similarity of reports and epidemics related articles is measured using the extracted information as features and extending them along predefined hierarchies. The similarity is computed in the Analytics layer in terms of hierarchical neighbors using fast streaming data structures called Hierarchical Neighborhood Trees (HNT).

The *neighbor* concept is used to measure the similarity of categorical data. If two entities (incident report and news article) have the same value for a categorical variable (e.g. both have fever for a symptom variable), then they are neighbors and will be put into a same node in the HNT. If they have different values (e.g., vomiting and diarrhea) then they will be put into different nodes in the HNT and thus they are neighbors only at the level of their lowest common ancestor (e.g., stomach upset). Based on HNT, we thus can specify a scale based neighborhood relationship which measures the level in the HNT tree of the closest common ancestor of the tags. The hierarchical neighborhoods are critical to finding correlations between the reports and the news items. For example, assume a report doesn't mention fever by name but indicates that the patient has chills and a news article talks about high body temperature. The chills belong to a hierarchy where one of its ancestors is fever, and similarly high body temperature belongs to the hierarchy that contains fever as an ancestor. In this case, the report and the news article are neighbors at the level of fever. As a result we not only learn that these are neighbors but also that they are related through "fever". Once correlations are obtained, relevance scores are derived by factoring in the reliability scores computed before. For details on HNTs the interested reader is referred to [10].

The steps described above are done in two phases. The first one, corresponding to the very top part in Figure 5, is off-line and is domain-specific; it is where models are learned to extract information from documents and to classify them. This phase is preceded by a specification step where the user defines through a GUI the entities to be extracted from the documents and other relevant domain information like document categories of interest. We have developed flexible models that learn regularities in the textual context of the entities to be extracted while allowing some degree of variability. The models are provided with knobs to tune according to quality requirements, for example, to tradeoff accuracy for performance.

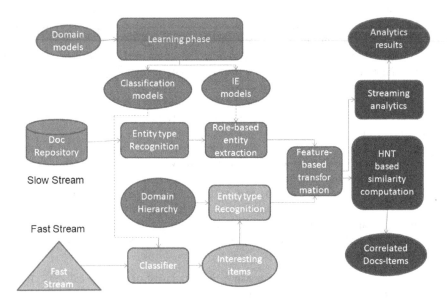

Fig. 5. IE & Analytics layers interaction

In the second phase, corresponding to the rest of Figure 5, the models learned in the previous phase are applied to classify documents and extract information from them. This can be done either off-line for a static document collection like a contracts repository or on-line for slow text streams like the contracts stream and for fast text streams like news feeds tweets or blogs. To speed up the search for role-based entities as well as to improve the precision of the results, the IE layer first uses entity recognizers to identify appropriate entity types. For example, to extract the date that symptoms appeared from a patient report it first recognizes entities of type date in the report and only on the textual neighborhood (surrounding words) of these entities it applies the extraction models learned in the previous phase to recognize the symptoms onset date. For the external fast stream the documents are first classified to discard those that are irrelevant (i.e, those that are not about epidemics) to avoid wasting time in attempting to extract information from them.

Once the information is extracted in terms of concepts (i.e., role-based entities) from the incident reports and the news items, the Analytics layer finds correlations using HNTs. Each concept belongs to one or more concept hierarchies and an HNT (Hierarchical Neighborhood Tree) is a tree based structure that represents all these hierarchies. Every document in the slow stream is converted to a feature vector (i.e., descriptor) where every feature is one of the extracted concepts. This document now coded as a multi-dimensional point is inserted into the HNTs for all the dimensions in the predefined hierarchies. For example suppose a report contains the concept "fever" and "Mexico" (e.g., the person visited Mexico). Then, it is coded as a two dimensional vector, where fever belongs to the "Symptoms" hierarchy and "Mexico", belongs to the "Location" hierarchy. In other words for the dimension "Symptoms" the value is "fever" and for the dimension "Location", the value is "Mexico". As a result of this process the reports in the slow stream are stored as multidimensional points in the HNTs. Likewise, when an "interesting" (i.e., "epidemics) news article comes in, it is similarly converted to a multidimensional point and inserted into the HNTs. The reports in each level of the hierarchy (for each of the dimensions) are the neighbors of the news item. For example, assume there is a news article n_1, which contains the concept "Honduras". The report (slow stream document) containing the concept "Mexico" is a neighbor at the level of "political region" to n_1. There are several advantages to using HNTs: (i) Mathematically, it allows us to overcome the problem of defining distances over categorical data such as regions, (ii) Inserts and deletes are very fast with the help of simple indexing schemes, (iii) It facilitates finding correlations at different abstraction levels, thus we can use the levels to quantify the "nearness", meaning points that are in the same node at a higher level are "less near" than the points which are in the same node at a lower level in the hierarchy.

The "nearness" is measured by the level of the lowest common ancestor. If news item n_i and a patient incident report c_k are neighbors in multiple HNTs they are more similar. This can be quantified with the following equation: $S_{i,k} = \prod_{j=1}^{d} s_{i,k,j}$, Where $S_{i,k}$ is the similarity between news item n_i and a report c_k and $s_{i,k,j}$ is the scale at which news item n_i and a report c_k are neighbors in HNT d. Here d corresponds to the dimension. This whole process is efficient as explained in [10].

Once we have computed the scale based similarity using the equation above, we compute the 'top k' reports that are most correlated to the news. These reports represent the best candidates of patients affected by the epidemics (they match characteristics of the epidemic in terms of symptoms, place of origin, pre-conditions, etc). They make possible to derive information about the spread and effect that the epidemic onset is having in the area. With this valuable information, the hospital gains awareness of how the situation is developing and can get prepared accordingly.

In a nutshell, the goal of the IE and analytics layers of the platform is to reduce the time and effort to build data flows that integrate structured and unstructured data, slow and fast streams, and analyze (through correlation in our epidemics example) this combination of data in near-real-time to provide awareness of situations with potential impact on the enterprise business operations.

As part of this effort we are developing algorithms for different functions, including information extraction and analytics, to be wrapped as operators of a library used to build data flows. In addition, we are developing techniques to optimize these flows with respect to different quality metrics in addition to performance (we call this, QoX-based optimization) [11].

5 Related Work

Efforts have been made in this space in both industry and academia. There are OLTP database systems like Oracle, document management systems such as Documentum™, workflow management systems like FileNet™, Text Mining products such as PolyAnalyst™ and Attensity™, Extract-Transform-Load products like Informatica™. However we have found no solution that encompasses all of these functionalities in a single platform that manages structured and unstructured data, integrating them seamlessly to provide actionable insight for optimizing business operations.

Information extraction at the high semantic level done here is very challenging and requires training learning models. Although many proposals for such models exist [4], our experiments revealed that they do not work well for large documents nor they are flexible enough to cope well with variability in the text surrounding the information to be extracted (even if they claim the opposite). At the end, a lot of manual tuning needs to be done. This led us to decide creating our own algorithms for role-based entity extraction. In contrast, for plain entity extraction there are commercial and open-source frameworks and suites with recognizers well trained on vast collections of text or even manually created like ThingFinder [12] which is used in the IE layer for plain entity recognition.

Distance based clustering approaches are most relevant to our streaming correlation work. For high dimension data, distance computation can be expensive. In such cases a subspace projection approach can be used. Aggarwal [13] uses this approach for outlier detection. However, none of the approaches are designed for streaming data. A recent approach [14] computes distance based outliers over streaming windowed data. However, they do not consider the case of heterogeneity of fast streams and slow streams. In this paper we propose a novel and efficient solution with HNT to measure the correlation among the patients incident reports in a slow stream and news items in a fast stream based on categorical distances.

6 Conclusions

We have presented our operational business intelligence platform where information extraction, real-time processing and analytics are the core functionalities to achieve the goal of providing actionable insight in a timely manner for improving business operations and at the same time to extend the scope of transactional processes with data from unstructured sources. Our work is unique in that it integrates seamlessly all the necessary components into a single platform, making it easy to develop applications that benefit from the insight obtained. We have shown an application where unstructured data from patients incident reports and from streaming news articles is extracted and analyzed in real-time to enable situational awareness of epidemic onsets for hospitals to be proactive in being prepared to deal with the epidemic.

Currently some components of the architecture have been implemented but we still have a long way to complete the development of algorithms and techniques for the rest of the platform components.

References

[1] Halevy, A.: Why your Data Won't Mix. Association for Computing Machinery Queue Magazine (October 2005)

[2] Castellanos, M., Dayal, U.: FACTS: An Approach to Unearth Legacy Contracts. In: Proc. First International Workshop on Electronic Contracting (WEC 2004), San Diego, CA (July 2004)

[3] Inmon, W.H., Strauss, D., Neushloss, G.: DW 2.0: The Architecture for the Next Generation for Data Warehousing. Morgan Kauffman, Burlington (2008)

[4] Sarawagi, S.: Information Extraction. Foundations and Trends in Databases 1(3), 261–377 (2008)

[5] Soderland, S.: Learning Information Extraction Rules for Semi-Structured and Free Text. Machine Learning 34(1-3), 233–272 (1999)

[6] Freitag, A.M.: Information Extraction with HMM Structures Learned by Stochastic Optimization. In: Proc. 17th National Conference on Artificial Intelligence. AI Press (2000)

[7] Peng, F., McCallum, A.: Accurate Information Extraction from Research Papers using Conditional Random Fields. In: HLT-NAACL, pp. 329–336 (2004)

[8] Michalewicz, Z.: Genetic Algorithms + Data Structures = Evolution Programs, 3rd edn. Springer, Heidelberg (1996)

[9] Inmon, W.H., Nesavich, A.: Tapping into Unstructured Data: Integrating Unstructured Data and Textual Analytics into Business Intelligence. Morgan Kaufmann, San Francisco (2007)

[10] Castellanos, M., Gupta, C., Wang, S., Dayal, U.: Leveraging Web Streams for Contractual Situational Awareness in Operational BI. To appear in Proc. EDBT Workshops, BEWEB 2010 (2010)

[11] Dayal, U., Castellanos, M., Simitsis, A., Wilkinson, K.: Data Integration Flows for Business Intelligence. In: Proc. EDBT (2009)

[12] Business Objects Thing Finder Language Guide and Reference. Business Objects an SAP Company (2009)

[13] Aggarwal, C., Han, J., Yu, P.S.: A framework for projected clustering of high dimensional data streams. In: Proceedings of the 30th VLDB Conference (2004)

[14] Angiulli, F., Fassetti, F.: Detecting distance-based outliers in streams of data. In: CIKM, pp. 811–820 (2007)

On Realizing Quick Compensation Transactions in Cloud Computing

Shinji Kikuchi

School of Computer Science and Engineering, University of Aizu,
Ikki-machi, Aizu-Wakamatsu City, Fukushima 965-8580, Japan
d8111106@u-aizu.ac.jp

Abstract. The Cloud Computing paradigm influences various items including principles and architectures in the domain of the transaction processing. It might be predictable that the number of processing will exponentially increase. Although the domain of the transaction processing has evolved and a number of implementations have also been developed according to the various requirements over the ages, there are still some remaining engineering issues in practice within SOA (Service Oriented Architecture) area. Therefore, it is predictable that developers will face a lot of difficulties to realize implementations and scalability. In this paper, we show a tentative proposal on a regulated framework and an abstract model in which a compensation transaction plays the central role. Further we propose a more effective transaction processing which rely on the new possible features by scalable environment.

Keywords: Long Live Transaction, Compensation, Optimization.

1 Introduction

The Cloud Computing paradigm influences various items including also principles and architectures. The transaction processing has its long history in the computer era [1]; however it is not an exception and will also suffer from this paradigm shifting to the next evolution. The Cloud Computing paradigm consists of various elemental notions, thus it is almost impossible to define it exactly. However, it might be predictable that a number of processing and its complicacy will increase.

Until now, the domain of the transaction processing has evolved and a number of implementations have also been developed according to the various requirements of each age, for instance the long live transaction which is dominant in the B2B (business-to-business), SOA area. However there are still some remaining engineering issues in practice due to a lack of consensus among the notions in these areas. Therefore, it might also be predictable that we will face a lot of difficulties to realize implementations with scalability.

In this paper, we show a tentative proposal on a regulated framework and an abstract model in which a compensation transaction plays the central role. Then based on this framework we clarify the crucial points to realize the more effective transaction processing, and also propose potential candidates for tackling these points, which rely on the new possible features due to the scalable environment.

S. Kikuchi, S. Sachdeva, and S. Bhalla (Eds.): DNIS 2010, LNCS 5999, pp. 46–64, 2010.

The remainder of this paper is organized as follows; in section.2 we describe the background and issues. In this section, after clarifying the potential requirements caused by the Cloud Computing environment and current status on the compensation transaction, we will identify our approach. In section.3 we define the framework in order to regulate the disordered notions in regards to compensations. In section.4 we continuously describe potential candidates in order to realize the quick compensation transaction which is our main proposal in this paper. In section.5 the evaluation in regards to these candidates is mentioned. As it is currently difficult to demonstrate these quick compensation transactions due to the conceptual level of discussion, the evaluation will be done by comparing to the existing approaches. And we conclude and discuss about future's works in section.6.

2 Background and Issues

Here, we will clarify the structure of the issues and focus on our main targeted one through three items. The first item is to identify the issues including the potential requirements. The second item is to identify the current status in aimed area. Then as the third item we will finally organize a new suitable approach as the solutions.

For the first item, we will focus on the primary requirement. Gartner research predicts the emergence of a new stage of transaction processing due to raising the number of transactions, as the software as a service gains the central role of processing. According to their explanation, the notion of this stage is named 'extreme transaction processing' [2]. The main characteristic features are; (1) applying SOA notions, (2) increasing the number of transactions. These matters are related to the multiple and combined factors, so that we carefully need to analyze relationships among these factors. However, it is definite that realizing scalable processing is the most potential and crucial requirement.

In the case of the second item, we might retain the current approach under the insufficient status in the SOA transaction area. The current dominant approach has been influenced by multiple factors, and the design methodology of the transaction processing might tend to be complicated. A history of the transaction processing is explained in detail by T.Wang and et al, [1]. Under the current status of SOA, not only the ACID properties, but the long live transactions are also charged with the big roles. Currently the most potential methodology and the specification to realize these long live transactions might be Web Service Business Activity as the OASIS standard [3]. In this standard, the compensation transaction is crucial to maintain the consistency, although the notion of the compensation transaction used to be proposed in the movement of the advanced transaction models in the late of 80's.

Various new independent notions and methodologies had been considered in the B2B area. For instance the distributed workflows, Grid architecture applied for businesses and dependability with fault tolerance for the Grid have emerged [4]. However, identifying the relationship among the basic notions and solving the disorders in regards to the compensation transactions had definitely remained insufficient. For instance, according to P.Greenfield [5], there are some conceptual variations on the compensation. In the narrow sense, the terminology of the compensation means a compensation transaction in the sense of a long live transaction, especially in 'Saga' case computation. Whereas, in the wide sense, that terminology could imply a part of the fault tolerance. Furthermore, according to M.Chessell et al. [6], in a wider sense, the procedures in the compensation can include the sub-procedures which are not related to the ACID transaction.

Under the current status, it is impossible to develop more suitable methodology for the engineering of the combined long live transactions. Furthermore, it is impossible to define more efficient protocol oriented to the scalable solutions. Therefore, new issues are regulating the notions of the current status of the compensation transactions, and identifying suitable approaches to realize the high efficiency of processing with focusing on the potential requirements. It is identified as the third item, here.

In the B2B and SOA areas, there had been some prior studies done by B.Limthanmaphon and Y.Zhang [7], P.Hrastnik and W.Winiwarter [8]. However, when considering the new potential requirements, one might need to think about a more comprehensive precise approach and solution. And they might not be in the same level which remains under single and specialized views.

In order to respond to the above new needs, we might be required to view the traditional transaction model with a wider scope and to make more abstract models for synthesizing the traditional elemental models. According to our observation, to do that we have only a short history as we have just started, for example defining taxonomy between transaction and workflow [9]. There are insufficient numbers of fragments of the complete model. In this paper, we show a tentative proposal on a regulated framework and an abstract model in which the compensation transaction has the central role. Then based on this framework, we aim to realize the more effective transaction processing for tackling the scalable environment.

3   Definition of the Framework

3.1   Outline

Due to our general experience, the following methodologies and notations sound suitable when reorganizing the regulated framework and the abstract model, especially to explain about the fundamental notions.

  (1)Function layer model.
  (2)Class organization as a static model.

The function layer model is popular to regulate the architectural requirements as we can see Open Systems Interconnection/Basic Reference Model as a typical example [10]. Whereas the class organization is important in modeling to grasp the structural, configuration relationships. Based on the regulated notions on the compensations expressed in these models, we clarify the crucial points to realize the more effective transaction processing in the section.4.

3.2   Function Layer Model

Considering the concept of the compensation according to the classification by P.Greenfield [5], the compensation transaction might be a sub-element. The proposed function layer model includes the extended notions from the original compensation transaction. Consequently, five sub-function layers are proposed. They are suitable in the sense of corresponding software modularity. However, when comparing with another framework for the fault tolerance in workflow, there seems to be a room to improve the structure of the layers. In this section, we explain about the five sub-function layers and features on the software components. Additionally, we mention the remaining issues caused by a comparison to concepts of the fault tolerance in workflow.

Fig.1 shows the outline of our function layer model consisting of the five sub-function layers as mentioned. This also shows the correspondence between sub-function layers and granulites of activities' network. Here, we assume that the four layers from the bottom will automatically be executed along the specified workflows, whereas the fifth layer at the top must be executed by inter-mediation of the human operations.

(1) The first layer is 'Service Invocation Layer', which corresponds to invoked elemental activities as services. As we have often encountered some divergences in expression due to treating wide domains, we will use 'service' and 'activity' within the synonyms.

(2) The second layer is 'ACID Transaction Layer', which corresponds to ACID transactions grouping the set of elemental activities, and guaranteeing the ACID properties. The typical example of this behavior is the defined as X/Open transaction model.

(3) The third layer is 'Compensation Transaction Layer', which corresponds to the instance of the long live transaction like 'Saga' [11]. That uses the elemental ACID transactions including the compensation transactions. However, we define the obvious constrained features to this layer, which differ from the fourth layer. That is complete-recovering (Or equivalent-recovering in looser cases) to the original status at the passed safe point mentioned later whenever a fault happens. Furthermore, the following four properties 'visibility', 'permanence', 'recover' and 'consistency' are naturally important in the this layer instead of the traditional ACID properties according to P.K.Chrysanthis [12].

(4) The fourth layer is 'Autonomic Compensation Layer', in which the compensation procedures will not be limited within transactional manners, but a more generalized and non-transaction manner as well. In this layer, there is an obvious feature, which is 'never faults and absolutely succeeds'. This feature is originally pointed out in [23]. In particular, the fourth layer should be treated with all of the possible procedures featured as fault tolerant instead of the transactional features.

(5) The fifth layer at the top is 'Compensational Operation Layer', which must be run by an inter-mediation of human operations. In this layer, semantically equivalent operations for the fault tolerant will be executed under the operators' judgments.

As for the relationship with the granulites of the activities' network, the Autonomic compensation layer should be defined including all paths of processes between safe points. The definition of the safe point is the same as one in [13]. Therefore, a path between safe points corresponds to an instance of workflow on the Compensational transaction layer. If there is an alternative path having the same start and end points, another instance of workflow on the Compensational transaction layer must be invoked. Any part of the above path consisting of set of element activities should be mapped as an instance on the ACID transaction layer, and elemental invocation should correspond to an invocation on the Service invocation layer.

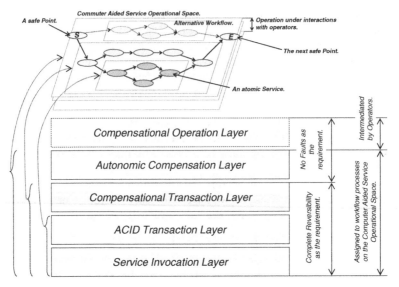

Fig. 1. Outline of the function layer model for compensation transactions. This also shows the correspondence between elements of layers and granulites of activities' network.

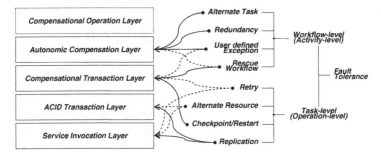

Fig. 2. The correspondence between elements of layers and fault tolerance methods defined by J.Yu and R.Buyya [4]

Fig.2 shows our correspondence between layers and methods for fault tolerance defined by J.Yu and R.Buyya [4]. So far based on our analysis, this correspondence tends to be ambiguous and includes multiple mappings. This might be caused not only by existing ambiguities on the defined granulites in the framework of fault tolerance methods, but also by no explicit definitions in regards to the resources in our function layer model. In this paper we don't intend to analyze these multiple mappings more precisely because that activity requires us to define the common Ontology sharing notions over both of the compensation procedures and the fault tolerance methods. Furthermore we need to re-organize the models of them individually. For example, P.Grefen and J.Vonk report the taxonomy of transactional workflows through analyzing their conceptual relationship and plotting all cases on coordinates consisting of concepts and architectures [9]. According to their report, the integrated concept model is applied in order to compare a case to others. More precise analysis would require us to apply the similar methodology if we would try to clarify the relationship between compensation procedures and the fault tolerance methods. However as the fault tolerance contains the heterogeneous categories like replica of data, the fault tolerance might naturally be related to wider domains than merely workflow.

This suggests to us that the number of classes included in the targeted Ontology might be huge. Unfortunately, there are obviously no limits when modeling our world, and our current model remains as explained in the next section. Thus, it might be insufficient for analyzing the correspondence more precisely. By the sense mentioned above, our function layer model remains as a tentative version. However it is sufficiently explanatory for evaluating the quick compensation transactions as our current target.

3.3   Class Organization as a Static Model

In the traditional modeling of transaction processing, most of the cases have usually been modeled in the two notations. The first is a sequence chart which can deal with dynamical interactions between entities within time passing. The second is a graphical notation for general transaction models defined by J.Gray and A.Reuter which can express transitions of entities' states [14]. And there are not many notation instances, which can express the static and structural aspects of the transaction processing. However when thinking about the framework which will deal with the relationship between a transaction processing and its corresponding compensations through abstracted models, that might closely mean evaluating instances of that relationship from the point of view of structural aspects.

Therefore, it is mandatory to adopt the notation, which has properties corresponding to the class chart. In previous works, there are the following instances which treat the notions in regards to the compensation transactions. The first instance is the metamodel expression in XML proposed by P.Hrastnik and W.Winiwarter [8]. The second one is UTML (Unified Transaction Modeling Language) proposed by N.Gioldasis and S.Christodoulakis [15], [16]. In particular UTML has the notation, which are extended from the UML class chart, and constraint rule expression for instances. Therefore we might feel potentiality and some advantages as a meta-language on this UTML. Here we define our targeted class organization as a static model by inheriting and extending this UTML.

In this UTML, 'Activity' and 'Operation' classes which correspond to the definitions of the management units are defined, and 'Activity Instance' and 'Operation Instance' classes as execution instances refer to them. Difference between 'Activity' and 'Operation' classes relies on the atomic property, and the minimum atomic unit is 'Operation'. This 'Operation' class corresponds to 'Service Invocation Layer' of the function layer model expressed in Fig.1, and the 'Activity' class is mapped to 'ACID Transaction Layer' and 'Compensation Transaction Layer' of the function layer model. Furthermore the 'Activity' class can express a hierarchical invocation relationship not only by referring 'ActivitySet' class but also by being referred by that. And 'ManagementOperation' which expresses the abstraction of procedures at an initiation and a termination is explicitly drawn.

However according to our review of the class organization of the original UTML from point of view of integrating multiple models and realizing a quick compensation transactions as mentioned later, there are the following five items to be enhanced.

(1) Within our recognition the 'propertySet' class seems to focus on the ACID properties. However, it might be insufficient for our assumed cases, as there seem to be no suitable classes for the properties featured at execution time. As a typical instance, we often need to specify the order of procedures 'Only in sequence' or 'In parallel'. An opportunity of this specifying obviously arises in the case of using OASIS/WS-BPEL(Web Service Business Process Execution Language) [17], which is a standard for the distributed workflows, the instances of which consists of a set of 'Activities'.

(2) In regards to the properties, there seem to be no corresponding notions to save points and safe points. These notions are definitely important to decide of a scope for compensation transactions.

(3) Weimin Du and et al categorize the compensations and define the method of Lazy Compensation [18]. In this sense, there is a characteristic property for the notion of the compensation. And the comprehensive performance of the compensations might lie on that property. In particular, in our function layer model drawn in Fig.1 there is the clear border between the 'Compensation Transaction

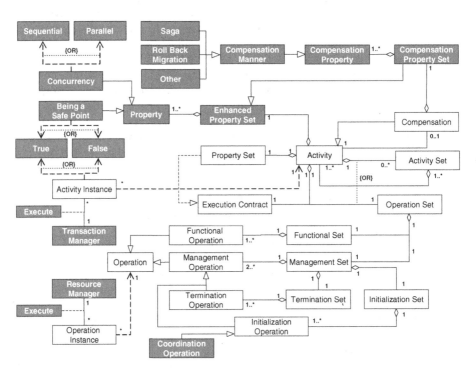

Fig. 3. Enhanced class structure by using notions defined the UTML: Unified Transaction Modeling Language defined by N.Giodasis and S.Chritodoulakis [15], [16]

Layer' aiming for maintaining a transaction consistency and 'Autonomic Compensation Layer' lying above the 'Compensation Transaction Layer'. Therefore, the characteristic property describing the compensation itself is implicitly important. In our extended model, we need to specify this property.

(4) It is important to decompose the class 'ManagementOperation' like initialization more explicitly. These procedures are merely costs for the whole of transactions. Therefore, it is naturally required to optimize and compress these procedures to be more effective. In order to identify the cost structure of the transaction, we need to decompose the overhead operations.

(5) It is obviously important to grasp the explicit relationships between the execution entities like a transaction manager, a coordinator and a resource manager.

According to the above items, our enhanced class organization based on the UTML is shown as Fig.3. In this figure, the aforementioned properties are expressed as gray rectangles with white letters.

4   Potential Approaches for the Quick Compensations

4.1   Outline

The following items might have potentialities, when identifying the crucial points to realize more effective transaction processing and quick compensations according to the function layer model as Fig.1 and the class organization as Fig.3.

(1) Shrinking the safe point intervals and making management operations more effective.

(2) Identifying potential phases to be compressed.

In this paper, we explain about outlines of them more. Then due to limited space of this paper, with omitting the first item, we will newly mention the enhanced rollback migration protocol as one of potential candidates to the above second item in section 4.3.

4.2   Background for the Two Candidates

The first item mentioned above includes two related points and making balance of both is the most important matter here. In general, there might be both an effect and a side effect in regards to shrinking the safe point intervals. As the effect of shrinking them, we could realize a quick and short term recovering when a fault would happen. Whereas, it could lead more overhead cost for the initial coordination.

'ManagementOperation' class in Fig.3 definitely remains as an overhead from the point of view of efficiency of the transaction processing. Therefore increasing the number of instances of 'ManagementOperation' class means less efficiency to the whole of the transaction processing. Furthermore, from the point of view of ACID properties, too short safe point intervals in the long live transactions are substantially equivalent with managements by ACID transactions. Thus, managing by long live transactions would lose their value and meaning. Therefore, it would become an important issue to design the suitable granulites of safe point intervals according to the probability of occurrences of faults.

When thinking about the second item mentioned in the previous section with Fig.1, it is important how to compress the procedures on the 'Compensation Transaction Layer', in other words the procedures for recovering to the original status at the passed safe point. The reason for this is because the function on 'Autonomic Compensation Layer' remains merely as a function to specify the alternative path, whereas the main responsibility of the function of 'Compensation Transaction Layer' is managing the consistency of the long live transactions.

In the case where an activity belonging to an instance of a long live transaction has a complicated hierarchical invocation relationship and is applied with the normal Saga for its compensation transaction, the invocation order of the compensation transaction to the safe point as the start point must usually be inverse to the order of the original long live transaction. The concrete algorithm for this is mentioned by P.Grefen, et al. [13] and the concrete requirements to maintain the consistency are explained by Weimin Du, et al. [18].

In Fig.3 there are no explicit expressions. But if an instance of 'Compensation Manner' class as a property attached to the 'Compensation' class is specified as 'Saga', that instance of the 'Compensation' class must be bound to the instance of the 'Concurrency' class as a part of 'EnhancedPropertySet' class which is also a part of the original 'Activity' class. Thus there are certain possibilities that all of the compensation procedures must be carried out in sequence in the worst case where the activity has a complicated hierarchical invocation relationship. Therefore it becomes important as the second item as to how the compressed procedure for the compensation should be established.

4.3 Enhanced Rollback Migration Protocol

Rollback Migration is proposed by J.Vonk and P.Gefen, and it might be allowable to maintain the consistency of the transactions with some time lags [19]. With our more consideration, we could mention that Rollback Migration is the method to move the management responsibility on the processed transaction to the outside requester once. Consequently, it might be postponed more to maintain the consistency inside of the scope of the transactions compared to the normal Saga. As mentioned before, in order to compress the procedure in the compensation, we have considered a new method named 'Enhanced Rollback Migration protocol' by adding new elements into the original Rollback Migration. The features of this protocol could be described as following;

(1) The super huge scale of data management will be realized in the Cloud Computing era. Therefore, facilities of temporal data management will become more popular than the previous and current conditions. Therefore here, we rely on these facilities existing in the near future.

(2) The execution of the compensation transactions will be done under other transaction context from the original one. Therefore, it is possible to assume the recombination of the compensation transactions regardless of their original orders and contents.

(3) The sequential inverse constrained by Saga must be replaced with more parallel compensation processing, if there are few negative influences and few dependencies, which are mentioned by Weimin Du [18]. In particular, there are some

dependencies on data, if the procedures are changed according to the values of data processed before completing the procedures. However, if we were to apply the temporal data management and it could easily provide the previous data status at any time, we could also decrease those dependencies at least in recovering phase. In these cases, the constraints due to Saga might become meaningless.

(4) There are definitely constraints to access the resources under the long live transactions, due to maintaining security for them, autonomy of them and the current dominant architecture consisting of three tiers. However, if some of the resources managed by ACID transactions on the edges would be disclosurable to outside transaction managers, all of the compensation transactions to them must be invoked directly by the aforementioned outside transaction manager in the ACID manner. Otherwise, in impossible cases, the compensation transactions managed under the long live transactions will be invoked.

In order to explain about the procedure of this Enhanced Rollback Migration protocol, we will assume the following simplified model drawn in Fig.4. This model consists of four tiers as a special case and contains the notational symbols mentioned in following individual explanation. We can naturally extend this into the more general model with finite number of tiers.

(1) The first tier is the client application, which is both of the requester and the beneficiary of the whole of a long live transaction instance.

(2) The second tier corresponds to the superior transaction manager which manages the instances of the long live transaction with multiple subordinate transaction managers. Here we will express this superior transaction manager with the symbol of $TM_{(11)}$.

(3) The third tier corresponds to multiple subordinate transaction managers each of which manages the instances of the ACID transaction with multiple subordinate resource managers. Here we will express the subordinate transaction manager assigned at k_{th} with the symbol of $TM_{(2k)}$.

(4) The fourth tier corresponds to multiple subordinate resource managers. Here we will express the subordinate resource manager assigned at m_{th} under the aforementioned subordinate transaction manager $TM_{(2k)}$ with the symbol $RM_{(2km)}$.

The set of tiers from the first tier to the third one composes a scope of a long live transaction, and each subordinate transaction manager at the third tier individually composes a subordinate scope of a long live transaction. Furthermore inside of these subordinate scopes of the long live transactions, the set of the ACID transactions is defined according to all of resource managers. In particular some of the multiple subordinate transaction managers at the third tier which can be executed in parallel under the same transaction contexts compose a block of procedure to be invoked at a time. And this block contains all of the subordinate resource managers under the set of the subordinate transaction managers. Then the set of the blocks of procedure will sequentially be invoked by the superior transaction manager at the second tier. Furthermore in Fig.4 we will regard the prior status of the invoking the subordinate transaction manager $TM_{(21)}$ as the first safe point. Then we will also regard the after status of invoking the subordinate transaction manager $TM_{(2n)}$ as the next safe point.

Now we assume that a failure would happen during processing at the resource manager $RM_{(2k2)}$ as Fig.4 shows. In this case, what the system should do as the first reaction is to identify the failure and the scope of the compensation transaction. In the case of Fig.4, the area surrounded by the bold dotted line until the latest safe point becomes the scope of the compensation transaction.

Fig.5 and Fig.6 demonstrate the heart of the procedures of the Enhanced Rollback Migration protocol. The resource manager $RM_{(2k2)}$, which detects the failure during updating and prepare phases of the two phase commitment, will inform about the failure to the subordinate transaction manager $TM_{(2k)}$ which marshals the ACID transaction and long live transaction managing the aforementioned ACID transaction. If the subordinate transaction manager $TM_{(2k)}$ doesn't find out the necessity to execute the compensation transaction within its own scope, this subordinate transaction manager $TM_{(2k)}$ will also inform about the failure to the superior transaction manager $TM_{(11)}$ as step.1. In the case of the normal Saga, the superior transaction manager $TM_{(11)}$ might invoke the instance of the compensation transaction to the prior subordinate transaction manager $TM_{(2k-1)}$. However in the case of the Enhanced Rollback Migration protocol, the superior transaction manager $TM_{(11)}$ may once inform about the failure to the client application and close the failed transaction as step.2. Then the compensation process will be started under another transaction context. Naturally this invocation of another is not mandatory, so that the superior transaction manager $TM_{(11)}$ can continue to execute the failed instance of the long transaction.

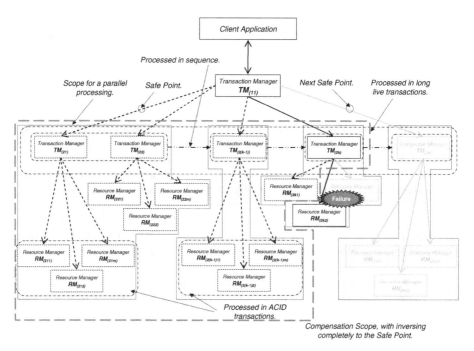

Fig. 4. Instance model containing the compensation transactions executers. This model has four layers for the entire of transaction processing.

Regardless of the manners, the superior transaction manager $TM_{(11)}$ will send queries in parallel to all of the subordinate transaction managers inside the scope of the compensation transaction as step.3 in order to confirm the following two conditions.

(1) Whether the version management is applied in order to record the transition of data.

(2) Whether it is possible to permit the superior transaction manager to access all of subordinate resource managers directly in the ACID transaction manner.

After receiving all of the results of queries, the superior transaction manager $TM_{(11)}$ will invoke a new ACID compensation transaction to revive the previous status before the safe point. This ACID compensation transaction will be executed under another transaction context with a set of resource managers marshaled by the subordinate transaction managers, all of which positively respond to both of the aforementioned two conditions. Fig.6 shows the situation of this invocation. According to this Fig.6, the subordinate transaction manager $TM_{(22)}$ cannot satisfy the first condition and the subordinate transaction manager $TM_{(2k-1)}$ cannot satisfy the second condition. Both of them will be compensated by another long live transaction after the aforementioned ACID compensation transaction. In this case, it will also be done in parallel unless there are constraints and negative influences in regards to the order of invocations.

The pseudo-code described in Appendix. A expresses the algorithm in the case of another invocation in the above explanation. The most characteristic parts are around step#4.1, step#5. These parts correspond to the procedures shown in Fig.5 and Fig.6 and include the process in the two phase commitment. However the failed resource managers will repeatedly try to do reviving the specified status until allowable times. On the other hand successful resource managers will fall out from repeating procedure. Then, the ordinary compensation transaction will be carried out with aforementioned subordinate transaction managers $TM_{(22)}$, and $TM_{(2k-1)}$ around step#8.

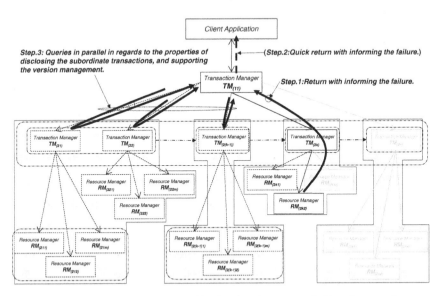

Fig. 5. Phase.2 action of the enhanced rollback migration protocol. Here the superior transaction manager interacts with subordinate transaction managers in order to seek the possibilities of the compressed compensations.

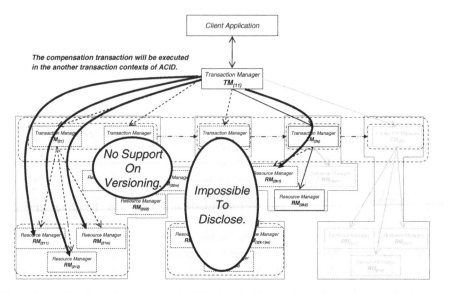

Fig. 6. Phase.3 action of the enhanced rollback migration protocol. Here the superior transaction manager will start a new compensation transaction in ACID, if the resource managers support temporal or version management functions, and the subordinate transaction managers can allow the superior transaction manager to access the subordinate resource managers.

5   Comparisons to the Relative Works

5.1   Outline

In this section, we consider the meaning of our approach and the potential position of the proposed 'Enhanced rollback migration protocol' by comparing to the existing related works. As categories of the existing related works, at least the following items should be mentioned; however the degree of comprehension on recovery will be omitted due to limited space.

(1) Evaluation on the framework.
(2) Optimization.
(3) Degree of comprehension on recovery.

5.2   Evaluation on the Framework

There are not many studies on the framework, which treat the compensation as the central topic with explicit demonstration in regards to the function layers and class organization. In general, the framework in this area might be explained implicitly as a part of the architecture for workflow management and its formalization. The reason of this trend might be related to the strong dependency of the compensations on the features of the application logics. In particular, the evaluation by the formalization is one of the important approaches in this area due to the big issue on how to maintain the consistency. The following are the existing and known studies with the formalization.

(1) H-transaction.
(2) ACTA.
(3) ConTract.
(4) StAC and related approaches.

H-transaction is the description framework proposed by L.Mancini et al, which is applicable to Saga [20]. In this study, the layered architecture model, behaviors of elemental components, and hierarchical primitives named as 'basic primitives' and 'new primitives' are explained. And it also explains its ability to express the various advanced transaction models. In this study the function layers are explicitly defined, however, the main contribution of this study might be to define the high level linguistic framework. Thus the issues on the compensations are not necessarily treated as the main concerns. Therefore, their motivation might be different from ours as one of our issues is to clarify how the function layers for the compensation should be organized.

ACTA is a comprehensive transaction framework to verify effects from the extended transactions and it is demonstrated in applying it to the constructible Saga [12], [21]. This formalizes the constraints from the dependencies in the transaction processing. Furthermore, the aforementioned UMTL also explicitly defines the constraints on the defined classes [16]. Although we consider the function of the compensations with a broader perspective in this paper, there is no sufficient formalization in regards to the constraints within the individual function layer. However, there tends to be neither sufficient decomposed formula into the individual function layer in the existing studies. This might suggest that the bidirectional approach between defining function layers comprehensively and decomposing formulas into the individual function layer should be important and required more.

ConTracts defines the categories of the compensations in formulas which are based on the predefined basic elemental formal models [22], [23]. In particular, the following two concepts are defined in the formal approach; the first is the comprehensive compensation which treats multiple execution steps in its procedure. The second is the partial compensation which can change the execution models. On the other hand, there are certain conceptual differences from ours in regards to decomposing function layers in particular between 'Compensational Transaction Layer' and 'Autonomic Compensation Layer' in Fig.1. As mentioned before, our 'Compensational Transaction Layer' should be defined to maintain the consistency, whereas our 'Autonomic Compensation Layer' corresponds to realizing high reliability. In this sense, we do not decompose both functions in the same view with ConTracts.

As for StAC and the new trend of the formalization, the detailed explanations are mentioned in references [24], [25], [26]. StAC is a business process modeling language the scope of which includes the compensations and this is the latest research in the formalization area to the best of our knowledge. In this research the various styles of Saga have been defined. As there are some related matters with the optimization in StAC, it will briefly be mentioned in the next section.

By using the formalization we can verify the appropriateness of the compensations and definitions of the constraints to maintain the consistency. However the aspect in regards to the composition might be ambiguous because this might be treated indirectly. Therefore it might cause some ambiguities in defining the function sets of the compensations. Thus it might become a crucial factor causing some difficulties in the practical operational designing. In order to solve these difficulties, we need to redefine the framework consisting of the function layers and also need to positively formalize the elemental functions and behaviors within the individual function layer. This is the main argument of our consideration.

5.3 Optimization

The aforementioned 'Enhanced rollback migration protocol' which is aimed for realizing more efficiency is also related to the optimization. Here, we mention the relationships to the existing optimization techniques. The following list shows the existing and known techniques of the transaction optimization.
(1) Presume abort and Presume commit.
(2) One-phase.
(3) Read-Only.
(4) Integrated way with optimizing queries to replica.
(5) Lazy Compensation.
(6) Parallel Sagas.
Presume abort, Presume commit, One-phase and Read-Only are traditional and well-known approaches explained in [27], [28]. Therefore, the explanation about them should be omitted here. The integrated way with optimizing queries to replica is proposed by A.Helal et al. [29]. This proposes a method of allocating a replica for being referred under integration with optimizing queries. These five methods are applicable in both of the traditional distributed environments and long live transaction environments. On the other hand, Lazy Compensation is proposed by Weimin Du because the cost of compensation procedure is expensive [18]. This method includes postponing a compensation to be applied as late as possible and judging them after checking the changed status. These methods are not directly related to our 'Enhanced rollback migration protocol'.

The study which our protocol directly has a relationship to is Parallel Sagas proposed by R.Bruni [24]. This gives the formalized model for decomposing sequential Sagas into a set of concurrent processing. And a related implementation is reported in [30]. Comparing our protocol to this Parallel Sagas, our proposal regards that the complete reverse order in compensations might become meaningless due to following three items; the first is emergence of the data management on a huge scale and increasing popularity of the temporal data management. The second is the easy revitalization of data at any specified time due to the aforementioned temporal data management. And the third is less dependencies in the recovering phase. Furthermore, we assume the new model with compressing compensation procedures under the disclosed resources although we need more to verify this hypothesis. That is an issue which remains for future's work.

6 Summary and Conclusions

In this research study, an attempt has been made to make the notion of the compensation transaction clearer. The paper aims to identify the crucial points in order to prepare for the potential demand caused by 'extreme transaction processing'. In particular the study also explains about the possibility of the 'Enhanced rollback migration protocol'which executes the compensation processing in parallel. Therefore we conclude it has the sufficient potentiality to be more effective for the scalable environment.

Further, there are several items of studies to be done as follows; the first is to prove how 'Enhanced rollback migration protocol' has advantages in the performance through the numerical evaluation. The second is to define the formal semantic model for the 'Enhanced rollback migration protocol' strictly. And the third is to consider the optimized coordination processes.

References

1. Wang, T., Vonk, J., Kratz, B., Grefen, P.: A survey on the history of transaction management: from flat to grid transactions. Distributed and Parallel Databases 23(3) (2008)
2. Pezzini, M., et al.: Etreme Transaction Processing. Technologies to Watch, Gartner Inc.: ID Number: G00146107 (2007)
3. OASIS Standard Web Services Business Activity (WS-BusinessActivity): http://docs.oasis-open.org/ws-tx/wsba/2006/06
4. Yu, J., Buyya, R.: A Taxonomy of Workflow Management Systems for Grid Computing. Journal of Grid Computing 3(3-4) (2005)
5. Greenfield, P., et al.: Compensation is Not Enough. In: Proceedings of the 7th IEEE Intl. Enterprise Distributed Object Computing Conference (2003)
6. Chessell, M., Butler, M., Ferreira, C., et al.: Extending the concept of transaction compensation. IBM Systems Journal 41(4) (2002)
7. Limthanmaphon, B., Zhang, Y.: Web Service Composition Transaction Management. In: Proceedings of 15th Australasian Database Conference ADC 2004, vol. 27 (2004)
8. Hrastnik, P., Winiwarter, W.: Using Advanced Transaction Meta-Models for Creating Transaction-Aware Web Service Environments. International Journal of Web Information Systems 1(2) (2005)
9. Grefen, P., Vonk, J.: A Taxonomy of Transactional Workflow Support. International Journal of Cooperative Information Systems 15(1) (March 2006)
10. ISO/IEC 7498-1:1994: Open Systems Interconnection – Basic Reference Model. The Basic Model (1994)
11. Garcia-Molina, H., Gawlick, D., Klein, J., Kleissner, K., Salem, K.: Modeling long-running activities as nested sagas. IEEE Data Engineering archive 14(1) (1991)
12. Chrysanthis, P.K., Ramamritham, K.: ACTA: A Framework for Specifying and Reasoning about Transaction Structure and Behavior. In: Proceedings of the 1990 ACM SIGMOD international conference on Management of data (1990)
13. Grefen, P., Vonk, J., Apers, P.: Global transaction support for workflow management systems: from formal specification to practical implementation. The VLDB Journal 10, 316–333 (2001)
14. Gray, J., Reuter, A.: Transaction Processing: Concepts and Techniques, p. 184. Morgan Kaufmann Publishers, San Francisco (1993)
15. Gioldasis, N., Christodoulakis, S.: UTML: Unified Transaction Modeling Language, http://www.cse.ust.hk/vldb2002/VLDB2002-proceedings/papers/S34P02.pdf
16. Gioldasis, N., Christodoulakis, S.: UTML: Unified Transaction Modeling Language. In: Proceedings of the Third International Conference on Web Information System Engineering, WISE 2002 (2002)
17. OASIS Standard Web Services Business Process Execution Language Version 2.0, http://docs.oasis-open.org/wsbpel/2.0/OS/wsbpel-v2.0-OS.html

18. Du, W., et al.: Flexible Compensation of Workflow Processes (1997), http://www.hpl.hp.com/techreports/96/HPL-96-72r1.html
19. Vonk, J., Grefen, P.: Cross-Organizational Transaction Support for E-Services in Virtual Enterprises. Distributed and Parallel Databases (2003)
20. Mancini, L., Ray, I., Jajodia, S., Bertino, E.: Flexibile Commit Protocols For Advanced Transaction Processing. Advanced Transaction Models and Architectures, 91–124 (1997)
21. Chrysanthis, P.K., Ramamritham, K.: ACTA: The SAGA Continues. Database Transaction Models For Advanced Applications (1992)
22. Wachter, H., Reuter, A.: The ConTract model. In: Elmagarmid, A. (ed.) Database Transaction Model for Advanced Applications. Morgan Kaufmann Publishers, San Francisco (1992)
23. Reuter, A., Schneider, K., Schwenkreis, F.: ConTracts Revisited. Advanced Transaction Models and Architectures, 127–151 (1997)
24. Bruni, R., Melgratti, H., Montanari, U.: Theoretical foundations for compensations in flow composition languages. In: Proceedings of POPL 2005, 32nd ACM SIGPLAN-SIGACT Symposium on Principles of Programming Languages, ACM Press, New York (2005)
25. Butler, M., Ferreira, C.: An operational semantics for StAC, a language for modelling long-running business transactions. In: De Nicola, R., Ferrari, G.-L., Meredith, G. (eds.) COORDINATION 2004. LNCS, vol. 2949. Springer, Heidelberg (2004)
26. Bruni, R., Butler, M., Ferreira, C., Hoare, T., Melgratti, H., Montanari, U.: Comparing two approaches to compensable flow composition. In: Abadi, M., de Alfaro, L. (eds.) CONCUR 2005. LNCS, vol. 3653, pp. 383–397. Springer, Heidelberg (2005)
27. Mohan, C., Lindsay, B.: Efficient commit protocols for the tree of processes model of distributed transactions. ACM SIGOPS Operating Systems Review archive 19(2) (April 1985)
28. JBoss Transaction Service 4.2.3.: Web Service Transactions Programmers Guide. JBoss (2006)
29. Helal, A., Kim, Y.S., Nodine, M.H., Elmagarmid, A.K., Heddaya, A.A.: Transaction Optimi-zation Techniques. Advanced Transaction Models and Architectures, 238–255 (1997)
30. Bruni, R., Ferrari, G., Melgratti, H., Montanari, U., Strollo, D., Tuosto, E.: From Theory to Practice in Transactional Composition of Web Service. In: Bravetti, M., Kloul, L., Zavattaro, G. (eds.) EPEW/WS-EM 2005. LNCS, vol. 3670, pp. 272–286. Springer, Heidelberg (2005)

Appendix A

```
compensationProcess(abortedTransactionContext){
  transactionContext = getNewTransactionContext(); //#1
  invokeSuperiorTransactionManager(transactionContext);   //#2
  List<subordinateTransactionManager> =
    identifyListOfSubordinateTransactionManager(
    abortedTransactionContext);   //#3
  FOR All of List<subordinateTransactionManager>{
    //#4 in parallel
    List<disclosurableResourceManager> =
      listenCompensationProprty(
        subordinateTransactionManager); //#4.1
    IF List<disclosurableResourceManager> != EMPTY){
      List<qCDisclosurableResourceManager> =
        addListForQuickCompensation(
          List<disclosurableResourceManager>); //#4.2.1
```

```
    }ELSE{
      List<normalSubordinateTransactionManager> =
        addListForNormalCompensation(
          subordinateTransactionManager); //#4.2.2
    }
  }
  setLoopLimit(ITERATION);
  setCount(counter=0);
  WHILE(qCDisclosurableResourceManager != EMPTY &&
    counter < ITERATION){ //#5
    FOR All of List<qCDisclosurableResourceManager>{
      //#5.1 invoked in parallel
       doCoordination(qCDisclosurableResourceManager);  //#5.1.1
      result = requestRevive(
        qCDisclosurableResourceManager,
        abortedTransactionContext);  //#5.1.2
      IF(result == 'Success'){
        List<2PCResourceManager> = addEnlist(
          qCDisclosurableResourceManager); #5.1.3.1
      }
    }
    FOR All of List<2PCResourceManager>{
      //#5.2 invoked in parallel
       result=do2PCPreperationPhase(2PCResourceManager);//#5.2.1
      IF(result != 'Success'){
        List<2PCResourceManager> = removeFromList(
          2PCResourceManager);//#5.2.2.1
      }
    }
    FOR All of List<2PCResourceManager> {
      do2PCCommitmentPhase(2PCResourceManager); //#5.2.3
      List<qCDisclosurableResourceManager>=
        removeFromList(match(2PCResourceManager,
          List<qCDisclosurableResourceManager>)); //#5.2.4
    }
     counter++;
  }
  IF(counter==ITERATION){
    ThrowError( ); //#7.1
  }
  FOR All of List<normalSubordinateTransactionManager>{
    //#8 invoked in parallel
    doCoordination(normalSubordinateTransactionManager,
      transactionContext); //#8.1
    result = doCompensation(
      normalSubordinateTransactionManager,
       transactionContext,abortedTransactionContext); //#8.2
    IF(result!='Success'){
      Throw Error (  );
    }
  }
}//#End
```

```
doCompensation(normalSubordinateTransactionManager,
  transactionContext,abortedTransactionContext){
  List<resourceManager> = invokeResourceManager(
    transactionContext);//#10
  FOR All of List <resourceManager> {
    doCoordination(resourceManager,transactionContext); //#11.1
  }
  setLoopLimit(ITERATION);
  setCount(counter=0);
  setStatusOfCompensation(
    statusOfCompensation='incompleted');
  WHILE(statusOfCompensation != 'completed'&&
    counter < ITERATION){ //#15
    FOR All of List<resourceManager> {
      //#15.1 invoked in parallel
      result = doCompensation (resourceManager,
        transactionContext,abortedTransactionContext); //#15.1.2
      IF(result == 'Success'){
        List<2PCResourceManager> = addEnlist(
          resourceManager);
      }ELSE{
        doAbort(List<resourceManager>,
          transactionContext); //#15.1.3
        List<2PCResourceManager> = removeFromList(ALL); //#15.1.4
        Break;
      }
    }
    counter++;
    IF(List<2PCResourceManager> != EMPTY){
      FOR All of List<2PCResourceManager>{
        //#15.2 invoked in parallel
        result = do2PCPreperationPhase(
          2PCResourceManager); //#15.2.1
        IF(result != 'Success'){
          doAbort(List<2PCresourceManager>,
            transactionContext);   //#15.2.2
          List<2PCResourceManager>=removeFromList(All); //#15.2.3
          Break;
        }
      }
      IF(List <2PCResourceManager> != EMPTY){
        FOR All of List<2PCResourceManager>{
          //#15.3 invoked in parallel
          do2PCCommitmentPhase(2PCResourceManager);
            //#15.3.1
        }
        setStatusOfCompensation(
          statusOfCompensation='completed'); //#15.4
      }
    }

  IF(counter == ITERATION){return('Failure'); //#16.1 }
  ELSE{return ('Success');//#16.2 }
}//#End.
```

Optimization of Query Processing with Cache Conscious Buffering Operator

Yoshishige Tsuji[1], Hideyuki Kawashima[2,3], and Ikuo Takeuchi[1]

[1] Graduate School of Information Science and Technology, The University of Tokyo
[2] Graduate School of Systems and Information Engineering, University of Tsukuba
[3] Center for Computational Sciences, University of Tsukuba

Abstract. The difference between CPU access costs and memory access costs incurs performance degradations on RDBMS. One of the reason why instruction cache misses occur is the size of footprint on RDBMS operations does not fit into a L1 instruction cache. To solve this problem Zhou proposed the buffering operator which changes the order of operation executions.

Although the buffering operator is effective, it cannot be applied for RDBMS in the real business. It is because Zhou does not show an algorithm for optimizer to select the buffering operator.

Thus we realized the CC-Optimizer which includes an algorithm to appropriately select the buffering operator. Our contributions are the design of new algorithm on an optimizer and its implementation to RDBMS.

For experimental RDBMS, we used PostgreSQL as Zhou did, and our machine environment was Linux Kernel 2.6.15, CPU Intel Pentium 4(2.40GHz). The result of preliminary experiments showed that CC-Optimizer was effective. The performance improvement measured by using OSDL DBT-3, was 73.7% in the greatest result, and 17.5% in all queries.

1 Introduction

The difference between the CPU and memory degrades relational database management system (RDBMS) performance. The traditional approach to this problem has been to reduce L2/L3 data cache misses [1,2,3]. In another approach, Zhou proposed to reduce the number of L1 instruction cache misses [4]. Zhou proposed a buffering operator, which buffers usual operators such as index scan, sequential scan, aggregation, and nested loop join.

The buffering operator reduces the number of L1 instruction cache misses under certain conditions, but it increases misses under other conditions. Beyond that, the positive and negative conditions were not clarified in the original work [4]. Also, [4] does not go far enough in the design of an optimizer with buffering operator. In designing such an optimizer, the positive and negative conditions should be clarified.

This paper investigates the effects of the buffering operator by re-implementing it. We adopted PostgreSQL-7.3.16, Linux Kernel 2.6.15, CPU Intel Pentium 4 (2.4 GHz). We propose a Cache Conscious query optimizer, the CC-optimizer with algorithms that judge the usage of buffering operators.

S. Kikuchi, S. Sachdeva, and S. Bhalla (Eds.): DNIS 2010, LNCS 5999, pp. 65–79, 2010.

The rest of the paper is organized as follows: Section 2 describes related work that reduces the number of CPU cache misses on RDBMS. We introduce the buffering operator paper [4] in detail. We then describe the flaws in [4] and formulate the problem. Section 3 describes re-implementation of the buffering operator and experimental investigations of its behavior. Section 4 describes implementation of the CC-optimizer. Section 5 describes experimental results of the CC-optimizer measured by a standard benchmark tool, OSDL DBT-3[5]. Section 6 concludes this paper.

2   Related Work and Problem Formulation

2.1   CPU Archichitecture and L2 Data Cache

CPU has a hierarchical memory architecture. CPU memory accesses are accelerated by cache memories. Cache memories are incorporated into the CPU die internally, taking into account both spatial locality and time locality. Figure 1 shows the architecture of the Pentium 4 cache memory[1].

The CPU first accesses an L1 cache to look for data or instructions. If it fails, then CPU accesses an L2 cache. If that fails, the CPU then accesses an L3 cache if it exists, or memory if an L3 cache does not exist. The access cost for memory is more than

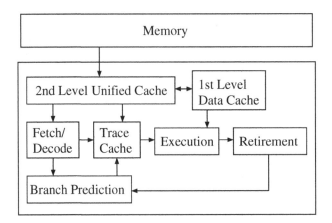

Fig. 1. Pentium 4 Cache Memory Architecture

SELECT COUNT(*) FROM item;

Fig. 2. A Query with Simple Aggregation

---

[1] Reference: Zhou, J. and Ross, K.A., Buffering Database Operations for Enhanced Instruction Cache Performance, Figure 2, Proc. SIGMOD'04.

10 times that for the L2/L3 cache [4]. That is why considerable work has focused on reducing L2/L3 cache misses [1][6][7][3].

2.2   L1 Instruction Cache: Buffering Operator

As shown above, some work focused on reducing the number of L2/L3 cache misses. On the other hand, a few work focused on reducing the number of L1 cache misses [2,4].

To the best of our knowledge, the [4] only focused on reducing L1 cache misses for general relational query processing. Although the penalty for an L1 cache miss is smaller than for an L2 cache miss, L1 cache misses outnumber L2 cache misses. It is therefore important to reduce L1 cache misses from the standpoint of DBMS performance.

[4] proposes a method that reduces L1 instruction cache misses on a real DBMS, PostgreSQL. The method is referred to as "buffering operator." Although the advantage of the buffering operator is clear, it should not be directly applied to real DBMSs. This is because the buffering operator improves performance in some situations, but it degrades performance in other situations. The following section describes the buffering operator in detail.

The total size of operators required for a query processing sometimes exceed capacity of an L1 instruction cache. When that happens, an L1 instruction cache miss occurs, degrading performance of the RDBMS. Let us look at the mechanism of an L1 instruction cache miss in detail.

First, an operator is a processing unit with a specific meaning in an RDBMS. PostgreSQL implements many operators including sort, join, aggregation, etc. We take into account the behavior of operators when processing the simple query shown in Figure 2. To process the query, a scan operator that reads the item relation and an aggregation operator that aggregates the result of the scan are necessary. On PostgreSQL, these operators are implemented as the TableScan operator and Agg operator. They construct a plan tree as shown in Figure 3.

On the plan tree, Agg is denoted as a parent operator (P) and TableScan is denoted as a child operator (C). For each aggregate processing on P, P fetches a tuple from C followed by the pipeline execution manner. This mechanism contains the sporadic L1

Fig. 3. Simple Plan Tree   **Fig. 4.** Buffering Operator for a Simple Plan Tree

cache miss occurrences. Assuming that cardinality of the item relation in Figure 3 is 8, operators are executed as follows:

$$PCPCPCPCPCPCPCPC$$

If the total footprint of P and C is larger than the L1 instruction cache, an execution of both operators causes an L1 instruction cache miss. This occurs because only one of P or C can fit into the L1 instruction cache; the other will be excluded from the cache.

To solve this problem, [4] proposed a novel technique, the buffering operator. The buffering operator buffers a pointer to data that is generated by child operator. Buffering operator can be implemented with a small footprint. With the buffering operator, usual operators can be executed continually, thereby reducing the number of L1 instruction cache misses. Let us look at an example. By applying buffering operator to the simple plan tree in Figure 3, a buffering operator is inserted into a link between two nodes, and a new plan tree is constructed as shown in Figure 4. If the buffering size of the buffering operator is 8, the execution order of parent/child operators can be changed as follows using the buffering operator.

$$PCCCCCCCCPPPPPPPP$$

The same operators are executed continually (C...C or P...P), thereby reducing the number of L1 cache misses. However, please note that the reduction is achieved only when the total footprint of P and C is larger than L1 instruction size.

[4] evaluated the buffering operator, reporting that the buffering operator reduced the number of L1 instruction cache misses by about 80%, and the buffering operator improved query performance by about 15%.

2.3   Problem Formulation

As Section 2.2 described, the buffering operator reduces the number of L1 instruction cache misses if total size of the parent/child operators is larger than the L1 cache. Otherwise, the buffering operator increases the number of L1 instruction cache misses because of the overhead produced by the buffering operator itself [4]. Therefore, the technique should be applied only when it is effective. In the final stage of query processing in a usual RDBMS, after several plan trees are generated by a planner module, a query optimizer chooses the best plan from among them. For the chosen plan tree, with the buffering operator, the optimizer should find links where the buffering operator should be inserted to improve performance further.

To realize such a query optimizer, the following two problems should be addressed.

1. Deep analysis of buffering operator behavior.
   The behavior of the buffering operator is not analyzed in depth by [4]. The paper does not present enough data for a query optimizer to find links in a plan tree to insert the buffering operator. To obtain the data, detailed investigation should include cases in which the buffering operator is appropriate and when it is inappropriate; and the boundaries should be clarified with several parameters.
2. Realization of a query optimizer with buffering operator.

After finishing the in-depth analysis of buffering operator behavior, an algorithm to find links, where the buffering operator is inserted, should be constructed based on the analysis.

We describe approaches to these problems in the remainder of this paper.

3   Analysis of Buffering Operator

3.1   Reimplementation of Buffering Operator

To analyze the behavior of the buffering operator, we re-implement it reading [4]. For implementation, we used PostgreSQL-7.3.16 since [4] used PostgreSQL-7.3.4; their basic architectures are the same.

We implemented the buffering operator as a new operator. Therefore, conventional operators are not modified with this implementation. In our implementation, the buffering operator is invoked by the executor module. The executor module indicates the position of the buffering operator in a plan tree. An example of the indication is shown in Figure 4. To achieve it, we modified the executor module so that it can deal with the buffering operator similarly to conventional operators.

We adopted the open-next-close interface used in PostgreSQL. The open function and close function allocate and release pointers to tuples generated by child operators, and the arrays that hold them. However, our implementation of arrays differs from [4]. The difference is that [4] used only fixed length arrays, while we allow variable length arrays. The length of an array is specified by our optimizer module. The next function performs as follows: (1) If the array is empty, then it executes child operators until the array is fulfilled or the final tuple is given; it then stores pointers to tuples, generated by a child operator, into the array. (2) If the array is not empty, then it provides pointers in the array to a parent operator. When a buffering operator is inserted, (1) if the array is empty, then a child operator is continually executed and thus the L1 instruction cache miss does not occur, and (2) if the array is not empty, then a parent operator is continually executed; thus the L1 instruction cache miss does not occur either.

As we described in Section 2.2, the buffering operator should be small enough to fit into the L1 instruction cache with another operator. The footprint of the buffering operator we implemented was about 700 Bytes while large types of PostgreSQL operators are more than 10 KBytes. In measuring the footprint, we used Intel VTune.

3.2   Experimental Analysis Outline

Hardware. To analyze the experiment, we used hardware with the specs given in Table 1. The hardware for [4] and our hard are the same, which is Pentium 4. Please note that "trace cache" is the same as the usual L1 instruction cache.

Relations. In our experiment, we used two simple relations, points and names. The points relation has two integer attributes, id and point. The names relation has two integer attributes, id and name.

Table 1. Hardware Environment

CPU	Pentium(R) 4 2.40GHz
OS	Fedora Core 5 (Linux 2.6.15)
RAM	1GB
Trace Cache	12K μops (8KB–16KB)
L1 Data Cache	8KB
L2 Cache	512KB
Compiler	GNU gcc 4.1.0

Investigation Parameters. Using the buffering operator, we observed its behavior through experiments. This subsubsection describes the result of experiments. On the experiments, we thought the following three parameters should be investigated.

1. Total footprint of parent and child operators
   When total foot print size of the parent and child operators is smaller than the L1 instruction cache size, then the cache miss does not occur. Otherwise it occurs. Therefore, the footprint of each operator should be investigated to know the pairs that cause cache misses.
2. Number of buffering operators in the L2 cache
   The L2 cache stores both buffering operators and child operators buffered by the buffering operators. When there are too many buffering operators, L2 cache capacity is exceeded and an L2 cache miss results. Therefore, the relationship between the number of buffering operators and L2 cache misses should be investigated.
3. Number of buffering operator executions
   Execution of the buffering operator requires excess cost. The cost, however, is small and its execution may reduce L1 cache misses. Therefore, the buffering operator improves the performance of query processing if the number of execution times of buffering operators crosses a specified threshold. Execution times are directly related to the cardinality of a relation. Therefore, the relationship between improving query processing performance and the cardinality of a relation should be investigated.

3.3   Result with (1) Total Footprint Size of Parent and Child Operators

Footprint size of each operator is investigated by [4], and this paper follows it. It should be noted that simply adding the sizes of two operators is not enough to estimate the total size of two operators. This is true because (1) operator sizes after compiling differ depending on a machine environment, and (2) even if the architectures are the same, different modules may share the same functions. Therefore, we executed a variety of queries and investigated conditions under which the buffering operator is appropriate.

Large Size Opetators. As described above, when total size of the parent operator and child operator exceeds L1 cache capacity, an L1 cache miss occurs. To confirm, we evaluated the performance of queries with large operators. As representatives, we selected the aggregation operator and sort operator. The following descrbes the result.

Table 2. Queries varying Aggregation Complexity

ID	SQL
A	SELECT COUNT(*) FROM points WHERE id < 5000000 AND point ≥ 0
B	SELECT COUNT(*), AVG(point) FROM points WHERE id < 5000000 AND point ≥ 0
C	SELECT COUNT(*), AVG(point), SUM(id) FROM points WHERE id < 5000000 AND point ≥ 0
D	SELECT COUNT(*), AVG(point), SUM(id), SUM(id/2) FROM points WHERE id < 5000000 AND point ≥ 0
E	SELECT COUNT(*), AVG(point), AVG(id), SUM(id/2) FROM points WHERE id < 5000000 AND point ≥ 0
F	SELECT COUNT(*), AVG(point), max(point/2), SUM(id/2) FROM points WHERE id < 5000000 AND point ≥ 0
G	SELECT COUNT(*), AVG(point), AVG(id), SUM(id/2), max(point/2) FROM points WHERE id < 5000000 AND point ≥ 0
H	SELECT COUNT(*), AVG(point), AVG(id), AVG(id/2), AVG(point/2), AVG(id/3) FROM points WHERE id < 5000000 AND point ≥ 0

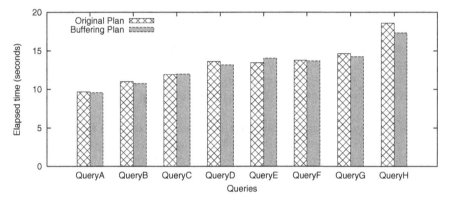

Fig. 5. Result of Aggregate Queries

Aggregation Operator. The aggregation operator comprises a base operator and each functional operator [4]. The size of the aggregation operator differs depending on query complexity. Varying the complexity of aggregation operators, we measured query execution time. The queries are shown in Figure 2. The result of experiments is shown in Figure 5. The following summarizes the result of experiments. Please note that all the queries included buffering operators. (1) When more than three kinds of aggregations, including AVG and SUM, were used, performance improved. (2) When two kinds of aggregations, including AVG and SUM, were used, performance sometimes improved and sometimes degraded. (3) When only AVG was used and the number of AVGs was more than 5, performance improved. (4) When only SUM was used and the number of SUMs was more than 5, performance improved.

From the above result, we judged that the buffering operator should be applied when more than three kinds of aggregations, including AVG and SUM, were used.

Sort Operator. Next, we investigated effectiveness of the buffering operator with the sort operator through experiments. For the experiments, we used queries shown in Figure 6.

SELECT * FROM points WHERE id < 5000000 AND point ≥ 0 ORDER BY point DESC

Fig. 6. Query with Sort

Table 3. Execution Time with Sort

Original Plan	Buffered Plan	Improvement by Buffering
91.63 sec	90.93 sec	0.76%

Table 3 shows the results of experiments. When a sort operator was parent, the buffered plan slightly (0.76%)improved performance. On the other hand, when a sort operator was child, the buffered plan degrades performance. This occurs because the sort operator simply generates sorted tuples; therefore, its footprint is small.

In summary, the buffering operator performed effectively when the total size of operators was too large to fit into the L1 instruction cache.

Small Size Operators. Third and last, we measured the execution times of queries when buffering operators are applied to small-sized operators. The query is shown in Figure 7. The buffering operator was applied only to sub queries.

The results of experiments are given in Table 4. The result shows that the buffering operator slightly degrades performance (1.25%).

3.4   Result with (2) the Number of Buffering Operators in L2 Cache

As the number of buffering operators increases, the number of L1 cache misses decreases. However, if the buffering operators do not fit in the L2 cache, L2 cache misses occur. We investigated appropriate sizes for buffering operators through the following two experiments.

Single Buffering Operator into a Plan Tree. In the first experiment, we inserted only one buffering operator into a plan tree. For the query, we included four aggregations so that operators exceed the L1 cache size. Results of the experiment are shown in Figure 8. The results show that execution times improve when buffering size is from 100 to

SELECT * FROM names n, (SELECT COUNT(*) FROM points GROUP BY point) AS t
WHERE t.count = n.id

Fig. 7. Small Size Operator

Table 4. Execution Time with Small Size Operator

Original Plan	Buffered Plan	Improvement by Buffering
184.6 sec	186.9 sec	-1.25%

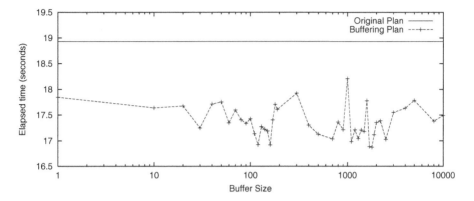

Fig. 8. Execution Time with Single Buffering Operator

2500. In particular, when buffering size is 1800, the execution time improved 10.86% comparing with the original plan.

Further, we measured the number of L1 cache misses and L2 cache misses. The results are shown in Figure 9. With the L1 cache, as buffering size increases, the number of cache misses decreases. With the L2 cache, when buffering size is from 10 to 1000, the number of L2 cache misses is relatively small, and when buffering size is more than 1000, the number of cache misses rapidly increases.

From the above results, with this query, appropriate buffering sizes are from 100 to 1800.

Multiple Buffering Operators into a Plan Tree. Second, we conducted an experiment with multiple buffering operators. For the experiment, we used a query with two join operators. The result of the experiment is shown in Figure 10. When buffering size is 700, performance improvemes 19.7%. Similar to aggregation operators, when the buffering size of a buffering operator was more than 100, large improvements were observed.

3.5 Result with (3) Number of Buffering Operator Executions

Even if the buffering operator is effective, the invocation of buffering operator requires excess cost. Thus, for a query processing to improve performance, the reduction in L1 cache misses should at least complement the excess cost. We measured the relationship

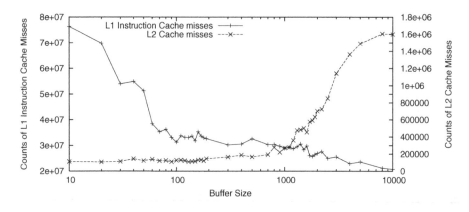

Fig. 9. L1 Cache Misses with Single Buffering Operator

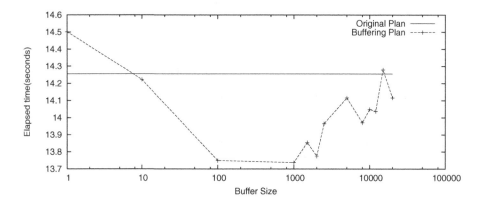

Fig. 10. Execution Time with Multiple Buffering Operators

between the complement and the cardinality of a relation. For the experiment, we used the queries in Figure 2.

The result of experiments are shown in Figure 11 and 12. The results show that when cardinality is about 600, the original plan lost stability and degraded performance.

We also measured the relationship between cardinality and the number of cache misses. The result is shown in Figure 12. For the L1 cache, when cardinality is more than 300, the buffered plan showed better performance than the original plan. For the L2 cache, when cardinality is more than 500, the buffered plan showed better performance than the original plan. Therefore, when cardinality is more than 500, the buffered plan is better than the original plan with both the L1 cache and the L2 cache.

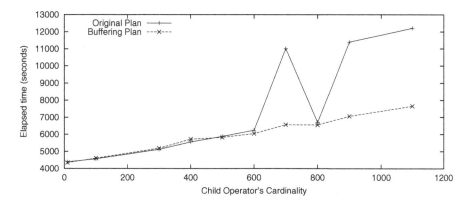

Fig. 11. Execution Time with Cardinality

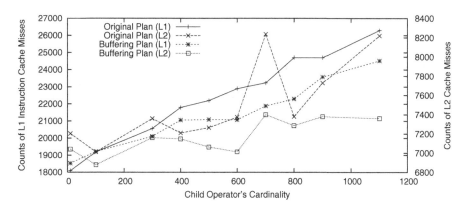

Fig. 12. The Number of Cache Misses with Cardinality

4   Proposal of a Cache Conscious Query Optimizer

This section proposes a cache conscious query optimizer, the CC-optimizer. This section describes algorithms to judge buffering operator insertion in the CC-optimizer, and implementation of the CC-optimizer.

4.1   Algorithm to Insert Buffering Operator

From the experimental results described in Section 3, we summarize conditions under which the buffering operator improves performance in Figure 13. Only when the two conditions in Figure 13 are satisfied, buffering operator is invoked. Since the conditions are simple, execution cost for them is almost for free, which does not incur performance degradation with query optimization processing.

1. At least one of the following should be satisfied.
   (a) Parent node includes more than three kinds of aggregations, and child node is a main module except for sort.
   (b) Parent node is a join operator and child nodes are main modules except for sort.
   (c) Parent node is a sort operator and child node is a main module.
2. The number of tuples to be processed is more than 500.

Fig. 13. Conditions to Invoke Buffering Operators

4.2 Distribution of Buffer into Multiple Links

When multiple links in a plan tree satisfy conditions in Figure 13, an optimizer should coordinate the number of buffering operators for each link. Two things should be considered. The detailed procedure is described in Section 4.3.

– The L2 cache size limit of buffering operators in all links
  The total size of buffering operators in the L2 cache is computed as: (*the size of a tuple obtained from child operator*) × (*the buffer size of each buffering operator*). This size should be smaller than the L2 cache size. Please note that the L2 cache is not dedicated for buffering operators, so the usable area is smaller than the total L2 cache size. In our experiments, we set the usable area as 14 KBytes of 512 KBytes.
– Minimum number of buffering operators in a link
  From Figure 11 and 12, the number of tuples to be processed should be more than 100 to improve query processing performance when the buffering operator is applied.

4.3 Implementation of Optimizer

In PostgreSQL, plan trees are implemented using list structures that correspond to operators. The judgment of buffering operator insertion is conducted at the final stage of optimizer execution. For the judgment, the conventional optimizer module generates a plan tree with the best cost, and the root node of the plan tree is provided for the judgment. We implemented CC-optimizer through the following three steps.

1. Searching places to which buffering operators can be inserted
   In the first step, the CC-optimizer searches places in which The buffering operators can be inserted. A PostgreSQL server generates primitive logical plan trees when a query arrives. The conventional PostgreSQL optimizer recursively reads the plan trees, assigns minimum cost operators, and generates the execution plan. For example, when join conditions are included in a query, PostgreSQL chooses the fastest operator from nested loop join, hash join, and merge join.
   Based on the algorithm, the CC-optimizer judges whether a buffering operator can be inserted when aggregation, join, and sort operators are selected following the conditions in Figure 13.

2. Calculating size for each buffering operator insertion point
   In the second step after finishing the first step, buffering size for each buffering operator insertion point is calculated following Figure 13.
   The calculation is as follows: First, for each insertion candidate point, a buffering size of 100 is given because the buffering operator is effective as shown in Figure 9 and 10. Second, if L2 cache capacity remains, then the remaining capacity is equally divided and assigned to each insertion candidate point.
3. Insertion of buffering operator
   Finally, buffering operators are inserted into candidate points.

5 Evaluation

5.1 Measurement of Cache Misses with Performance Counter

Most recent CPUs, including Pentium 4, hold registers dedicated to collect specific events for performance analysis. We obtained the events to evaluate performance of the CC-optimizer. The detailed hardware environment for experiments is shown in Table 1.

To obtain the events we used Perfctr, a monitor driver. To use Perfctr, we reconstructed the Linux-2.6 kernel after applying Perfctr patches. We then used PAPI (Performance Application Programming Interface) to conduct the measurement easily. PAPI provides an integrated interface to view events collected by Perfctr. PAPI provides the C language interface. So we implemented the measurement procedures in PostgreSQL internal. The procedures are start point and end point. The start point is where the PostgreSQL server receives a query. The end point is where the socket connection is closed.

We collected four events. They are (1) Trace cache misses, (2) L2 total cache misses, (3) Conditional branch instructions miss-predicted, and (4) Total cycles. We used Pentium 4 in our experiment. The Pentium 4 cpu implements a trace cache instead of an L1 instruction cache. Therefore, we treated trace cache misses as L1 instruction cache misses.

5.2 Benchmark

For benchmarking, we used OSDL DBT-3 [5]. DBT-3 is a database benchmark software based on the TPC-H benchmark. DBT-3 provides 22 complex queries to measure the performance of decision support systems.

In our experiment, we set two parameters of DBT-3 as follows: The first parameter was the scale factor to express the scale of data for the experiment. When the scale factor is 1, then 1 GByte text is generated, and more than a 4 GByte database cluster is generated. We set the scale factor as 1. The second parameter was the number of streams to express the number of concurrently executed transactions on the throughput measurement experiment. We set the parameter as 1 so that context switches will not likely occur. The remaining parameters were default values.

Table 5. Experimental Result (Cardinality=500)

Query	Total Execution Time Improvement (%)	Process Execution Time Improvement (%)	Trace Cache Miss Improvement (%)	L2 Cache Misses Improvement (%)
Q4	73.65	8.036	7.607	25.52
Q18	39.26	0.03162	6.942	76.26
Q20	34.74	5.274	1.771	-2.464
Q10	20.88	2.096	33.39	-14.74
Q7	17.47	2.326	17.39	16.79
Q6	11.36	2.547	36.87	-2.266
Q13	6.356	-3.848	-83.95	-7.666
Q16	5.255	4.637	40.96	-7.072
Q2	5.066	5.094	23.27	18.54
Q14	2.528	4.147	37.87	8.395
Q11	0.7313	-0.7235	36.00	-35.76
Q12	0.1540	-0.9972	28.05	9.089
Q22	-0.05150	-0.04429	-27.34	-3.397
Q19	-0.6108	4.600	2.716	28.22
Q1	-0.7540	1.457	-13.68	-16.69
Q9	-1.042	3.326	4.562	4.664
Q21	-1.084	2.330	-2.256	10.35
Q17	-1.482	-1.542	12.31	6.670
Q3	-5.073	2.434	15.47	-1.632
Q15	-13.22	3.220	38.07	25.58
Q5	-13.43	4.871	18.03	17.22
Q8	-24.57	2.577	19.13	8.761

5.3   Result

When L2 cache capacity is set to 14 KB and the cardinality threshold is 500, we obtained the result of the DBT-3 benchmark as given in Table 5. In the experiment, the CC-optimizer applied the buffering operator to 16 of 22 queries.

Table 5 shows that for 9 of 16 queries, execution times were improved. Q4 improved most, and the ratio was 73.6%. Further, most queries that achieved large execution time improvement also improved trace cache misses (L1 cache misses).

On the other hand, 7 of 16 queries degraded performance. The maximum degradation ratio was 24.5% by Q8.

The buffering operator was not applied to Q2, Q6, Q9, Q17, Q19, and Q20. For these cases, excess cost of the CC-optimizer potentially degrades performance. However, the degradation was observed only in Q17, and it was just 1.5%. Therefore, the implementation cost of CC-optimizer algorithms is considered small.

In all, a 17.5% execution time improvement was achieved by the CC-optimizer.

It should be noted that interesting results are also observed in Table 5. Negative correlation between cache misses and execution time is observed in Queries 13, 19, 9,

17, 15, 5, and 8. It may be caused by disk accesses, but the detailed analysis is left in the future work.

6 Conclusions and Future Work

This paper proposes algorithms that apply buffering operators to query processing. To construct the algorithm, we re-implemented the buffering operator proposed in [4], and conducted several experiments and analyzed behavior of the buffering operator. We added a new module to a conventional query optimizer module so that buffering operators can appropriately be inserted at the final stage of query optimization. We proposed the CC-optimizer, a modified cache conscious query optimizer. The CC-optimizer was implemented on PostgreSQL-7.3.16 and evaluated using the OSDL DBT-3 benchmark suite. The result of experiments showed that the CC-optimizer improved execution time of a query 73.6% in the maximum case; it improved total execution time 17.5%. We therefore conclude that this work provides algorithms that appropriately apply the buffering operator to the query optimizer module.

The algorithms we presented can be applied only for a specific machine. This is because parameters in algorithms must be investigated with a lot of human intervention. In future work, we will investigate the behavior of PostgreSQL more in detail to analyze the reason of negative correlation between cache misses and execution time, and sophisticate the proposed algorithms.

Acknowledgements

This work is partially supported by the Exploratory Software Project Youth Program of the Information-Technology Promotion Agency, Japan (IPA), and by the Grant-in-Aid for Scientific Research from JSPS (#20700078).

References

1. Boncz, P., Manegold, S., Kersten, M.L.: Database architecture optimized for the new bottleneck: Memory access. In: VLDB (1999)
2. Harizopoulos, S., Ailamaki, A.: Improving instruction cache performance in oltp. ACM Trans. Database Syst. 31(3), 887–920 (2006)
3. Cieslewicz, J., Mee, W., Ross, K.A.: Cache-conscious buffering for database operators with state. In: DaMoN, pp. 43–51 (2009)
4. Zhou, J., Ross, K.A.: Buffering database operations for enhanced instruction cache performance. In: SIGMOD, pp. 191–202 (2004)
5. OSDL DBT-3, http://osdldbt.sourceforge.net
6. Ailamaki, A., DeWitt, D.J., Hill, M.D., Skounakis, M.: Weaving relations for cache performance. In: VLDB (2001)
7. Liu, L., Li, E., Zhang, Y., Tang, Z.: Optimization of frequent itemset mining on multiple-core processor. In: VLDB, pp. 1275–1285 (2007)

Usability Confinement of Server Reactions: Maintaining Inference-Proof Client Views by Controlled Interaction Execution

Joachim Biskup

Technische Universität Dortmund, Dortmund, Germany
`biskup@ls6.cs.uni-dortmund.de`

Abstract. We survey the motivation, the main insight and the perspective of our approach to policy-driven inference control of server-client interactions for a logic-oriented information system. Basically, our approach aims to confine the usability of the data transmitted by the server to a client. The confinement is achieved by enforcing an invariant that, at any point in time, a client's view on the actual information system is kept inference-proof: the information content of the data available to the client does not violate any protection requirement expressed by a declarative confidentiality policy. In this context, the information content of data and, accordingly, the inference-proofness of such data crucially depend on the client's a priori knowledge, general reasoning capabilities and awareness of the control mechanism. We identify various parameters of the approach, outline control mechanisms to enforce the goals, and sketch the methods employed for a formal verification.

1   Introduction

We consider a scenario of a *server-client* information system where a server manages data for assisting clients to share dedicated parts of information, as roughly illustrated in Fig. 1. Basically, on the one hand, the server maintains a current *instance* of a *schema*, provides *answers* to query requests and *reacts* to view update requests, and might *update* the instance itself and then send *refreshments* to the clients; and on the other hand, a client issues *query* requests, constructs its own *view* on the instance and might submit *view update* requests. Moreover, we assume that the information system is *logic-based* in the sense that the semantics of interaction requests, i.e., *query evaluation* and *update processing* including preservation of integrity constraints, are founded on the well-defined notions of validity (truth) and implication (entailment) of some logic. This includes, e.g., a relational database system based on first-order logic, where a query expressed in the relational calculus is evaluated by determining the set of those tuples that match the query format and can be seen as a sentence that is valid (*true*) in the interpretation constituted by the current instance.

The logical basis provides a clear distinction between the raw *data* stored by the server or transferred between the server and a client and the actual

S. Kikuchi, S. Sachdeva, and S. Bhalla (Eds.): DNIS 2010, LNCS 5999, pp. 80–106, 2010.

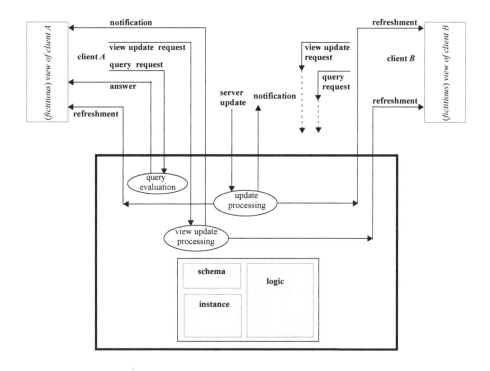

Fig. 1. Functionality of a logic-based server-client information system

information represented by that data. In particular, each individual piece of data contains some immediate information, e.g., a tuple contained in a query answer expresses that the corresponding sentence is valid in the pertinent interpretation. However, by means of logical *implication* (entailment), several pieces of data together might comprise additional information that is not represented solely by any of the individual pieces.

While the server and the clients operate on raw data, the human agents on behalf of whom these components are running are mainly interested in the information content. Considering information as a fundamental asset, the human agents might want to *discretionarily share* some pieces of information while *hiding* other pieces. In our scenario, we assume that, for each of the clients (or for each of some suitably formed and mutually disjoint groups of clients), all such interests in availability on the one hand and confidentiality on the other hand are suitably declared by a specific *availability policy* and a specific *confidentiality policy*, respectively, that are held and enforced by the server. Accordingly, the server is responsible for controlling any interaction with a client with regard to the policies declared for that client (or the client's group).

Distinguishing between data and information, such a control should aim at suitably *confining* the information conveyed by the server's reactions shown to clients while executing interactions. In more general terms, the control should inspect the *usability* – rather than the actual usage – of shown data by a client, i.e., the *options* for a *potential* exploitation of that data by a client. In particular, the control has to consider the information that a client can possibly obtain by *logical inferences*, applied not only to the individual pieces of data arising in a current interaction but rather to the collection of all data supposed to be available to that client. This includes the data that represent the postulated a priori knowledge or result from previous interactions. Additionally, the control should also be *self-reflective* in the sense that a client's awareness of the control mechanism does not provide any advantage for gaining hidden information.

The control mechanisms surveyed in this work *statically inspect* beforehand or *dynamically intercept* and thereby effectively control transmitted data regarding declared confidentiality and availability policies. The fundamental features of our approach include *content-dependent* and *history-dependent* control, *resource assumptions* on a client's knowledge and kind of reasoning, *theorem proving* for relating data on the level of the underlying information system, and *meta-inferences* for dealing with the client's awareness of the control functionality. The basic control methods distort potentially harmful information by *refusal* (e.g., notifying the client that a submitted query will not be answered), *lying* (e.g., returning the negation of the correct answer to a closed query), or a suitable *combination* thereof. The achievements of the methods are formally verified by establishing an *indistinguishability* property. This property roughly expresses that for any sequence of interactions of the server with some client, from that client's point of view, any sentence of a declared confidentiality policy might be false in the actual instance(s); in other words, based on the client's view, an actual instance of the information system is indistinguishable from another possible instance that makes the sentence considered false. We call such a view *inference-proof* (with respect to the confidentiality policy and the assumptions about the client).

2   Inference-Usability Confinement

In this section, we introduce our concept of *usability confinement* already mentioned above. This concept is first expressed in general terms of accesses to some resource and exemplified for read accesses, and then specialized to *inference-usability confinement* regarding *options to infer information* to be kept secret.

Consider a request of a subject (client) to perform an operation (to interact with the server) on an object (regarding some data managed by the information system). Pure discretionary access control (DAC) decides on such a request on the basis of previously granted permissions and prohibitions (positive or negative access rights). These access rights are supposed to reflect both the *trustworthiness* of the subject and the *sensitivity* of performing the operation on the object. Once a positive decision is taken, there is *no further* control and, accordingly, the requesting subject could *unrestrictedly* do anything with the result of the executed operation, in particular inferring additional information that is not immediately contained in the result. In principle, the decision maker just has to *trust* the requesting subject to further use the results as anticipated when granting the rights. Similarly, like most of the cryptographic mechanisms, encryption relies on appropriately distributed secret keys, and the key distributor just has to trust the receiver of a decryption key to behave as anticipated later on.

Many and diverse mechanisms for further controlling *local usage* and *proliferation* of a read result (and, correspondingly, of the result of any other object access) have been proposed, and a specific application might demand sophisticated combinations of them (see, e.g., [29,4,24]). Among others, the proposals include state-dependent DAC and workflow control [1], obligation control [2] and, more generally, usage control [36,37], mandatory access control (MAC) [32], inference(-usability) control [27] (see also, e.g., [44,23,21,22,45,35,31,34,40]) and (cryptographic, secure) multiparty computations [25,28]. We inspect some approaches in Table 1: for each selected approach, we sketch the basic mechanism to come up with a (kind of) read permission, and we roughly describe the options afterwards left to the recipient for local usage or proliferation of the read result.

As indicated by the table, in general such mechanisms rely on *assumptions* about the recipient, e.g.:

– state-dependent DAC: objects cannot leave workflow control;
– obligation/usage control: objects are monitored by a shielding wrapper;
– MAC: objects cannot leave MAC enforcement;
– inference(-usability) control: resource assumptions are appropriate.

Though inevitably relying on *some* assumption, we would like to favor an approach that

– enables security enforcement *solely by the provider* (in our specific context: the server) of the (data) objects (and the represented information) to be protected, and
– does *not* require any dedicated protection mechanism on the recipient's side (in our specific context: a client).

We refer to such an approach as *usability confinement*: The owner (server) *customizes* the crucial items in such a way that the manipulated items are still useful for the recipient (supporting availability) but – under some assumptions – show a purposely confined usability; ideally, the latter property does not offer any possibility to the recipient (client) for a misuse of the item. The following approaches to customization appear to be most promising:

– multiparty computations: a crucial object is never released but only a suitably prepared encrypted version is forwarded to cooperating agents;
– inference(-usability) control: crucial information is eliminated beforehand.

While holding the former approach to open a fascinating world of astonishing new ways of securing information with an arithmetic flavor, in our own research surveyed in this paper, we are elaborating the latter approach for information with a logical flavor, characterizing it as *inference-usability confinement*. Both approaches are particularly recommendable for distributed (information) systems, where an agent cannot directly control another agent at a remote site.

Table 1. Some approaches to control of read access

Approach	Mechanism	Recipient's option for local usage	Recipient's option for proliferation
discretionary access control	grant read privilege for object to subject; permit access iff privilege exhibited	unrestricted	unrestricted
state-dependent discretionary access control	grant read privilege for object to subject dependent on some "workflow state"; permit access iff subject in workflow state	unrestricted	within workflow control: as declared outside workflow control: unrestricted
discretionary access control with obligations	grant read privilege with obligations for object to subject; permit access and attach obligation iff privilege exhibited	under "usage control": restricted according to attached obligation outside "usage control": unrestricted	under "usage control": restricted according to attached obligation outside "usage control": unrestricted
mandatory access control	assign classification to object and clearance to subject; permit access iff clearance dominates classification	inside MAC: unrestricted computation; writing only in "upwards" containers outside MAC: unrestricted	inside MAC: "upwards" outside MAC: unrestricted
encryption	distribute decryption key; encrypt object; permit unconditional access to encrypted object	without decryption key: "meaningless" with decryption key: unrestricted	without decryption key: "meaningless" with decryption key: unrestricted
multiparty computation	generate keys jointly; provide encrypted object as input; participate in joint computation	unrestricted, but *only* for goal of joint computation "meaningful"	"meaningless"
inference control (controlled query evaluation, k-anonymity, ...)	customize objects to make them "inference-proof" according to "user assumption"; permit access to customized objects	unrestricted, but while "user assumption" is valid: *actually confined* by customization	unrestricted

3   An Example and a Notion of Inference-Proofness

To preliminarily illustrate the task of inference-usability confinement, we will present a very simple example. The server maintains an *instance* of a propositional information system, currently only containing the health record of some patient Lisa, represented as a set of (propositional) atoms, understood to denote the *"true* part" of an interpretation:

{brokenArm, brokenLeg, lowWorkload, highCosts}.

To protect poor Lisa against a client curious to learn her working abilities, the security officer responsible for the server declares a *confidentiality policy*:

{highCosts ∧ lowWorkload}.

In doing so, the officer intends to only hide the full consequences of Lisa's bad state but still permits to learn something about the underlying diseases and one part of the consequences – if individually considered.
Furthermore, the security officer reasonably assumes that the agent on the client side has some *a priori knowledge* by common sense:

{brokenArm ⇒ lowWorkload, brokenLeg ⇒ highCosts}.

Let now the client issue a sequence of *queries*, indicated by "?", explicitly asking firstly whether or not Lisa suffers from a broken arm and secondly whether or not Lisa suffers from a broken leg:

⟨brokenArm(?), brokenLeg(?)⟩.

The answer to the first query is still harmless. The policy does not prohibit the correct answer, brokenArm: neither directly, since brokenArm is not an element of the policy, nor indirectly, since the sole (strongest) sentence entailed by the correct answer together with the a priori knowledge, lowWorkload, is a not an element of the policy either. However, the provider now has to update the *client view*, which is initialized by the a priori knowledge, by inserting the answer, brokenArm and, at least conceptually, the entailed sentence, lowWorkload:

{brokenArm ⇒ lowWorkload, brokenLeg ⇒ highCosts,
          brokenArm, lowWorkload}.

Inspecting the second query, the server will detect the following: The correct answer, brokenLeg, and the current client view entail highCosts; adding the answer and, conceptually, the entailed sentence to the current client view, which already contains lowWorkload, would entail the sentence highCosts ∧ lowWorkload. However, that sentence is declared to be kept secret. Now the server has two straightforward options to react by a distorted answer: either *refusing* to give an informative answer, e.g., just returning a special value mum, or *lying*, e.g., returning ¬brokenLeg. Clearly, both options should be handled with great care, in particular: a refusal should not enable a *meta-inference* based on the information conveyed by the refusal notification, see Sect. 6; a lie should not lead to an *inconsistency* in the future, see Sect. 7.

More generally and now also including all kinds of interactions, but still informally speaking, we would like to always ensure a declarative *confidentiality* property expressed in terms of a client being unable to distinguish whether or not a sentence to be kept secret actually holds; in other words, for that client it should always appear to be possible that the sentence considered does not hold:

Definition. *A control mechanism for inference-usability confinement maintains an* inference-proof client view *iff the following property holds:*
For all situations that are possible for the underlying information system and the control mechanism, i.e.,
    for all instances of the information system satisfying the integrity constraints,
    for all a priori knowledges of a client compatible with the instance,
    for all confidentiality policies,
    for all interactions sequences,
for each of the policy elements there exists an alternative possible situation,
in particular an alternative sequence of instances such that
    – from the point of view of the respective client –

1. *the actual and the alternative situation are indistinguishable, in particular all reactions of the server visible to that client are the same, but*
2. *the alternative instances do not make the policy element considered true.*

Depending on the concrete instantiation of the parameters that we will discuss in Sect. 5, this informal definition of inference-proofness has got a specific formalization, as reported in the respective work.

The definition given above is adapted to a *dynamic* mode of inference-usability control, which considers interaction requests step by step. Alternatively, using a *static* mode of inspection, we could generate an alternative instance in advance that subsequently can be queried without any further control. For our example, by lying regarding either illness, we could generate one of the following instances before any user interaction:

$$\{\texttt{brokenArm}, \neg\texttt{brokenLeg}, \texttt{lowWorkload}, \neg\texttt{highCosts}\},$$

$$\{\neg\texttt{brokenArm}, \texttt{brokenLeg}, \neg\texttt{lowWorkload}, \texttt{highCosts}\}.$$

For the specific query sequence of our example, the first version would be expected to trigger the same answers, whereas the second version would return a different reaction already for the first query. This comparison indicates the power of the dynamic approach to be most informative by only introducing a distortion as a "last-minute" action in dependence of the actual knowledge the client has obtained before. On the other hand, however, the static mode does not require us to perform the subtle considerations exemplified above at runtime. The resulting tradeoff situation is apparent, e.g., seen from the point of view of a client, either favoring availability of information or preferring computational efficiency and thus short response times.

4   Logical Approach and Basic Features

In general, any control mechanism for effective inference-usability confinement must necessarily comprise the following features:

- The semantics of the information system underlying the control must provide means to distinguish between *data* and the *information* represented by the data.
- The distinction to be made must be founded on some logic in which a suitable notion of *logical implication* (entailment) can be expressed; the logic employed must capture the pertinent universe of discourse, correctly and completely modeling all relevant aspects of an application.
- Given some pieces of data, the control has to be able to *effectively determine* the represented information; in particular, the control has to be able to decide whether the information represented by a further piece of data is *comprised* (logically implied, entailed) by the information of the given pieces.
- The control has to be *content-dependent*, i.e., to take an access decision, the content of the requested object – seen as a kind of data *container* – must be inspected.
- The control has to be *history-dependent*, i.e., to take an access decision, all previously permitted accesses must be considered, and thus the mechanism has to keep a *log* of all relevant events.
- A security officer supervising the control has to postulate and to declare a reasonable *resource assumption* on a client's *a priori knowledge* (available data and the information represented) and *reasoning capabilities* (available options to actually determine the information represented by some data).
- Additionally, the security officer has to detect and to evaluate options of several clients to collude; a set of clients suspect to *collusion* should be treated like one (group) client.
- On the basis of the logged history and the declared resource assumption, the control has to model a *client view*, which constitutes the *control's expectation* (which might be inappropriate, since the control cannot intrude the client) about the *client's belief* (which might be erroneous for several reasons, including accepting data that has purposely been distorted by the server's control) about the actual instance(s) of the information system.
- Moreover, the control has to maintain an invariant expressing that the client view *does not imply* any information considered harmful for the client according to the pertinent *confidentiality policy*.
- Accordingly, the control must algorithmically solve *implication problems*, and thus the control in general needs means for theorem proving.
- To react on a potential violation of the confidentiality policy, the control has to block *meta-inferences* regarding refusal; basically, this can be achieved by additional refusals for the sake of indistinguishability.
- To react on a potential violation of the confidentiality policy, the control has to avoid "hopeless situations" regarding lying caused by an inevitably arising *inconsistency*; basically, this can be achieved by additionally protecting all disjunctions of harmful information.

On the one hand, the goals of inference-usability confinement appear to be widely *socially accepted*. In fact, e.g., in many countries these goals are part of the legislation aiming at privacy protection: data becomes "personal" and thus subject to protection, if the pertinent human individual is identified or *identifiable*, i.e., can be inferred from other data available to some client.

But obviously, on the other hand, the listed requirements on effective control mechanisms are *extremely challenging*. These requirements both impose a high burden on the security officer to model the clients appropriately, and might demand an infeasible computational overhead of resources, space for storing the log and time for solving implication problems. In fact, e.g., decision problems regarding logical notions like (un)satisfiability or implication (entailment) tend to be of high computational complexity or even computationally unsolvable.

Like in other fields of computing engineering, the apparent discrepancy can be resolved by interpreting the situation as a *tradeoff*, describing which *amount* of computational costs is demanded by/enables which *degree* of inference-usability confinement. So far, we don't have fully elaborated the tradeoff. Rather, we inspected various parameters that might have an impact on the tradeoff and designed concrete mechanisms for several parameter choices.

5 Parameters and Components

In our work on inference-usability confinement we have identified various parameters, and we are studying their impact on the concrete forms of the overall approach, whose dynamic version is illustrated by Fig. 2. In the following, we sketch the parameter choices that we have considered, which potentially span a large space of possibilities by taking all combinations. We emphasize, however, that not all combinations are meaningful, and so far, as indicated by Table 2, we considered only a small fraction of the potentially useful combinations.

Logic underlying the information system. Choosing a logic, we have to find a suitable compromise between expressiveness and computational feasibility.

- In the basic conceptual work [3,5,7,6], ignoring computational aspects, we refer to an *abstract logic* by only specifying some necessary properties: the propositional connectives include negation and disjunction without any restriction, with classical semantics (and thus all other connectives are expressible as well); we have the notion of an interpretation, which can make a (set of) sentence(s) *true* (valid) or *false*; then the semantic relationship of logical implication (entailment) is defined in the standard way; and we assume the logic to be compact.
- To take care of computational aspects but to avoid subtle details [18,17,42,9,14,11], e.g., regarding infinite domains of an interpretation, in some work we focus on *propositional logic*, then implicitly postulating that we would be able to describe a complex situation by an appropriate propositionalization.

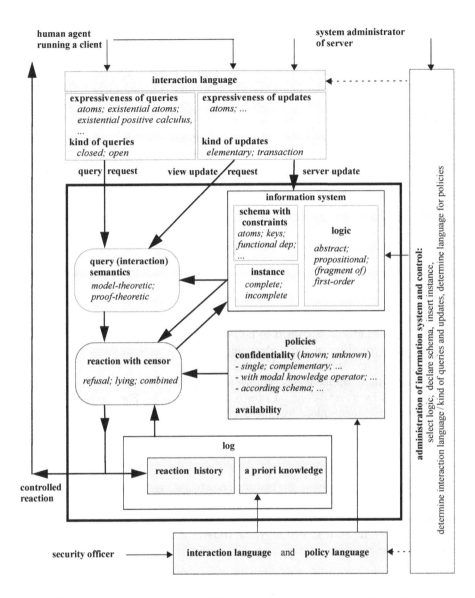

Fig. 2. Main agents and overall architecture of a dynamic control mechanism for inference-usability confinement, also indicating some parameter choices

— To deal with relational databases, we use *first-order logic* where in each case a suitable fragment is selected in order to allow a computational treatment of the decision problems arising in the specific context or an optimization we are looking for. For example, to deal with open queries and the completeness sentences needed to express closed world assumptions [8], we restrict to the Bernays-Schönfinkel prefix class $\forall^*\exists^*$, which has a decidable (finite) universal validity problem; for propositionalizing incomplete instances [15],we require a finite vocabulary; for statically generating an inference-free instance [43,19], we restrict to either universal sentences or existential sentences or a dedicated class of mixed sentences; for optimization to get rid of the log [12,10,13], we basically restrict to atomic sentences which might have an existential prefix. In these cases, except for the propositionalization, we only consider interpretations that have a fixed infinite set of constant symbols as (Herbrand) domain, interpret the constants by themselves and allow only finitely many positive interpretations of ground atoms (and thus correspond to database relations). Notably, these conventions lead to the notions of DB-interpretations and, accordingly, of DB-implication, which slightly deviates from the classical semantics. For example, a sentence stating that for some predicate there exists a negatively interpreted ground atom is a tautology.

Instances of the information system. We distinguish two cases regarding whether or not an instance is expressive enough to enable us to always give a definite result.

– In most of our work, we consider only *complete* instances formed by a finite set of ground atoms, assumed to be positively interpreted, that can serve as a full interpretation of the underlying logic, and thus to evaluate the truth (validity) of any sentence of that logic.
– However, many applications require to deal with *incomplete* instances that are formed by a consistent set of arbitrary sentences. For the case of propositional logic, incomplete instances are studied in [17,42]; in the work [15], under some essential restrictions, incomplete first-order logic instances are reduced to propositional instances.

Constraints on instances. An instance has to satisfy the integrity constraints (semantic constraints), that are declared in the schema. Moreover, any update processing has to maintain these constraints as an invariant. We suppose the schema to be public and, accordingly, the constraints are seen as part of a client's a priori knowledge.

– Dealing with relational databases on the basis of first-order logic, keys and foreign keys, more generally equality generating and tuple generating dependencies, play a crucial role to syntactically define a schema to be in a *normal form*, e.g., Boyce-Codd normal form, BCNF. Semantically, normal forms can be characterized by guaranteeing that all instances are in some sense "redundancy-free", roughly saying that there will never be a tuple in an instance such that we can infer the tuple's membership from the remaining part of the instance. This characterization suggests that for normalized schemas and under suitable further restrictions, any immediate information returned to a client just stands for its own: the client will never be able to infer any further information beyond the immediate one. Based on this observation, inference-usability confinement could possibly be achieved by only considering the harmfulness of individual pieces of information; then we could do without keeping a log file and thus operate a substantially optimized control. In fact, this idea has been successfully elaborated for the special but practically relevant case of a schema in object normal form (complying with BCNF and having *unique keys*) [10] and is currently further investigated.

– In general, constraints declared in the schema are seen as part of a client's *a priori knowledge*. Accordingly, if we impose a restriction on that knowledge, we also restrict the kind of constraints the control can take care of. For example, we might restrict the constraints for a relational instance to *functional dependencies* [12,10,13] – together with further restrictions, again in order to enable an optimized control without keeping a log file. In a simpler setting based on propositional logic, a restriction to express constraints only by *atoms* leads to a substantial optimization of the control of refreshments triggered by server updates [11].

Semantics of query evaluation. Basically, the semantics depends on the expressiveness of instances.

– In the case of completeness, i.e., if instances are full interpretations of the underlying logic, we employ the *model-theoretic* approach, which is based on evaluating the truth (validity) of sentences with respect to an interpretation, and thus always returns either *true* or *false*.
– In the case of incompleteness [17,42,15], we rely on the *proof-theoretic* approach, which is based on evaluating whether an instance, seen as a set of sentences, logically implies (entails) a sentence: besides *true* (yes, implied) and *false* (no, negation implied), such an evaluation might return a third value *undefined* (not known to the instance). The additional value provides new flexibility to distort a harmful reaction by modification.

Kind of queries. A query request is specified by a formula that complies with the syntax of the underlying logic.

– If the query formula does not contain (occurrences of) free variables and thus is a *closed* formula, i.e., a sentence, then the expected answer is either *true* (yes, the sentence is implied by the instance/valid in the instance) or *false* (no, the negation of the sentence is implied by the instance/valid in the instance) or *undefined* (neither the sentence nor its negation are implied by the instance). Such closed queries can arise in all kinds of the underlying logic, and accordingly we consider them in all our work; the propositional logic only allows closed queries.
– If the query formula has free (occurrences of) variables and thus is an *open* formula, then so far we only treat first-order logic [8,12,10,13,43,19], and the expected answer is formed by considering all ground substitutions of these variables over the fixed infinite (Herbrand) domain: The positive part is built

from those substitutions such that the resulting sentence is implied by the instance/valid in the instance; the negative part by those substitutions such that the negation of the resulting sentence is implied by the instance/valid in the instance; for incomplete instances there might be a third part consisting of the substitutions where the result is undefined. Since we aim at getting a finitely represented answer, we have to impose some restrictions: The open query formula must be *safe* such that the positive part is guaranteed to be always finite. The complement of the positive part is then captured by a *completeness sentence* (*closed-world statement*) as a first-order sentence, expressing that all substitutions except the finitely many positive ones are negative or undefined. Clearly, for a complete instance, there are no undefined substitutions. Notably, a completeness sentence added to the positive part might be harmful regarding the confidentiality policy, and thus returning such a sentence must be controlled with great care.

Kind of updates. The processing of a request to update the current instance has to maintain the integrity constraints declared in the schema and to notify the issuer about acceptance or rejection.

- If the update is *elementary*, requiring to insert or delete just one item [14], then the integrity constraints must be valid just after the modification. The issuer is notified whether or not the update of the single item stated in the request has been accepted/committed or rejected/aborted; this notification corresponds to an answer for a query about satisfaction of the constraints after handling the single item. Accordingly, a sequence of elementary update returns a sequence of such answers, to be controlled like any other reaction of the server.
- If the update refers to a *transaction*, requiring to modify a sequence of items [11], then the integrity constraints must be valid only after the last modification, but may be violated in between. Again, the issuer is notified whether or not the updates of the whole transaction requested have been accepted/committed or rejected/aborted at the end. This notification corresponds to just one answer for a query about satisfaction of the constraints after the whole transaction. In general, this answer will provide less information than a sequence of answers, one for each element listed in the transaction.

Issuer of updates. Though the current instance is kept by the server, the request for an update can be issued by any of the agents.

- If (the agent running) the *server* requests an update [11], then all direct reactions, i.e., notifications about acceptance or rejection according to the preservation of integrity constraints, can be freely communicated to the issuer, who is supposed to have full access to the information system anyway. In case of an acceptance, however, the server has to send a refreshment to each of the clients, to inform them about the aging of information previously released. Clearly, such refreshments have to be carefully controlled. In particular, the control has to confine the potential information gain from comparing the reactions received for the old instance with the refreshed reactions for the new instance, where all kinds of reactions might be distorted.

- An update request issued by a *client* [14] is handled as a view update that has to be translated into an update of the instance maintained by the server. A crucial point for the control is to take into consideration that the client's view referred to by the request might be already distorted. Moreover, if the server actually modifies the instance, refreshments for the other clients must be generated.

Confidentiality policies. Declaring (an instance of) a confidentiality policy by means of the underlying logic, we express which information we want to be kept secret to a client. The harmful information should be denoted explicitly, but the protection should apply to any appearance of the information, in particular it should be independent of any specific syntactic representation (basically, protection should refer to some suitably formed equivalence classes).

- In the simplest case, used in most of our work, we just specify a set of *single* sentences, in this context called *potential secrets*. The intuitive meaning of a sentence as a potential secret is the following: If the sentence is *true* in the instance(s) considered, then the client should never be able to infer this fact; otherwise, if the sentence is *false*, then it is considered harmless when the client will infer that fact. In other terms, from the point of view of the client, it must always be possible that a potential secret is *false* in the instance(s) considered. In order to enable an optimized control, in particular to get rid of the log file, we have to restrict the class of sentences employed as potential secrets, e.g., allowing only sentences expressing a so-called "fact" in an ONF schema [10], or possibly existentially quantified atomic sentences [12,13].
- Alternatively and restricting to complete instances [3,7,6], we can take pairs of *complementary* sentences, in this context called *secrecies*, as elements of a policy. Then the intuitive meaning of such a pair is the following: From the point of view of the client, it must always be possible that either sentence of the pair is *true* (equivalently: is *false*), i.e., there are always two different instances both seen to be possible by the client, such that one instance makes one of the sentences in the secrecy *true*, and the other instance makes the complementary sentence *true*.
- For incomplete instances, where a sentence can be assigned any of the three different values *true*, *false* and *undefined*, we are studying the generalized option to freely specify which sets of these values the client is not allowed to infer [16,42,15]. To do so, we convert a specified set into a so-called *epistemic potential secret*, expressed in *modal logic*, where the value *undefined* for some sentence is represented by intuitively saying "the system does not know that the sentence is *true* and the system does not know that the sentence is *false*".

In fact, beyond that conversion, epistemic potential secrets offer powerful and flexible means to specify policies.

Client knowledge of instance of the confidentiality policy. As usually recommended for security, we always assume that a client is fully aware of both the overall approach of inference-usability confinement and the specific instantiation in operation. However, we distinguish whether or not a client knows the actual instance of the confidentiality policy (and could do so also for the instance of an availability policy).

- In most cases, we follow the conservative assumption that the policy instance is *known* by the client, i.e., the client is aware about the dedicated sentences to be kept secret to it.
- A more relaxed assumption is that the policy instance is *unknown* to the client [3,7,6,42], such that the client's uncertainty about the actual instance of the information system also extends to the policy. Accordingly, in reasoning about a situation, the client has to consider more possibilities, which might offer additional options for the control to hide the secret part of the actual situation. In fact, under the assumption of an unknown policy instance, the necessary control tends to be essentially less restrictive; basically, the control has to consider only those parts of a policy instance that are *true* in the instance of the information system, whereas the remaining *false* parts can be neglected.

Availability requirements. A server could trivially enforce any confidentiality policy by always providing "no information" at all to a client. For example, the server could never react to a request, or, less intuitively, the server could always generate a reaction that is correct with respect to a fixed public instance, independently of the actual (hidden) instance. We should avoid such trivial solutions and, more generally, we have to balance interests in confidentiality with legitimate interests in the availability of information needed.

- As a basic *implicit* rule, in all our work, we first of all aim at correctly providing the information represented by the instance(s) of the information system to each of the clients as requested. As a kind of exception, the basic availability rule may be violated *only if* this is strictly necessary (for a given setting) to enforce a confidentiality policy declared for a client.
- In some work [9,17,43,41], we additionally consider *explicit* availability policies, which are specified as a set of sentences that should not be distorted. Clearly, a confidentiality policy and an explicit availability policy might be in conflict. Accordingly, we have to require that conflicts are resolved beforehand, or we must include some means to rank conflicting requirements.

Representation of a priori knowledge and history. To represent the current knowledge of a client in a log file, we basically need the expressiveness of the logic underlying the information system, possibly restricted according to the kind of constraints, queries and updates that may be employed. However, there are two specific observations. For complete instances, we do not have to store a refused reaction explicitly [38,5]. For incomplete instances [16,17,42], however, refusals might enable some information gain and thus should be logged. Moreover, we need to represent a returned answer value *undefined*, roughly saying that the "system does neither know that the query sentence is implied nor that its negation is implied". This goal can be achieved by using *modal logic*.

Assumed reasoning capabilities of a client. Postulating the reasoning capabilities is a crucial issue for the overall success of inference-usability confinement. In principle, we are faced with a large and, to the best of our experience, unstructured and widely unexplored set of options for modeling a client.

- Throughout our work, we follow the extreme option that considers a client to be potentially a *perfect attacker*. In particular, a client is a rational and omnipotent reasoner; it keeps the full interaction history; it employs some common or personal a priori knowledge; and the client is fully aware of the control mechanism of the server it is interacting with.
- Another extreme option is used in supplementary work on incomplete, propositional instances for querying, contained in [42], where a client is seen as *plain*, taking any reaction truthfully for granted. Astonishing enough, control mechanisms defending against the perfect attacker tend to be suitable also against a plain client.
- Unfortunately, so far we see no obviously distinguished candidate models to be explored in the future. Rather, we plan to successively study some selected examples, taking from the rich field of knowledge engineering.

Languages for schema declarations, interactions and policies. As explained above, we might want or are forced to restrict the kind of integrity constraints, of interactions offered to the server and the clients, and of confidentiality and availability policies. Such restrictions can be implemented by configuring the interface for the respective agents.

- The *schema declaration language* (data definition language), offered to a (conceptual) system administrator of the server, defines the employable integrity constraints.
- The *interaction language* (data manipulation language), offered to the (human) agents running the clients and to the system administrator, comprises a *query language* and an *update language*. For the query language, the *expressiveness* is defined in terms of the syntactic structure of the formulas to be used as queries, and the *kind* indicates whether besides closed queries also open queries can be issued. Similarly, for the update language, the *expressiveness* is defined in terms of the syntactic structure of the sentences to be inserted or deleted, and the *kind* indicates whether requests are elementary or transactions.

- The *policy languages* for confidentiality and availability policies, offered to a (conceptual) security officer, defines the expressiveness in terms of the syntactic structure of policy elements.

Reaction on harmful actions. Whenever the server has to send a reaction to a client, regarding a *single sentence* that has been found to be harmful, there are two basic options:

- The server might explicitly *refuse* to return a meaningful reaction, denoted by a special reaction mum. This option is further discussed in Sect. 6.
- The server might (implicitly, clearly without a notification) *lie*, i.e., return a reaction that deviates from the correct semantics. In the complete case, such a lie is formed by just switching the correct truth value, i.e., *true* into *false*, and vice versa; the incomplete case, offering three possibilities for reactions, allows for more flexibility to form a modification. These options are further discussed in Sect. 7.

We can employ either refusals or lies *uniformly*, distorting a harmful reaction always by the same option, or we can aim at deciding case by case on the option, leading to the *combined* approach, which is further discussed in Sect. 8.

Mode of inference-usability control. The control can be performed at different points in time and, accordingly, might or might not depend on the actual behavior of a client.

- In the mode of *static* inspection beforehand [18,43,19,41], for each of the clients considered, the server generates an alternative inference-proof instance that the client can freely query afterwards. Clearly, the alternative instance must be still useful: it should comply with explicit availability requirements and only minimally deviate from the correct instance. For propositional logic, these goals can be accomplished by a branch-and-bound search based on SAT-solving [18,41]; for first-order logic, such a search can be extended by determining violations as answers to specific safe queries expressed in the relational calculus [43,19].
- In the mode of *dynamic* interception and control, favored in most of our work for the sake of maximizing availability, the control follows a "last-minute" distortion strategy. Keeping a log file of the preceding interactions, to react on a specific request, the control considers the actual situation case by case. In general, dynamic control enables us to return more correct information than static inspection, however at the price of keeping a log file and spending the runtime overhead. For special cases, however, the dynamic mode can be optimized to get rid of the log file [10,12,13].

To treat the apparent tradeoff between static inspection and dynamic control, we suggest to design suitable combinations, roughly aiming at "compiling" all distortions necessary under all possible behaviors of the client beforehand and then treating only the remaining ones dynamically.

Table 2. Selected parameter combinations studied so far

paper	logic	semantics/ completeness	query/ interaction	confidentiality policy	user image/ log file	modus	reaction/ censor	avail-ability
DKE00 [3]	abstract	model-theoretic	closed queries	known/unknown, complementary	abstract	dynamic	refusal/lying	implicit
DKE01 [5]	abstract	model-theoretic	closed queries	known, single	abstract	dynamic	refusal/lying	implicit
AMAI04 [7]	abstract	model-theoretic	closed queries	known/unknown, single/complementary	abstract	dynamic	combination	implicit
IJIS04 [6]	abstract	model-theoretic	closed queries	known/unknown, single/complementary	abstract	dynamic	refusal/lying/ combination	implicit
DBSEC06, JCS08 [18]	propositional	model-theoretic	closed queries	known, single	not necessary	static	lying	implicit/ explicit
AMAI07 [8]	first-order	model-theoretic	open queries, existential positive calculus	known, single	first-order suit-able for Bernays-Schönfinkel class	dynamic	refusal/lying/ combination	implicit
IJIS08 [17]	propositional	proof-theoretic, incomplete	closed queries	known, single	modal	dynamic	refusal/lying/ combination	implicit
DBSEC07 [16]	propositional	proof-theoretic, incomplete	closed queries	known, modal	modal	dynamic	refusal/lying/ combination	implicit
IPL08 [10]	first-order	model-theoretic	closed queries, only existential atoms	known, single, only facts for ONF schemas	not necessary	dynamic	refusal	implicit
ISC07 [12]	first-order	model-theoretic	closed queries, only atoms	known, single, only existential atoms	not necessary	dynamic	refusal	implicit
ICISS07 [9]	propositional	model-theoretic	closed queries	known, single	not necessary	static	lying	explicit
SEC09 [13]	first-order	model-theoretic	closed queries, only existential atoms/conjunction	known, single, only existential atoms/ disjunction, open	not necessary	dynamic	refusal	implicit
DBSEC09 [14]	propositional	model-theoretic	closed queries/ view updates	known, single	propositional	dynamic	lying	implicit
ISC09 [19]	first-order	model-theoretic	open/closed queries	known, single	not necessary	static	lying	(explicit)
ESORICS 09 [11]	propositional	model-theoretic	closed queries; view/ provider updates	known, single	propositional	dynamic	lying	implicit
FOIKS10 [15]	first-order	proof-theoretic, incomplete	closed queries	known, modal	modal	dynamic	refusal	implicit

6   Refusal Approach

The refusal approach originates from the seminal proposal made for secrecies by Sicherman, de Jonge and van de Riet [38], later rephrased by Biskup [3], adapted for potential secrets by Biskup and Bonatti [5], and further elaborated as indicated below. The refusal approach follows a quite straightforward and at first glance attractive idea: If the provider detects that executing a client's request correctly would violate the confidentiality policy, then the provider explicitly refuses the execution and returns a corresponding notification to the client; otherwise, the request is executed as demanded and the correct return value(s) is (are) sent to the client. Accordingly, the client's view on the actual instance of the information is always correct, and the client is fully aware where its knowledge has remained fragmentary due to refusals. However, two issues immediately arise, which will lead to a tradeoff situation in general:

- Regarding a malicious agent running the client, can a refusal notification carry information that leads to a violation of confidentiality?
- Given an honest agent running the client, does the fragmentary knowledge suffice to comply with the agent's obligations, i.e., can we ensure sufficient availability?

Assuming that the client *knows* the instance of the confidentiality policy, refusal notifications must not be sent carelessly, i.e., *not only* in the case of an immediately recognized violation. For otherwise, the client could exploit a *meta-inference* of the following kind:

1. working under the refusal approach, the client receives an explicit refusal notification;
2. simulating the (supposedly known) behavior of the provider, the client probes all executions that are possible by design;
3. knowing the policy instance, the client identifies those executions whose visible reactions lead to an immediately recognized violation;
4. postulating careless refusals, the client infers that one of the violating executions is the correct one (if only one possible execution is harmful, the client obtains full knowledge about the situation).

Obviously, if we fix the conditions of the first three points, to disable such meta-inferences, in general we have to carefully add further refusals [38,3,5,17,42]. Doing so, however, decreases the availability of information, and might require further adjustments. For example, consider *closed queries* to a *complete instance* under *known potential secrets*. If the correct answer – together with the current client view – is harmful and thus must be refused in any case, then the only option for an *additional refusal* is to treat the negation of the correct answer as harmful as well. There are two immediate consequences. First, the control then works *instance-independent* (possibly causing unacceptably many refusals). Second, controlling a query whose answer is already known to the client – according to the current client view – would always be refused (since tentatively adding the negation would produce an inconsistency), unless we *improve* the refusal procedure by first checking whether the query result is already known [7].

The outline of meta-inferences given above suggests to alternatively relax the conditions of the first three points. Relaxing the first condition immediately introduces *lies* as a possible reaction. Relaxing the second condition would mean either to hide the details of the control (but "security by obscurity" is *not* advisable in general) or to protect against *non-perfect* attackers (but still only preliminarily explored [42]). Relaxing the third condition, now assuming that the client does *not know* the policy instance, turns out to be helpful (but might be questionable). Basically, the refusal approach can then work instance-dependent by considering only those policy elements that are actually *true* in the instance of the information system; moreover, there is no need to employ "additional refusals" [38,6,42].

The refusal approach appears to suffer from some further essential shortcomings. First, so far, when additionally dealing with *open queries* (for complete instances under potential secrets) we only succeeded by being rather restrictive (basically repeatedly checking the harmfulness of a completeness sentence after considering each individual ground substitution), thereby causing a lot of refusals [8]. We conjecture that we cannot improve the refusal approach for open queries in general. Second, employing refusals for the mode of *static inspection* applied to a complete instance would necessarily destroy the completeness; so far we have not explored the consequences in detail. Third, dealing with *update* requests is known (from the work on polyinstantiation) to be incompatible in general with notifying refusals: It might happen that, on the one side, accepting the request is impossible since it would invalidate an integrity constraint and, on the other side, notifying a refusal is forbidden since it would reveal a secret.

Finally, the refusal approach behaves like classical *discretionary access control*, which either executes an access request or explicitly denies it. While pure access control can be efficiently implemented, inference-usability confinement tends to be of high computational complexity. We might aim at reducing that complexity for special cases ending up with an acceptable overhead needed for access control anyway. This goal has been achieved for some cases of relational databases reported in [10,12,13]. In these cases the client's a priori knowledge only includes *functional dependencies*, declared as integrity constraints in the schema, and confidentiality policies and queries may not be freely formed. We are currently looking for further promising cases that could narrow the computational gap between refusal-based inference-usability confinement and access control.

7   Lying Approach

The lying approach originates from the challenging proposal made by Bonatti, Kraus and Subrahmanian [20], later rephrased by Biskup [3], adapted for potential secrets by Biskup and Bonatti [5], and further elaborated as indicated below. Again, the lying approach follows a quite straightforward though in this case not necessarily intuitively acceptable idea: If the provider detects that executing a client's request correctly would violate or endanger the confidentiality policy, then the provider deviates from the expected, correct behavior (for short: the provider lies), without notifying the client about the distortions made; otherwise, the request is executed as demanded and the correct return value(s) is (are) sent to the client. Accordingly, the client's view on the actual instance of the information might be incorrect, and the client is aware that this can happen. Clearly, the client should never be able to find out any concrete distortion. Two main issues immediately arise:

- Regarding a malicious agent running the client, can we prevent that purposely modified (lied) reactions as returned to the client nevertheless carry information that could lead to a violation of confidentiality? In particular, can we always guarantee that we can modify reactions consistently, never running into trouble in the future?
- Given an honest agent running the client, does the potentially unreliable knowledge suffice to comply with the agent's obligations, i.e., can we ensure sufficient availability? Moreover, can the agent running the provider take the responsibility – pragmatically justified or ethically founded – to provide incorrect and thus potentially misleading information?

Somehow astonishing, the first issue has a simple solution, at least in our framework. We just have to avoid a careless treatment of disjunctive knowledge, which *in future* might lead to a "hopeless situation" in the following sense: both the correct reaction and every possible distorted reaction would violate the confidentiality policy. More specifically, assume the following:

1. As an alternative reaction, *only* lying in the form of returning the negation of the correct answer is possible.
2. The current knowledge of the client includes the *disjunctive* sentence $\Psi_1 \vee \Psi_2$, for example.

Further assume that the individual sentences Ψ_1 and Ψ_2 are to be protected. Let the client then issue the sequence of queries $\langle \Psi_1, \Psi_2 \rangle$. Both answers have to be distorted. As the final result, the client's view will comprise the sentences $\Psi_1 \vee \Psi_2$, $\neg\Psi_1$ and $\neg\Psi_2$ and thus will become inconsistent, which clearly should be strictly avoided. Describing the example alternatively, we observe that the provider faces the following situation when processing the first (next to the last) query. The client already knows $\Psi_1 \vee \Psi_2$. Returning the correct answer Ψ_1 would violate the protection of that sentence; returning the lied answer $\neg\Psi_1$ would enable the entailment of (the last query) Ψ_2 and thus violate the protection of the other sentence: a "hopeless situation".

If we fix the first point, to avoid hopeless situations, we have to ensure that a client never acquires knowledge about a disjunction of information to be protected. In fact, if we require that individual sentences $\Psi_1, \ldots, \Psi_k$ should be kept secret, then we must additionally take care that the disjunction $\Psi_1 \vee \ldots \vee \Psi_k$ is always kept secret [20,3,5,17,42,18,43]. In other words, we can only handle confidentiality policies that are (explicitly or implicitly) closed under disjunction.

It has been shown that, basically, doing so also suffices to avoid troublesome inconsistencies: The security invariant that the client view does not entail the disjunction of the policy sentences is always maintained when a needed lie is returned; more intuitively expressed, if the correct answer is harmful, the negated answer is guaranteed to be harmless. Notably, such results indicate that there is no need to additionally protect the negation of a correct answer, as identified to be necessary for refusals.

The two assumptions listed above suggest to alternatively relax the first point, i.e., to offer also *refusals* as a possible reaction. In the next section, we will show that this alternative can be successful indeed.

The second issue raised, whether or not we can responsibly apply lying, is much more subtle and should be answered from different points of view and depending on the requirements of a concrete application.

First, for some good reasons, in some situations we might want or be obliged to strictly enforce that only correct information is disseminated, without any exception, and thus decide to employ only refusals. Then, however, comparing refusals and lying, we see a tradeoff regarding availability:

- The refusal approach has to additionally inspect negated answers, whereas the lying approach has not.
- The lying approach has to additionally protect disjunctions of information to be kept secret, whereas the refusal approach has not.

Consequently, banning lying and favoring refusals instead might decrease availability. In the special case of explicitly declaring a policy that is closed under disjunctions, the decrease is apparent. In other cases, the disadvantages and advantages of the two approaches regarding availability are difficult to exactly compare, due to the dynamic "last-minute" distortion strategy [5]. While the refusal approach tends to decline for open queries producing (too) many refusals, the lying approach works suitably cooperatively also for open queries [8]. In fact, we were able to design two variants to deal with the problem of returning a convincing *completeness sentence* (closed-world statement) about the negative part of the controlled answer. Roughly summarized, the first variant produces such a completeness sentence after having fully inspected the positive part of the correct answer and in general even many more tuples as well, whereas the second variant generates such a completeness sentence in advance, at the price of potentially needing more lies for the positive part of the correct answer. Finally, the lying approach also profits from assuming that the client does *not* know the policy instance; basically, then only disjunctions of policy elements that are actually *true* in the instance of the information system must be protected [6].

Second, we might wonder whether a client is able to find out by itself whether or not a reaction is *unreliable*, i.e., roughly described, *potentially* lied. Astonishing enough, as proved in [5] for *closed queries* to a *complete instance* under *known potential secrets*, assuming a disjunctively closed policy, the abstract *information content* of returned answers under the refusal approach and under the lying approach is exactly the same: From the point of view of the client, two instances of the information system are indistinguishable for that client under refusals iff they are indistinguishable for it under lying. Moreover, employing the a priori knowledge postulated by the server to run the control mechanism, the client itself can simulate the instance-independent refusal behavior, and then gets the following guarantee under the lying approach [5]: a received answer is *potentially lied* iff it is refused under the refusal simulation; equivalently, a received answer is *definitely correct* iff it is not refused under the refusal simulation. Thus, essentially, a main difference between lying and refusals can be expressed in terms of *computational complexity*: under refusals the client gets an assurance about correct information "for free" just by design, whereas under lying the client has to take the computational burden of the refusal simulation.

Third, we can complement the confidentiality policy by an explicit *availability policy*, and then enforce that all sentences of the latter policy are never distorted. If the availability policy appropriately covers the obligations of the agent running a client and that agent behaves *honestly* in the sense that he employs the client *only* for serving his obligation, he will always receive only correct information; otherwise, if he will be *curious* beyond his duties, seeing some lies would not affect the "real goals" of his permission to access the information system. Basically, this argument applies for a situation where a security officer carefully examines the application and is able to consistently separate apart the information needed by the client and the information to be kept secret to him by establishing the pertinent conflict-free policies. Under conflict-free policies, a simple means to implement an availability policy is to treat it like a priori knowledge. This approach has been particularly elaborated for the mode of *static inspection* under various conditions [9,18,43,41]. The basic approach is applicable for the dynamic mode as well.

Fourth, as already mentioned in Sect. 6, dealing with *update* requests might require us to employ lying anyway, since by refusals alone we sometimes cannot resolve the conflict between maintaining integrity constraints and keeping secrets. Accordingly, in our first steps towards dealing with updates we employ the lying approach [14,11], and the alternative reactions designed in that work can be considered as some *discretionary* variant of *polyinstantiation*.

Finally, we note that the practical behavior of *query modification* to process open queries [39], as used in, e.g., *mandatory access control* [32] but also by Oracle's concept of "virtual private databases" [33], might be interpreted as a kind of lying. Basically, the correct result relation is filtered using some availability predicate (as an availability policy) describing which tuples are permitted to be shown or some confidentiality predicate (as a confidentiality policy) describing which tuples are to be suppressed, and then only the remaining tuples are returned to the client. If the client is not aware of the filtering and believes to work under a *closed world assumption*, seeing the non-appearance of a tuple in the output as denoting that the corresponding fact is *false*, then any suppressed tuple constitutes a lie for that client.

8   Combined Approach

The combined approach has been invented by Biskup and Bonatti [7], and further elaborated in [6,16,8,17,42], to resolve the tradeoff situation between refusals and lying described and discussed in Sect. 7. Deviating from using one of these approaches uniformly, the combined approach to control the result of a closed query to a complete instance under a *known* policy basically works as follows [7]: If the correct answer – together with the current client view – is harmless in the sense that none of the individual elements of the confidentiality policy is entailed, then the correct answer is returned. Otherwise, if the correct answer is harmful and thus there is an immediate and inevitable need to apply a distortion, the negation of the correct answer is inspected whether it will be harmless – again together with the current client view: if it harmless, this lie is returned; if this lie is harmful, then the answer is explicitly refused. Notably, we need neither to inspect the negated answer in all cases, as in the uniform refusal approach, nor to consider disjunctions of policy elements, as in the uniform lying approach. Consequently, the combined approach tends to be at least as responsive as the uniform approaches, as least regarding the first occurrence of a distortion [7].

We wondered whether the combined approach could be made even more cooperative, i.e., adapted to produce even less distortions, if we assume that the client does *not* know the policy instance. Unfortunately, we were not able to find a decisive result. We conjecture that the combined approach behaves in a "kind of a best possible way", but we were not even able to make such a statement precise [7].

The combined approach can be extended to work suitably cooperatively also for open queries [8]. This extension generates an appropriate completeness sentence in advance, and thus follows the second variant employed for lying. Notably, for the combined approach, we could not design a variant that produces a completeness sentence only after the inspection of the correct result is completed.

9   Perspectives

Controlled interaction execution, as surveyed in this paper, instantiates the basic features listed in Sect. 4 with the parameter variations sketched in Sect. 5. We restricted to a *possibilistic* approach, which merely ensures the existence of at least one "harmless" alternative instance, rather than following a refined *probabilistic* approach, where additionally probabilities of "harmless" alternative instances are considered, e.g., [35,30,26]. Though clearly worthwhile in general, dealing with probabilities raises the difficult problem of determining realistic probability distributions. However, if for simplification only equal distributions over a finite domain are inspected, then probabilistic information gain tends to show characterizations in pure combinatorial terms and the difference between the two approaches might vanish. Ideally, we would like to see results that are similar to the famous characterization of perfect (information-theoretic) encryption, roughly saying that any a priori distribution of "harmful information" is preserved by the controlled reactions and thus will coincide with the corresponding a posteriori distribution. Moreover, like in cryptography, we would like to take also computational feasibility into account and then to deal with *complexity-theoretic* confidentiality depending on some security parameter (like the length of a key), roughly attempting to keep the a priori distributions and the corresponding a posteriori distributions *computationally indistinguishable.*

Our approach allows application-specific control according to an *explicit* confidentiality policy, which can be extensionally declared with very *fine granularity.* Notably, we aim at enabling information flow as a general rule, but to confine it as an exception, only as far as strictly needed to enforce the specific policy instance. These requirements are best satisfied by the *dynamic mode* of control. We expect that our control mechanisms are somehow optimal; however, we still have to elaborate a formal model indicating a precise space of control options to rigorously prove such a claim.

Shortly summarizing the lessons we have learnt so far, we first of all emphasize that protection of *information* (rather than just some data) requires us to consider the features introduced in Sect. 4. Our work provides a *proof of concept* how to achieve the goal in principle, but the actual effectiveness is always subject to the *uncertainty* about the user behavior. The dynamic mode of control ensures high availability of information, but – inevitably – suffers from a high computational complexity in general. However, both the *static mode* for the general case and *optimizations* of the dynamic mode for special situations lead to affordable costs. Finally, we point out once again, as already indicated in Sect. 2, that information protection and usability confinement span a broad topic, which has been successfully treated in many complementary ways as well.

Acknowledgments. I would like to sincerely thank Piero Bonatti for the fruitful cooperation while laying the foundations of this work, and Dominique Burgard, David Embley, Christian Gogolin, Jan-Hendrik Lochner, Jens Seiler, Sebastian Sonntag, Cornelia Tadros, Torben Weibert, and Lena Wiese for their worthwhile contributions while jointly elaborating and extending the approach.

References

1. Bertino, E., Ferrari, E., Atluri, V.: The specification and enforcement of authorization constraints in workflow management systems. ACM Trans. Inf. Syst. Secur. 2(1), 65–104 (1999)
2. Bettini, C., Jajodia, S., Wang, X.S., Wijesekera, D.: Provisions and obligations in policy management and security applications. In: Very Large Data Bases, VLDB 2002, pp. 502–513. Morgan Kaufmann, San Francisco (2002)
3. Biskup, J.: For unknown secrecies refusal is better than lying. Data Knowl. Eng. 33(1), 1–23 (2000)
4. Biskup, J.: Security in Computing Systems - Challenges, Approaches and Solutions. Springer, Heidelberg (2009)
5. Biskup, J., Bonatti, P.A.: Lying versus refusal for known potential secrets. Data Knowl. Eng. 38(2), 199–222 (2001)
6. Biskup, J., Bonatti, P.A.: Controlled query evaluation for enforcing confidentiality in complete information systems. Int. J. Inf. Sec. 3(1), 14–27 (2004)

7. Biskup, J., Bonatti, P.A.: Controlled query evaluation for known policies by combining lying and refusal. Ann. Math. Artif. Intell. 40(1-2), 37–62 (2004)
8. Biskup, J., Bonatti, P.A.: Controlled query evaluation with open queries for a decidable relational submodel. Ann. Math. Artif. Intell. 50(1-2), 39–77 (2007)
9. Biskup, J., Burgard, D.M., Weibert, T., Wiese, L.: Inference control in logic databases as a constraint satisfaction problem. In: McDaniel, P., Gupta, S.K. (eds.) ICISS 2007. LNCS, vol. 4812, pp. 128–142. Springer, Heidelberg (2007)
10. Biskup, J., Embley, D.W., Lochner, J.-H.: Reducing inference control to access control for normalized database schemas. Inf. Process. Lett. 106(1), 8–12 (2008)
11. Biskup, J., Gogolin, C., Seiler, J., Weibert, T.: Requirements and protocols for inference-proof interactions in information systems. In: Backes, M., Ning, P. (eds.) ESORICS 2009. LNCS, vol. 5789, pp. 285–302. Springer, Heidelberg (2009)
12. Biskup, J., Lochner, J.-H.: Enforcing confidentiality in relational databases by reducing inference control to access control. In: Garay, J.A., Lenstra, A.K., Mambo, M., Peralta, R. (eds.) ISC 2007. LNCS, vol. 4779, pp. 407–422. Springer, Heidelberg (2007)
13. Biskup, J., Lochner, J.-H., Sonntag, S.: Optimization of the controlled evaluation of closed relational queries. In: Gritzalis, D., Lopez, J. (eds.) Emerging Challenges for Security, Privacy and Trust. IFIP AICT, vol. 297, pp. 214–225. Springer, Heidelberg (2009)
14. Biskup, J., Seiler, J., Weibert, T.: Controlled query evaluation and inference-free view updates. In: Gudes, E., Vaidya, J. (eds.) Data and Applications Security XXIII. LNCS, vol. 5645, pp. 1–16. Springer, Heidelberg (2009)
15. Biskup, J., Tadros, C., Wiese, L.: Towards controlled query evaluation for incomplete first-order databases. In: Link, S., Prade, H. (eds.) FOIKS 2010. LNCS, vol. 5956, pp. 230–247. Springer, Heidelberg (2010)
16. Biskup, J., Weibert, T.: Confidentiality policies for controlled query evaluation. In: Barker, S., Ahn, G.-J. (eds.) Data and Applications Security 2007. LNCS, vol. 4602, pp. 1–13. Springer, Heidelberg (2007)
17. Biskup, J., Weibert, T.: Keeping secrets in incomplete databases. Int. J. Inf. Sec. 7(3), 199–217 (2008)
18. Biskup, J., Wiese, L.: Preprocessing for controlled query evaluation with availability policy. Journal of Computer Security 16(4), 477–494 (2008)
19. Biskup, J., Wiese, L.: Combining consistency and confidentiality requirements in first-order databases. In: Samarati, P., Yung, M., Martinelli, F., Ardagna, C.A. (eds.) ISC 2009. LNCS, vol. 5735, pp. 121–134. Springer, Heidelberg (2009)
20. Bonatti, P.A., Kraus, S., Subrahmanian, V.S.: Foundations of secure deductive databases. IEEE Trans. Knowl. Data Eng. 7(3), 406–422 (1995)
21. Brodsky, A., Farkas, C., Jajodia, S.: Secure databases: Constraints, inference channels, and monitoring disclosures. IEEE Trans. Knowl. Data Eng. 12(6), 900–919 (2000)
22. Cuppens, F., Gabillon, A.: Cover story management. Data Knowl. Eng. 37(2), 177–201 (2001)
23. Dawson, S., De Capitani di Vimercati, S., Samarati, P.: Specification and enforcement of classification and inference constraints. In: IEEE Symposium on Security and Privacy, pp. 181–195. IEEE, Los Alamitos (1999)
24. De Capitani di Vimercati, S., Samarati, P., Jajodia, S.: Policies, models, and languages for access control. In: Bhalla, S. (ed.) DNIS 2005. LNCS, vol. 3433, pp. 225–237. Springer, Heidelberg (2005)

25. Du, W., Atallah, M.J.: Secure multi-party computation problems and their applications: a review and open problems. In: New Security Paradigms Workshop, NSPW 2001, pp. 13–22. ACM, New York (2001)
26. Evfimievski, A.V., Fagin, R., Woodruff, D.P.: Epistemic privacy. In: Principles of Database Systems, PODS 2008, pp. 171–180. ACM, New York (2008)
27. Farkas, C., Jajodia, S.: The inference problem: A survey. SIGKDD Explorations 4(2), 6–11 (2002)
28. Goldreich, O.: Foundations of Cryptography II – Basic Applications. Cambridge University Press, Cambridge (2004)
29. Gollmann, D.: Computer Security, 2nd edn. John Wiley and Sons, Chichester (2006)
30. Halpern, J.Y., O'Neill, K.R.: Secrecy in multiagent systems. ACM Trans. Inf. Syst. Secur. 12(1), 5.1–5.47 (2008)
31. Ishihara, Y., Morita, T., Seki, H., Ito, M.: An equational logic based approach to the security problem against inference attacks on object-oriented databases. J. Comput. Syst. Sci. 73(5), 788–817 (2007)
32. Lindgreen, E.R., Herschberg, I.S.: On the validity of the Bell-La Padula model. Computers & Security 13(4), 317–333 (1994)
33. Lorentz, D., et al.: Oracle Database SQL Language Reference, 11g Release 1 (11.1). B28286-03, Oracle Corporation (2008),
    `http://www.oracle.com/pls/db111/to_pdf?partno=b28286`
34. Machanavajjhala, A., Kifer, D., Gehrke, J., Venkitasubramaniam, M.: l-diversity: Privacy beyond k-anonymity. In: TKDD, vol. 1(1) (2007)
35. Miklau, G., Suciu, D.: A formal analysis of information disclosure in data exchange. J. Comput. Syst. Sci. 73(3), 507–534 (2007)
36. Pretschner, A., Hilty, M., Basin, D.A.: Distributed usage control. Commun. ACM 49(9), 39–44 (2006)
37. Pretschner, A., Hilty, M., Basin, D.A., Schaefer, C., Walter, T.: Mechanisms for usage control. In: Information, Computer and Communications Security, ASIACCS 2008, pp. 240–244. ACM, New York (2008)
38. Sicherman, G.L., de Jonge, W., van de Riet, R.P.: Answering queries without revealing secrets. ACM Trans. Database Syst. 8(1), 41–59 (1983)
39. Stonebraker, M., Wong, E.: Access control in a relational data base management system by query modification. In: ACM/CSC-ER Annual Conference, pp. 180–186. ACM, New York (1974)
40. Stouppa, P., Studer, T.: Data privacy for knowledge bases. In: Artemov, S., Nerode, A. (eds.) LFCS 2009. LNCS, vol. 5407, pp. 409–421. Springer, Heidelberg (2009)
41. Tadros, C., Wiese, L.: Using SAT solvers to compute inference-proof database instances. In: Fourth International Workshop on Data Privacy Management, DPM 2009 (2010) (to appear)
42. Weibert, T.: A Framework for Inference Control in Incomplete Logic Databases. PhD thesis, Technische Universität Dortmund (2008),
    `http://hdl.handle.net/2003/25116`
43. Wiese, L.: Preprocessing for Controlled Query Evaluation in Complete First-Order Databases. PhD thesis, Technische Universität Dortmund (2009),
    `http://hdl.handle.net/2003/26383`
44. Winslett, M., Smith, K., Qian, X.: Formal query languages for secure relational databases. ACM Trans. Database Syst. 19(4), 626–662 (1994)
45. Zhang, Z., Mendelzon, A.O.: Authorization views and conditional query containment. In: Eiter, T., Libkin, L. (eds.) ICDT 2005. LNCS, vol. 3363, pp. 259–273. Springer, Heidelberg (2004)

AccKW: An Efficient Access Control Scheme for Keyword-Based Search over RDBMS

Vikram Goyal, Ashish Sureka, and Sangeeta Lal

Indraprastha Institute of Information Technology Delhi
New Delhi, India
{vikram,ashish,sangeeta}@iiitd.ac.in

Abstract. Access control for relational databases is a well researched area. An SQL query is allowed or denied access to database according to the specified access control policy. On the other side, there has been a surge in research activities to provide keywords-based search interface over RDBMS. This has posed new challenges for access control enforcement as traditional solutions to access control will not be efficient for keyword-based search. This paper proposes a framework AccKW, which enforces access control policies on keyword-based search over RDBMS in the early phases of keywords based search process. The main contributions of this paper are twofold: (i) we have investigated the problem of access control in the domain of keyword-based search over relational databases, and (ii) we have implemented the framework AccKW, and found out that AccKW outperforms in terms of execution time as compared to the naive approach (brute force approach) in case of strict access control policy.

1 Introduction

Keyword-based search has become popular with the advent of Internet search engines. Although it may at times be challenging to end-users to specify a good set of keywords for their respective queries, due to ease in specification of query keyword-based search has become one of the mostly used search paradigm in many domains.

Recently, there has been a surge in research activity in the area of keyword-based search over relational databases [1,2,3,4,5,6,7]. The ability to search relational databases using keywords allows end-users to search relevant information without any knowledge of the database schema and SQL. BANKS [2,4,5], DbXPlorer [1] and DISCOVER [3] are few systems that have been proposed in the recent past. Based on our literature survey, we observed that the research on keyword-based search on structured or semi-structured databases has focused on aspects such as strategy for keyword based search, search effectiveness and efficiency.

This paper focuses on access control for keyword-based search in enterprise domain. Search in enterprise domain would be different from search in the Internet domain as the set of users and their information need may be known in enterprise domain. Further, access control is a critical aspect in enterprise domain which needs to be addressed to enable wide scale adoption and deployment of keyword-based search solutions on relational databases. Some examples of access control in enterprise domain are:

S. Kikuchi, S. Sachdeva, and S. Bhalla (Eds.): DNIS 2010, LNCS 5999, pp. 107–121, 2010.

- In a health organization, it may be desired to allow patients to see only their medical information in the medical database. On the other hand, a doctor should get access to all the medical information of the patients she is treating.
- In an academic institution information system, it may be desired that a student should be able to see only her grade. On the other hand, an instructor should be able to able to access all the grades for a course she has taught [8].
- In a bank database, a customer should be able to see only her bank account information. At the same time, a manager in the bank should be able to read transactions history of each bank account but not the personal information of customers.

Several techniques and models have been proposed as well as implemented in some of the commercial database management systems [9,8,10,11]. However, access control for relational databases is a well researched area, but, access control within the context of keyword-based search over relational databases poses certain unique challenges. According to our knowledge, access control for keyword-based search over relational databases in an enterprise domain is a relatively unexplored area. The work presented in this paper lies at the intersection of access control and keyword-based search on relational databases. Existing access control mechanisms on databases assume only SQL query interface. They analyze a user provided SQL statement's FROM clause and WHERE clause to decide the authorization. There are also some solutions which transform or rewrite an SQL query into a valid authorized SQL query if the input query is unauthorized [9,8].

In this paper, we have proposed a framework, AccKW, which integrates access control routines within the keyword-based search strategy. We identified that in some cases the naive approach would generate very large number of SQL queries which should then be processed through access control routines. Incorporation of access control in the early phases of search methodology drastically reduces the time to generate authorized queries. We define two performance metrics: (i) execution time to generate the queries and, (ii) number of authorized queries generated, to evaluate the performance of our framework. Experiments have been performed by varying number of input keywords, keywords selectivity, database size and access control policy. The results show that AccKW outperforms the naive approach when the access control policy is strict.

The key contributions of this paper are:

1. we have investigated the problem of access control in the domain of keyword-based search over relational databases, and have proposed a framework AccKW which enforces access control policies cost effectively in this domain.
2. we implemented the framework AccKW, and found out that AccKW outperforms in terms of execution time as compared to the naive approach in case of strict access control policy.

The rest of the paper is organized as follows. Section 2 describes related work. In Section 3, we describe the architecture of the AccKW framework. Section 4 presents our formal model of access control. Section 5 describes the algorithm used to enforce the access control policy in keyword-based search system. In Section 6 we describe the experimentation results. Section 7 concludes the paper.

2   Related Work

The work related to this paper can be discussed from two perspectives: access control and keyword-based search in RDBMS. We describe the work done in these two areas in this Section.

There has been extensive work in access control enforcement in relational databases [9,8,12,13,14]. The most closely related work is the Virtual Private Database (VPD) model of Oracle [9]. It uses the functions defined on relations by the administrator to enforce an access control policy. These functions return a predicate string for a given SQL query from a user, which is appended to the query. VPD is an excellent model to enforce access control policies but works on a single SQL query at a time. In case of keyword-based search system, this model will not be efficient as it would require all the SQL queries either authorized or unauthorized to be generated, whicshould then be analyzed by VPD model. Our proposed technique enforces access control policies during formation of SQL queries from input keywords and generates only valid SQL queries. We then use Oracle's VPD model to append predicates on the generated valid SQL queries to get a set of authorized SQL queries.

Similarly, the work proposed in [8,14] also considers a single SQL query and rewrites that SQL query into an authorized SQL query. These works focus on theoretical aspects of an SQL query after authorization rewriting. The work in [14] also discusses on optimal generation of safe plan of query execution to prevent leakage due to user defined functions having side effects. All these work are complementary to our work and can be used after generation of valid SQL queries. The work in [15] and the work in [13], both have described Cell-level authorization and its implementation techniques. However, there techniques are not general and are restricted to privacy policy enforcement. They also consider a single SQL query at a time and rewrites it or filter the query's result for enforcement of privacy. We have worked for general purpose access control and prune unauthorized SQL queries in the earlier phases.

Keyword-based search has been a focus of researchers in database community in the recent years [1,2,3,4,5,6,7]. They allow an application user to make a query using a set of keywords. It has made the task of application users easier as neither they should know and remember the SQL query semantics to search the information nor should they have knowledge about the database schema. Users input a set of keywords, and they get a set of ranked results. The search using keywords over databases [1,5] is different from search over Internet, as keywords are spread among a set of relations due to normalization of database. All the keyword-based search techniques over structured databases focus on aspects such as strategy for keyword-based search, search effectiveness and efficiency. None of this work addresses access control enforcement issues.

However, there have been proposed many different strategies for keyword-based search over relational databases. Our proposed access control enforcement strategy AccKW can be applied in conjunction with all of these techniques [3,5] in a cost effective way. For this paper, we choose the technique proposed in DbXplorer [1] as an example. DbXplorer uses two phases for reporting the result. These two phases are called publish phase and search phase respectively. In the publish phase, an inverted index is created using cell values in the database tables. Each entry of inverted index also called master index or symbol table has three values, i.e. keyword, relation-id and an attribute. The pair *<relation-id, attribute>* can be seen as a pointer to the table's attribute having the keyword as its cell value. The database schema graph is created using entities as nodes of the graph and primary key and foreign key relationship as an edge. This schema graph is then used in the search phase, in which input keywords of a user are annotated on the graph using the inverted index. Then different join trees containing all the input keywords are generated. Each join tree is finally translated to an SQL query and is executed on the database to get a partial result. In this paper, we adapt this technique to our needs. We call the adapted version of this technique as VarDbXplorer.

3   System Architecture

Figure 1 presents the system architecture of the proposed AccKW framework. The system architecture consists of modules already present in the DBXplorer system and additional modules introduced by us. The modules added by us are differentiated from existing modules by applying shading in the system architecture block diagram presented in Figure 1. As illustrated in Figure 1, we extend the existing DBXplorer system by introducing three functions: Keyword Tuples Filter, Access Control Policy and Schema Graph Modifier. The functionality of each of the module is described in the following paragraphs.

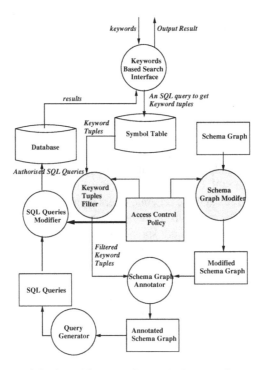

Fig. 1. Architecture of AccKW Framework

The Keywords Based Search Interface (KBSI) is the front-end module through which the application user interacts with the system. The end-user provides query keywords based on his or her information needs through the KBSI module. The search results obtained from the underlying Database is presented to the user through the same KBSI module. Typically, this module can be either a web-based interface or a form-based interface. The KBSI module is responsible for constructing an SQL query based on the supplied input keywords. The generated SQL query is then executed on the Symbol Table (a master index table implemented in the database) to get the keyword tuples. The resultant keyword tuples are passed on to the Keyword Tuples Filter (KTF) module.

The KTF module applies access control policy rules on these tuples and filters out keyword tuples having either unauthorized relations or unauthorized attributes. The filtered keyword tuples are eliminated and is not passed on to the next module in the pipeline for further processing. For example, consider a fine-grained access control rule which says that if the value of the age variable or field in a record is less than 30, then such a record should not be accessible to a specific user or role. In such a case, any keyword tuple consisting of age attribute and having the age value less than 30 will be filtered out from further processing. The module uses the application users context information to decide the access control rules for that user. Context information may include attributes such as role-id, user-id, time and location information.

The Schema Graph Modifier (SGM) module prunes the schema graph for a specific user using the access control rules applicable to the user. The pruning consists of operations such as relation node deletion, edge deletion, and specific node attributes deletion (i.e., node label modification). This module generates a Modified Schema Graph for the specific user.

The Schema Graph Annotator module takes filtered keyword tuples and modified schema graph as the inputs and produces an Annotated Schema Graph. This module annotates the keywords from the keyword tuples on the nodes of the schema graph. The Query Generator module generates all the valid join trees as described in [1] and forms the set of valid SQL queries. These valid SQL queries are then processed by the SQL Queries modifier (SQLM) module. The SQLM module takes each SQL query and the access control policy as inputs. It appends the predicates to the query by analyzing the relation and attribute information present in both the SQL query and the access control policy. For example, consider a rule wherein an access to a relation is authorized only if the user is accessing his or her information. In such a situation, a predicate checking for the match in user-id will be appended to the query. The queries generated by this module are called authorized SQL queries and are executed on the database. The result set of each query is forwarded to the KBSI module for final display to the application user.

4   Formal Model of Access Control

We assume that the access control policy has been defined as a set of rules. Each rule is a quad-tuple $(S, A, O, Auth)$. A rule defines an authorization decision $Auth$ (allow, deny) for a subject S for an object O (a set of tuples of a table in database) for an action A. For example, consider a rule $(manager, select, project, allow)$. The rule defines that a *manager* is allowed to access *project* table tuples. Subjects represent end-users as well as groups of users (wherein a group of users can be represented as a role). In general, an action on databases can be any common operation such as select, insert, update, and delete. In this paper, we investigate the select operation as it is the most widely used operation in keyword-based search over RDBMS.

The object component of the rule is specified as a pair: *(table_exp, predicate)*. The first tuple of the pair i.e., *table_exp* is a valid table expression which can be any valid projection of attributes of a table. The second tuple of the pair i.e., *predicate*, is either any valid WHERE clause condition of an SQL query as specified in [9] or a sequence of tables. If predicate is like a WHERE clause, then it can include exists sub-queries or conditions on more than one table attributes. For example, consider a rule *(salesman, select, (Employee.salary, salary >20000), deny)*. The rule denotes that a salesman cannot access salary of an employee if the salary of that employee is more than Rs 20,000. We specify following types of rules using above specified quad-tuple *(S, A, O, Auth)*:

1. **Entity Constraint.** An entity constraint defines authorization for a database relation for a subject. For an example, rule *(student, select, (faculty,), deny)* is an entity constraint and defines that a student is denied any type of access on faculty table. In other words, relation faculty can not occur in FROM clause of an SQL query issued by subject student.

2. **Context based Entity Constraint.** A context based entity constraint defines authorization for a relation by using predicates on attribute's value. As an example, rule *(student, select, (faculty,student.instructor_id != faculty.id), deny)* is a context based entity constraint and specifies that a student is not allowed to access faculty table tuples if faculty is not student's instructor.

3. **Entity-Attribute Constraint.** This constraint defines authorization for attributes of a database relation. As an example, rule *(student, select, (faculty.salary,), deny)* specifies that a student can not access any faculty's salary information. Further, salary attribute of faculty relation can not be used any where in the WHERE clause of an SQL query from student.

4. **Context-based Entity-Attribute Constraint.** This constraint uses predicates to define authorization for attributes of a relation. As an example, rule *(faculty, select, (student.marks, student.instructor_id != faculty.id), deny)* is a context-based entity-attribute constraint and specifies that a faculty can not access marks of a student if the faculty is not the instructor of that student.

5. **Entity Path Constraint.** An entity path constraint defines authorization on a set of relations in the database. For an example, rule *(student, select, (faculty, personal_detail), deny)* specifies that a student can not access faculty and personal_detail relation together in a single SQL query.

As described earlier in this Section, access control policy is a set of rules and we do not define any priority on the basis of ordering of rules in the policy specification. However, due to presence of both allow and deny decision there may occur a conflict for a query, in that case we give higher precedence to deny semantics.

5   Algorithm

We present an algorithm in this Section which enforces all the constraint types described in Section 4 in keyword-based search domain.

Algorithm 1. Algorithm to determine authorized SQL queries for a given set of keywords

---

Input: 1. Access control Policy, A
             2. Symbol Table, ST
             3. A schema graph, G
             4. User's Context Information, U
             5. Input keywords from a user U, $input\_keywords$

Output: Authorized SQL Queries set $output\_queries$

Method:
1. $output\_queries$ = []
2. key_tuples = []
3. **for all** $keyword$ in $input\_keywords$ **do**
4. k_tuples = get key_tuples from ST
5. key_tuples.extend(k_tuples)
6. **end for**
7. $filtered\_key\_tuples$ = prune-key-tuples(key_tuples, A, U)
8. **if** $\forall\, k \in keywords$, $k \notin key\_tuples$ **then**
9.   return $output\_queries$
10. **end if**
11. $pruned\_graph$ = prune-graph(G, A, U)
12. $annotated\_gr$ = annotate-graph($pruned\_graph$, $filtered\_key\_tuples$
13. $valid\_trees$ = get-valid-join-trees($annotated\_gr$)
14. $pruned\_trees$ = prune-trees($valid\_trees$)
15. $valid\_queries$ = get-valid-queries($pruned\_trees$)
16. $output\_queries$ = rewrite-queries($valid\_queries$, A, U)
17. return $output\_queries$

We now describe functions prune-key-tuples(key_tuples, A, U), prune-graph(G, A, U), prune-trees($valid\_trees$) and rewrite-queries($valid\_queries$, A, U) invoked in Algorithm 1.

- **prune-key-tuples**(key_tuples, A, U): In this function, all the keyword tuples which are not accessible to the user are removed. This function uses first four types of constraints to determine removable keyword tuples *(keyid,relid,attid)* in the following way:
  - A keyword tuple will be removed if it contains a relid value also present in object tuple of any entity constraint rule with deny authorization.
  - A keyword tuple will be removed if it contains a relid value also present in object tuple of context based entity constraint rule with deny authorization and the rule's predicate is true for keyid value.

- A rule of type entity attribute constraint would remove a keyword tuple if the rule has deny authorization and rule's relation and attribute ids match with the tuple's relid and attid.
  - Context based attribute constraint rule with deny authorization removes a keyword tuple if the tuple's keyid value makes rule's predicate true and also matches in relid and attid value of the rule's object tuple.
- **prune-graph**(G, A, U): This function prunes the schema graph and uses first and third type of constraints in the following way:
  - An entity constraint rule with deny authorization removes a node with relation label from the schema graph.
  - A rule of type entity attribute constraint removes attribute from the label of the relation node specified in the rule. It also removes an edge from the schema graph if the edge label has the attribute specified in rule.
- **prune-trees**($valid\_trees$): This function uses path constraint type rules to delete inaccessible join trees, i.e., if a join tree has relation nodes present in any of the path constraint rule then the join tree is deleted.
- **rewrite-queries**($valid\_queries$, A, U): In this function, valid queries are rewritten to get the authorized set of queries. This step is similar to Oracle's VPD approach and uses context based entity constraint, and context based entity attribute constraint types rules. These constraints are very similar to dynamic ACL association tuple of instance set in Oracle's VPD.

We illustrate Algorithm 1 through an example.

Illustrative Example

We present below an example to demonstrate our algorithm. The schema for example is given as below:

Employee(eid, name, age, gender, pan, salary, phone, address)
Project(pid, name, budget, start_date, duration, details)
Emp-Proj(eid, pid, join_date, role)
Sponsors(sid, name, phone, type, investment, did)
Complains(cid, pid, date, details)

The schema graph for this schema would be as given in Figure 2.
Let us further assume the following authorizations for a user on this database.

1. *(clerk, select, (Complains,), deny)* : A clerk can not execute any operation on Complain relation.
2. *(clerk, select, (Sponsors.investment,), deny)* :A clerk can not execute select query on sponsors investment attribute.
3. *(clerk, select, (Employee,Project.budget > 30 and Employee.eid=Emp-Proj.eid and Emp-Proj.pid=Project.pid), deny)* : A clerk can not access other employee information if employee is involved in at least one project with budget greater than 30 lakhs.

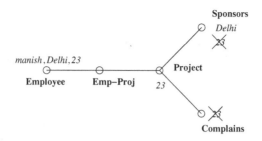

Fig. 2. Example Schema Graph

4. *(clerk,select, (Employee, Employee.age > 35), deny)*: A clerk can not access employee information if employee age is more than 35.

Let us now assume that the user with clerk role inputs Manish, Delhi and 23 keywords for information search. Keywords annotation in the schema graph for the above given authorization rules is shown in Figure 2. Figure shows crosses on relation Complains and keyword '23'. This is due to the reason that Complains relation and investment attribute of relation Sponsors are not accessible to the user. We have shown these crosses for the clarity purpose but in actual, these keywords will not be annotated to the schema graph due to filtering of keyword tuples and pruning of schema graph.

The following set of join trees as shown in Figure 3 will be generated in this case. The authorized SQL queries would be as given below:

– Select name, address, age from Employee where name = 'manish' and address = 'Delhi' and age = 23 and not (age > 35)
– Select name, age, address from Employee, Emp-Proj, Project, Sponsors where Employee.eid = Emp-Proj.eid and Emp-Proj.pid = Project.pid and

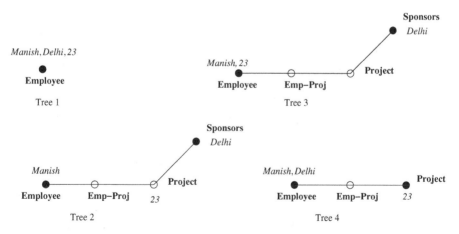

Fig. 3. Join Trees

Project.pid = Sponsors.pid and Employee.name = 'manish' and Employee.age = 23
and Sponsors.address = 'Delhi' and not (Project.budget > 20) and not (age >35)
- Select name, budget, address from Employee, Emp-Proj, Project, Sponsors where
  Employee.eid = Emp-Proj.eid and Emp-Proj.pid = Project.pid and Project.pid =
  Sponsors.pid and Employee.name = 'manish' and Project.budget = 23 and Spon-
  sors.address = 'Delhi' and not (Project.budget > 30) and not (age > 35)
- Select name, budget, address from Employee, Emp-Proj, Project where
  Employee.eid = Emp-Proj.eid and Emp-Proj.pid = Project.pid and Employee.name
  = 'manish' and Project.budget = 23 and Employee.address = 'Delhi' and not
  (Project.budget > 30) and not (age > 35)

6 Experimental Study

Our performance evaluation consists of experiments on our implementation of the VarD-
bXplorer (which is an adapted version of DBXplorer) and Access Control on TPC-H
[16] database on a HP desktop machine. The machine is a Intel(R) Core Due CPU
2.66GHz with 3GB of main memory. The experiment is implemented in Python version 2.5.2 and uses MySQL database system [17] at the backend.

We have used TPC-H database [16] for performance study as TPC-H database use is
common in researchers working on keywords based search on RDBMS [1,3]. The sizes
of the database are 10 MB, 100 MB, and 1GB generated using scaling factor of .01, .1
and 1 with dbgen utility of TPC-H benchmark.

We adapt DbXplorer keyword-based search strategy [1] for our scenarios and term
that as VarDBXplorer. The master index or symbol table is implemented as an inverted
index table in MySQL which is a column based approach [1]. We have written a python
program to make a master index for every value present in the database. Each value in
the database is stored with its occurrence information in the symbol table. A value is
a single word and is an alphanumeric expression which is obtained from an attribute
value after lexical analysis process. For an example, an address value is decomposed
into words using space as a delimiter and removing control characters. A value may
occur more than once in an attribute of a relation or in more than one tuple of a relation.
An entry of symbol table can be represented as a 3-tuple $< value, T_i, A_{ij} >$, where
$value$ is a keyword, T_i is a relation name and A_{ij} is j_{th} attribute in relation T_i. To
select keywords with a particular frequency, we created another table key_stats, which
is derived from master index table by executing a $group\ by$ SQL query on master index
table. Each tuple of this table is a (value, frequency) pair. Here, frequency specifies the
number of attributes in which the value occurs.

We measure the efficiency of applying access control policy in the early phases of
search in terms of performance parameters: time of execution (T_{exec}) in seconds and
number of generated queries (Num_{ge}). For that, we specified five types of access control policies which vary in terms of access restrictions for a user. Policy-1 has the least
restriction and Policy-5 has the maximum restriction among these 5 policies. Each policy rule is of the type defined in 4, i.e., entity constraint, context based entity constraint
etc. We considered different database sizes for evaluation, for that we have generated
data using different scaling factor of dbgen utility, i.e., 10 MB, 100 MB and 1000 MB

using scaling factor of .01, .1 and 1. We have also studied the effect on time and number of generated queries for variation in number of input keywords. For that we have used 2 to 6 keywords in a query. We have not used more than 6 keywords as it is quite uncommon that a user search has more than 6 keywords. We have also studied the effect of variation in keyword selectivity. Keyword selectivity of a keyword is defined as number of attributes in the database in which the keyword is present. The parameters table is given as Table 1

Table 1. Parameters Table

PARAMETER	DEFAULT VALUE	VARIATIONS
Access Control Policy Quantifier (λ)	3	1,2,4,5
Number of Keywords (n)	5	2,3,4,6
Selectivity of Keywords (F)	5	1 to 4, 6 to 10
Database Size (S)	1 GB	100 MB, 10 MB

Effect of Variation in Access Control Policy

The performance graphs of effect in variation of access control policy is shown in Figure 4a and Figure 4b. As explained earlier, the policies from policy-1 to policy-5 varies in terms of strictness. Policy-1 is the most relaxed one and policy-5 is the strictest one among the five. We see that as the access control policy becomes more restrictive for a user, it reduces the execution time T_{exec} to generate authorized queries as well as the number of authorized queries. Figure 4a shows that VarDBXplorer approach always takes more time as compared to AccKW. Figure 4b shows the effect of different access control policies on Num_{ge} and has log scale for Y-axis. It shows that the value of Num_{ge} goes on decreasing as the access control policy becomes more restrictive.

We find that this is due to pruning of schema graph size and authorized tuple set in the early phases. This reduction in the size of the schema graph results into less processing and hence reduction in both time T_{exec} and number of queries Num_{ge}.

Effect of Variations in the Number of Keywords

We did this experiment by selecting 2 to 6 keys at random from the master index which has selectivity (F) equal to 5.

The performance graphs of this experiment are shown in Figure 5a and Figure 5b. Figure 5b has log scale for Y-axis. The experiment shows that T_{exec} and Num_{ge} increases as the number of keywords (n) increases for a given access control policy option (default access control policy 3). This is due to generation of more number of tuples with the increase in number of keywords, that need to be annotated with the pruned schema graph. More keywords annotation to schema graph leads to more number of queries generation as well as more execution time.

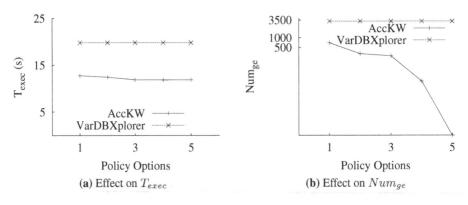

(a) Effect on T_{exec} (b) Effect on Num_{ge}

Fig. 4. Effect of variation in Access Control Policy

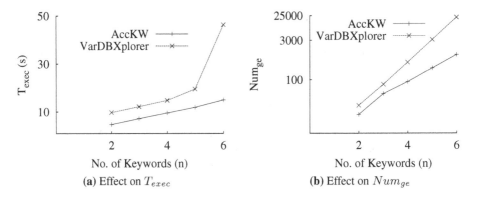

(a) Effect on T_{exec} (b) Effect on Num_{ge}

Fig. 5. Effect of Variation in Number of Keywords

Effect of Variations in Keyword Selectivity

The performance graphs of this experiment are shown in Figure 6a and Figure 6b (with log scale on Y-axis). The graphs show that the increase in the selectivity of keywords increases the T_{exec} as well as Num_{ge} for a given access control policy. The reason for this is similar to the previous experiment that the increase in selectivity of a keyword results into more tuples generation. This increase in tuples in the tuple set results in more number of keys to be annotated with the schema graph and hence increase in execution time Figure 6a and generated authorized queries Figure 6b.

Effect of Variations in the Database Size

Figure 7a and Figure 7b shows the performance results of variation in database size on T_{exec} and Num_{ge} respectively. Graph in Figure 7b uses log scale for Y-axis. As

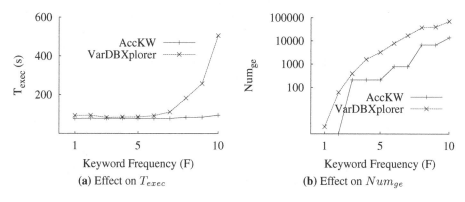

Fig. 6. Effect of variations in keyword selectivity

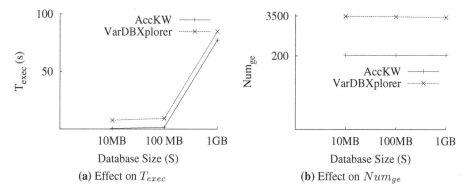

Fig. 7. Effect of variations in the database size

discussed earlier, we generated data of different sizes using scaling factor parameter of dbgen utility of TPC-H benchmark for this experiment.

The result shows a minimal effect of data sizes on execution time and number of queries for a given access control policy, keyword selectivity and number of keys. This constant difference is due to the difference in time to get the keyword-tuples from the master index. After keyword tuple selection, the process of authorized query generation will be same for each data size version.

7   Conclusions

This paper proposes a novel framework called AccKW which enforces access control in keyword-based search over RDBMS. Solutions for access control enforcement in the domain of keyword-based search over RDBMS is a relatively unexplored area and the work presented in this paper addresses the stated research gap. The paper proposes a solution framework and also discusses issues and challenges regarding access control in this domain. The proposed framework is implemented and emperically evaluated. The paper presents performance results from different perspectives such as efficiency and performance impact as a result of variation in number of keywords, access control policy, database size, and keyword selectivity. Based on the empirical evaluation and simulation results, we conclude that the proposed framework outperforms the naive approach in most of the cases.

References

1. Agrawal, S., Chaudhuri, S., Das, G.: Dbxplorer: A system for keyword-based search over relational databases. In: ICDE 2002: Proceedings of the 18th International Conference on Data Engineering, Washington, DC, USA, p. 5. IEEE Computer Society, Los Alamitos (2002)
2. Aditya, B., Bhalotia, G., Chakrabarti, S., Hulgeri, A., Nakhe, C., Parag, P., Sudarshan, S.: Banks: browsing and keyword searching in relational databases. In: VLDB 2002: Proceedings of the 28th international conference on Very Large Data Bases, VLDB Endowment, pp. 1083–1086 (2002)
3. Hristidis, V., Papakonstantinou, Y.: Discover: keyword search in relational databases. In: VLDB 2002: Proceedings of the 28th international conference on Very Large Data Bases, VLDB Endowment, pp. 670–681 (2002)
4. Hulgeri, A., Nakhe, C.: Keyword searching and browsing in databases using banks. In: ICDE 2002: Proceedings of the 18th International Conference on Data Engineering, Washington, DC, USA, p. 431. IEEE Computer Society, Los Alamitos (2002)
5. Kacholia, V., Pandit, S., Chakrabarti, S., Sudarshan, S., Desai, R., Karambelkar, H.: Bidirectional expansion for keyword search on graph databases. In: VLDB 2005: Proceedings of the 31st international conference on Very large data bases, VLDB Endowment, pp. 505–516 (2005)
6. Koutrika, G., Simitsis, A., Ioannidis, Y.: Précis: The essence of a query answer. In: International Conference on Data Engineering, vol. 0, p. 69 (2006)
7. Simitsis, A., Koutrika, G., Ioannidis, Y.: Précis: from unstructured keywords as queries to structured databases as answers. The VLDB Journal 17(1), 117–149 (2008)
8. Rizvi, S., Mendelzon, A., Sudarshan, S., Roy, P.: Extending query rewriting techniques for fine-grained access control. In: SIGMOD 2004: Proceedings of the 2004 ACM SIGMOD international conference on Management of data, pp. 551–562. ACM, New York (2004)
9. Murthy, R., Sedlar, E.: Flexible and efficient access control in oracle. In: SIGMOD 2007: Proceedings of the 2007 ACM SIGMOD international conference on Management of data, pp. 973–980. ACM, New York (2007)
10. Olson, L.E., Gunter, C.A., Madhusudan, P.: A formal framework for reflective database access control policies. In: CCS 2008: Proceedings of the 15th ACMconference on Computer and communications security, pp. 289–298. ACM, New York (2008)
11. Shin, H., Atluri, V.: Spatiotemporal access control enforcement under uncertain location estimates. In: Gudes, E., Vaidya, J. (eds.) Data and Applications Security XXIII. LNCS, vol. 5645, pp. 159–174. Springer, Heidelberg (2009)
12. Chaudhuri, S., Dutta, T., Sudarshan, S.: Fine grained authorization through predicated grants. In: ICDE 2007: Proceedings of the 23rd International Conference on Data Engineering, Istanbul, Turkey, pp. 1174–1183. IEEE, Los Alamitos (2007)
13. Agrawal, R., Bird, P., Grandison, T., Kiernan, J., Logan, S., Rjaibi, W.: Extending relational database systems to automatically enforce privacy policies. In: ICDE 2005: Proceedings of the 21st International Conference on Data Engineering, Washington, DC, USA, pp. 1013–1022. IEEE Computer Society, Los Alamitos (2005)

14. Kabra, G., Ramamurthy, R., Sudarshan, S.: Redundancy and information leakage in fine-grained access control. In: SIGMOD 2006: Proceedings of the 2006 ACM SIGMOD international conference on Management of data, pp. 133–144. ACM, New York (2006)
15. Lefevre, K., Agrawal, R., Ercegovac, V., Ramakrishnan, R., Xu, Y., DeWitt, D.: Limiting disclosure in hippocratic databases. In: VLDB, pp. 108–119 (2004)
16. TPC-H decision support benchmark (Transaction Processing Council),
http://www.tpc.org/
17. MySQL Database, http://www.mysql.com/

Social Intelligence Design for Knowledge Circulation

Toyoaki Nishida

Graduate School of Informatics, Kyoto University
Sakyo-ku, Kyoto 606-8501, Japan
nishida@i.kyoto-u.ac.jp

Abstract. Knowledge circulation is indispensable to improving the coverage and quality of knowledge shared in a community. In order for information and communication technologies to be successfully applied to realize knowledge circulation, social aspects need to be considered so that the technologies can be properly embedded into the society. This issue has been addressed in social intelligence design, a field of research aiming at understanding and augmenting social intelligence based on a bilateral definition of social intelligence as an individual's ability to better live in a social context and a group's ability to collectively solve problems and learn from experiences. In this paper, based on an overview of social intelligence design research, I present a generic framework of conversational knowledge circulation in which conversation is used as a primary means for communicating knowledge. I present attentive agents, autonomous interaction learner, situated knowledge management, self-organizing incremental memory, immersive conversation environment, as key technologies in conversational quantization for conversational knowledge circulation.

Keywords: Social Intelligence Design, Conversational Knowledge Circulation, Situated Knowledge Management.

1 Introduction

The advent of the information and communication technologies has significantly increased the amount of information available on the net. Kitsuregawa [1] called it information explosion. Information explosion brings about both negative and positive aspects. On the one hand, we often feel overloaded by the overwhelming amount of information, such as too many incoming e-mail messages including spams and unwanted ads. On the other hand, explosively increased information may also lead to a better support of our daily life. We can access not only public and infrastructure information such as the contact address of public service but also personal twitters and diaries that tell us how other people feel about perceived events.

Still we often run into problems which may be attributed to the lack and incompleteness of information and knowledge. From time to time, we are forced to waste long time to fix simple problems or loose critical moments due to the lack of timely information provision. After all, we are still suffering from unevenly distributed information

S. Kikuchi, S. Sachdeva, and S. Bhalla (Eds.): DNIS 2010, LNCS 5999, pp. 122–142, 2010.

and knowledge. In the industrial domain, uneven distribution of information can be thought of as a potential cause of various flaws in service such as incomplete manuals, ill-designed user interface, excessive functions, or even brittle machineries.

Information and knowledge need to be circulated so that demands for information can be communicated to information holders or providers, and information and knowledge can be communicated to those who need it. In the industrial domain, information and knowledge sharing among specialists in different sectors of service providers is considered to be a gold standard for service provision. In addition to it, communicating the engineers' intention underlying the products may lead the clients to better leverage the services and products; communicating demands of the clients may motivate service providers to design new services; communicating usage reports may force engineers to improve the quality of services and products; not only bug reports or criticisms but also reports on novel usage and proposals of new functions from skilled users might highly encourage new services and products.

In general, information and knowledge circulation is critical to ensure the coverage and quality of knowledge. Knowledge circulation may contribute to bringing about good coverage, by communicating information and knowledge demands among people. Knowledge circulation may help improve the quality of knowledge by collecting flaws, criticisms and proposals for products and services from people.

Although information and communication technologies are powerful, a simple deployment of they will not be enough. Social intelligence design [2, 3] is a field of research aiming at understanding and augmenting social intelligence based on a bilateral definition of social intelligence as an individual's ability to better live in a social context and a group's ability to collectively solve problems and learn from experiences. Issues in embedding information and communication technologies in the human society have been discussed in the context of social intelligence design.

In what follows, I first give an overview of social intelligence design research. Then, I present a generic framework of conversational knowledge circulation in which conversation is used as a primary means for communicating knowledge. Finally, I present attentive agents, autonomous interaction learner, situated knowledge management, self-organizing incremental memory, immersive conversation environment, as key technologies in conversational quantization for conversational knowledge circulation.

2 Social Intelligence Design

The central concern of social intelligence design research is the understanding and augmentation of social intelligence resulting from bilateral definitions of individual intelligence to coordinate her/his behavior with others' in a society and of collective intelligence to specify the discourse for the members to interact with each other. Social intelligence design can be discussed at the different levels of granularity. Social intelligence design on the macroscopic level is about social networking and knowledge circulation in a community. Social intelligence design on the mesoscopic level is about collaboration in a small groups and teams. Social intelligence design at the microscopic level is about fast social interactions in a social discourse.

2.1 The Idea of Social Intelligence Design -- Its Origin and Development

Social intelligence design research is based on bilateral definitions of social intelligence: social intelligence as an individual's ability to manage relationship with other agents and act wisely in a social situation, and social intelligence as an ability of a group of people to manage complexity and learn from experiences as a function of the well-designed social structure [2, 3]. Social intelligence is contrasted with problem solving intelligence / rational intelligence and emotional intelligence.

Social intelligence design research centers on five topics. The first is about theoretical aspects of social intelligence design, involving understanding group dynamics and consensus formation of knowledge creation, theory of common ground in language use, and social learning. The second is about methods of establishing the social context by such means as awareness of connectedness, circulating personal views, or sharing stories. The third is about embodied conversational agents for knowledge exchange, mediating discussions, or learning. The fourth is about collaboration design by integrating the physical space, electronic content and interaction. Multiagent systems might be used to help people in a complex situation. The fifth is about public discourse. Social intelligence design may be concerned with visualization, social awareness support, democratic participation, web mining and social network analysis [2].

Further topics, such as mediated communication and interaction [4], natural interaction [5], collaboration technology and multidisciplinary perspectives [6], evaluation and modeling [7], ambient intelligence [8], designing socially aware interaction [9], and situated and embodied interactions for symbolic and inclusive societies [10], have been added to the scope in subsequent workshops.

Social intelligence design is an interdisciplinary research area. Social intelligence design is discussed from conceptual, scientific and engineering viewpoint. Design is the most important feature to integrate scientific and engineering approach to achieve better social intelligence.

Social intelligence design is studied at three levels. Social intelligence design at the macroscopic level is concerned with networked interactions in community. Social intelligence design at the mesoscopic level focuses on structured social interactions in small groups. Social intelligence design at the microscopic level sheds light on fast interaction loops in the social discourse.

2.2 The Networked Interactions on the Macroscopic Level

Social intelligence design on the macroscopic level is concerned with understanding and supporting communities where knowledge evolves as a result of interaction among members. Major issues include community knowledge management, design and analysis of computer-mediated communication (CMC).

Community knowledge management is concerned with understanding and enabling organizational approach to identify, foster, and leverage insights and experiences shared in a community. It should recognize best practice in a community [11] and enhance the knowledge spiral between formal and tacit knowledge [12]. CMC tools should be amalgamated with organizational structure and process. Tacit knowledge might be better formalized into formal knowledge with CMC tools with face-to-face communication functions, while formal knowledge might be better internalized into tacit knowledge with anonymous communication means [13]. Caire [14] points out that conviviality contributes to promote values such as empathy, reciprocity, social cohesion, inclusiveness, and participation. Katai [15] introduces a framework of social improvisational acts towards communication aiming at creative and humanistic communities.

CMC tools support various phases of the knowledge process in a community. A corporate-wide meeting may not be possible without a powerful CMC tools. Faint-Pop [16] is designed to provide social awareness. Nakata [17] discusses a tool for raising social awareness through position-oriented discussions. Nijholt [18] discusses the design of virtual reality theater environment for a virtual community. At "World-Jam", the IBM's corporate-wide discussions held for three days and participated by over 53600 employees, a system called "Babble" was deployed which assisted synchronous and asynchronous text communications. Each participant was represented as a colored dot. The position of a dot within a visualization called "social proxy" was designed to allow each participant to grasp who else is present and which topics are being discussed [19, 20]. In the DEMOS project, Survey, Delphi and Mediation methods are combined to connect political representatives and citizens, experts and laymen. They are expected to strengthen the legitimacy and rationality of democratic decision making processes by using CMC tools to inspire and guide large scale political debates [21]. Public Opinion Channel was proposed as a CMC tool for circulating small talks in a community [22]. Kanshin was designed to allow for extracting social concern [23]. In order to cope with digital divide, the culture of the user need to be investigated with the greatest case and sensitivity [24].

CMC tools need to be analyzed in order to understand and bring about better community communication. In general, statistical or social network analysis may be applied to understand the structure and features of community communication [25]. Notsu [26] used the VAT (visual assessment of clustering tendency) to analyze the balance of the network modeling of conceptualization. Miura [27] found that medium-density congestion with a relevant topic might activate communication by experienced participants in online chat, and suggested the cognitive process in the course of communication congestion. Miura [28] suggested that information retrieval behaviors may vary depending on task-related domain specific knowledge in information retrieval. If the retriever has sufficient knowledge, s/he will cleverly limit the scope of retrieval and extract more exact information; otherwise, s/he will spend much efforts on comprehending the task-related domain for efficient retrieval. Matsumura [29] revealed that the dynamic mechanism of a popular online community is driven by two distinct causes: discussion and chitchat. ter Hofte [30] investigated placed-based presence (presence enhanced with concepts from the spatial model of interaction). The lessons learned include: place-based presence applications should be designed as an extension of existing PIM applications so that they may allow people to control the exchange of place-based presence information; place-based presence system should keep the user effort minimum, since the trust in presence status may be lowered otherwise; and wider presence and awareness scopes may be needed to allow people see each other since they will easily lose track of each other otherwise. Morio [31] made a cross-cultural examination online communities in US and Japan, and found that Japanese people would prefer to discuss or display their opinions when there is a lack of identifiability, while US people have a much lower rate of anonymous cowards. Furutani [32] investigated the effects of internet use on self efficacy. The results suggested that a belief of finding people with different social background may positively effect on self-efficacy (the cognition about one's capabilities to produce designated levels of performance), while staying in low-risk communication situation with homogeneous others might undermine self-efficacy. Moriyama [33] studied the relationship between self-efficacy and learning experiences in information education. They suggest that self-efficacy and abilities of information utilization may enhance each other. In addition, creativity and information utilization skills might promote self-efficacy.

2.3 The Structured Interactions on the Mesoscopic Level

Social intelligence design at the mesoscopic level is concerned with collaboration support in structured interactions of a group or team. Major issues include design and analysis of global teamwork, collaboration support tools, and meeting support and smart meeting rooms.

Design and analysis of global teamwork is a major concern in many industrial applications. Fruchter [34] proposed to characterize collaboration support systems for global teamwork in terms of bricks (physical spaces), bits (electronic content), and interaction (the way people communicate with each other). Fruchter [35] describes a methodology for analyzing discourse and workspace in distributed computer-mediated interaction. Fruchter [36] formalized the concept of reflection in interaction during communicative events among multiple project steakholders. The observed reflection in interaction is prototyped as TalkingPaper^{TM}.

Cornillon [37] investigated the conceptual design of a feedback advisor suggesting the knowledge co-construction aspect of a debate and noted that various aspects of social intelligence are coded in to the dialogue, such as repetitions encoding awareness of connectedness. Cornillon [38] analyzed how people work together at a distance using a collaborative argumentative graph. They found that the number of turning actions (those changing the structure of an argumentative graph) greatly varies between the face-to-face and remote condition, while that of building actions (those contributing new information on the screen) does not.

In the network era, workspaces are enhanced with information and communication technologies. In order to enable people to flexibly interact with one another in a hybrid workplace, communication in the real life workplace need to be analyzed in terms of physical space, communication space, and organizational space [39].

People's behavior in coping with multitasking and interruptions in the workplace has been studied in depth by Mark and her colleagues [40, 41, 42].

Various collaboration support tools have been proposed to facilitate collaboration from different angles. Martin [43] identified story telling as a vehicle for tacit-to-tacit knowledge transfer in architectural practice and proposed the Building Stories methodology. Fruchter [44] proposed RECALL, a multi-modal collaboration technology that supports global team work. Heylighen [45] presents DYNAMO (Dynamic Architectural Memory Online), an interactive platform to share ideas, knowledge and insights in the form of concrete building projects. Stock et al [46] presents a co-located interface for narration reconciliation in a conflict by making tangible the contributions and disagreements of participants and constraints imposed by the system to jointly perform some key actions on the story. Merckel et al [47] presents a framework for situated knowledge management. A low-cost three dimensional pointer is given to allow the user to associate information with arbitrary points on the surface of physical equipments. Analysis is as important as synthesis. Pumareja [48] studied the effects of long-term use of a groupware. The paradigm of social constructivism and the perspective of structuration was proposed as a framework of analysis. The finding from the case study suggests that collaboration technology can serve as a change agent in transforming the culture and structure of social interaction, through the various meanings people construct when interacting with technology and in benefiting from the structural properties of a system. Cavallin [49] investigated how subjective usability evaluation across applications can be affected by the conditions of evaluation and found that scenarios not only affect the task solving level, but also prime the subjective evaluation of an application.

Meeting support and smart meeting rooms have a large potential in application. Suzuki [50] discussed the social relation between the moderator and interviewees. Nijholt [51] describes a research on meeting rooms and its relevance to augmented reality meeting support and virtual reality generation of meeting. Reidsma [52] discussed three uses of Virtual Meeting Room: to improve remote meeting participation, to visualize multimedia data, as an instrument for research into social interaction in meetings. Rienks [53] presents an ambient intelligent system that uses a conflict management meeting assistant. Wizard of Oz experiments were used to determine the detailed specification of the acceptable behaviors of the meeting assistant, and obtain preliminary evaluation of the effect of the meeting assistant. Use of interaction media was studied by Mark [54] and Gill [55].

2.4 The Fast Interaction Loop on the Microscopic Level

Social intelligence design at the microscopic level is concerned with fast social interactions in the face to face interaction environment. Major issues include interactive social assistants, analysis of nonverbal social behaviors, social artifacts and multi agent systems.

Interactive social assistants help the user make social activities. S-Conart [56] supports conception and decision making of the user while online shopping. PLASIU [57] is designed to support job-hunter's decision making based on the observations from their actual job-hunting process. StoryTable [58] is a co-located cooperation enforcing interface, designed to facilitate collaboration and positive social interaction for children with autistic spectrum disorder.

Analysis of nonverbal social behaviors will provide insights needed to implement collaboration support systems or social artifacts. Yin [59] shows a method of extracting information from I-dialogue that captures knowledge generated during informal communicative events through dialogue, sketching and gestures in the form of unstructured digital design knowledge corpus. Biswas [60] presents a method for exploiting gestures as a knowledge indexing and retrieval tool for unstructured digital video data. Ohmoto [61] presents a method for measuring gaze direction and facial features to detect hidden intention.

Social artifacts aim at embodying social intelligence to interact with people or other social agents. Xu [62] presents a two-layered approach to enhance the robot's capability of involvement and engagement. Xu [63] describes a WOZ experiment setting that allows for observing and understanding the mutual adaptation procedure between humans. Mohammad [64] presents NaturalDraw that uses interactive perception to attenuate noise and unintended behaviors components of the sensor signals by creating a form of mutual alignment between the human and the robot. Mohammad [65] discusses combining autonomy and interactivity for social robots. Yamashita [66] evaluates how much a conversational form of presentation aids comprehension, for long sentences and when user had little knowledge about the topic, in particular. Poel [67] reports design and evaluation of iCat's gaze behavior. Nomura [68] studied negative attitudes towards robots.

Multi agent systems fully automate a computational theory of social agents. Roest [69] shows an interaction oriented agent architecture and language that makes use of an interaction pattern, such as escape/intervention. Rehm [70] integrates social group dynamics in the behavior modeling of multi agent systems. Mao [71] studied social judgment in multi agent systems. Pan [72] presents a multi-agent based framework for simulating human and social behavior during emergency evacuation. Cardon [73] argues that the emerging structure or the morphological agent organization reflects the meaning of the communications between the users.

2.5 Knowledge Circulation in the Context of Social Intelligence Design

Discussions in social intelligence design research may be applied to bring about better knowledge circulation.

At the macroscopic level, organizational design of knowledge circulation should be needed to make sure that a community knowledge process properly functions. A model of knowledge evolution need to be explicitly formulated which may specify when and how knowledge is created, how it is refined, how it is applied and evaluated, how it is archived, and how it is generalized for transfer. It is critical to identify contributors and consumers. The structure of participation need to be well-designed so that many people with different background and motivation can be motivated, participate in and contribute to knowledge circulation. Most importantly, it is critical to identify the structure of affordance and incentive of contribution. Consumers need to be provided enough affordance in order to apply knowledge. The social structure should be well-designed so that creators can find values in contribution in addition to affordance in order to create knowledge. Furthermore, trust and value of information and knowledge need to be addressed.

CMC tools for knowledge circulation should be designed so that they can enhance the affordance and incentive structure by reducing the overhead of knowledge circulation. They should be able to provide the user with cues for evaluating trust and dealing with a large amount of information. A method of evaluating CMC tools for knowledge circulation need to be established.

At the mesoscopic level, the central issue is to support the teamwork of steakholders who play a critical role in knowledge circulation. On the one hand, awareness need to be supported so that they can coordinate their behaviors with colleagues. Knowledge provision should be coordinated with team activities such as schedule maintenance. On the other hand, the complexity of the structure of workplace and multitask complexities must be considered, for knowledge workers are working simultaneously on multiple tasks by moving around multiple workplaces. The relationship among awareness, shared information, and privacies should be carefully analyzed by taking the structure and dynamics of participation into consideration. Collaboration support tools will augment a distributed team of steakholders. Smart meeting rooms will help co-located collaborative activities. Measurement and analysis of activities and the communication tools will be helpful in improving collaborations.

At the microscopic level, quick interaction loops using the combination of verbal and nonverbal behaviors need to be understood and supported. It should be noted that nonverbal behaviors not only control the discourse of communication, but also give additional meaning to verbal information. Understanding and leveraging quick interaction loops will help identify tacit information underlying the communication activities. Although it is challenging, developing social artifacts that can create and sustain interaction loops with people will significantly accelerate and improve the quality of knowledge circulation. Multi agent systems techniques might be used not only to control distributed systems but also to understand social systems as an accumulation of microscopic interactions among participating agents.

The above discussions may lead to a layered model of community knowledge process [74]. The first layer from the bottom is about context sharing. It accumulates the background information that serves as a common ground for a community. Thus layer remains often tacit in the sense that it is not explicitly and represented as a well-described static documents. Rather it is a dynamic collection of ongoing conversations among members or tacitly shared perception of the common ground. The second layer is information and knowledge explicitly shared in the community. The third layer is collaboration. Special interest groups are often formed to act to achieve a goal for a community. The fourth layer is discussion. Conflicting propositions are identified and discussed from various angles. The topmost layer is decision making. Resolutions to conflicting goals are determined at this level and disseminated to the community members. Knowledge circulation is indispensable to make knowledge process function effectively at each level.

3 Conversational Knowledge Circulation

Conversation is the most natural communication means for people. People are fluent in expressing ideas by combining verbal and nonverbal behaviors. People are skillful in interpreting communicative behaviors of other participants. People make nonverbal behaviors, iconic gestures for instance, not only to control the discourse but also to modulate the proposition conveyed by verbal utterances. Conversation is a heuristic process of knowledge creation by a group of people. Although it is pretty hard to express half-baked ideas, those filled with indeterminacies and inconsistencies, in a written language, vague thoughts often turn into clear ideas as a result of conversation on the spot by incorporating knowledge from participants and gaining better grounding on the subject. The discourse of conversation often allows participants to critically examine the subject from multiple angles, which may motivate further contributions from the participants. Conversational knowledge circulation centers on conversation, aiming at circulating knowledge in a conversational fashion by capturing information arising in conversations, organizing it into knowledge and applying knowledge to conversational situations. It focuses on communicating intuitive and subjective aspects of knowledge representation in a situated fashion.

3.1 Computational Framework of Conversational Knowledge Circulation

Conversational knowledge circulation depends on a method of capturing and presenting information at conversational situations. The result of conversation capture need to be packed into a some form of conversational content from which conversation will be reproduced. Design of the data structure of conversational content is critical to the design of conversational knowledge circulation. In general, the more sophisticated data structure is employed, the more flexible and reusable becomes conversational content, but the more complex algorithms may be required in implementation. Typically, conversational content may be implemented as an annotated audio-video segment. Although it is more useful if transcript of utterances or even semantic information is given as annotation, it will be more expensive and challenging to (semi-)automatically create high-quality annotated video clips. Basic elements of conversational knowledge circulation are the augmented conversational environment equipped with sensors and actuators, the conversational agent, and the conversational content server.

The augmented conversational environment is used for generating and presenting conversational content in the real world. Not only participants' conversational behaviors but also the objects and events referred to in conversation need to be captured. Although motion capture systems or eye trackers are useful devices for achieving the quality of data (such as accuracy or frequency), they may constrain the quality of interaction by compelling the participants to attach measurement devices or markers which may seriously distract natural conversations from time to time. The conversation capture may be enhanced by introducing conversational robots that may move around the environment to capture information at appropriate viewpoints or even to interview the participants to actively elicit knowledge. The key algorithms in smart environment and situated social artifacts are recognition of conversational environment and automated segmentation and annotation for captured conversation.

Conversational agents are used to interactively present conversational content. A conversational agent may be an embodied conversational agent that lives in a virtual world simulating the subject world. Alternatively, it may be a conversational robot that cohabits with people in a physical space. Although it is still a big challenge to build a conversational robot that can exhibit a proper conversational behaviors, conversational robots may embody strong presence as an independent agent in conversation once they are realized. In contrast, embodied conversational agents are portable over the net and versatile in expressing ideas without incurring by physical constraints, while their presence is often weaker than physical robots and their communicative expressions are usually bound to the two-dimensional display. The key algorithms for conversational agents are generating proper conversational behaviors and presentation of conversational content according to the conversation status.

Conversational content servers accumulate conversational content for distribution. Ideally, they may be equipped with a self-organization mechanism so that new conversation content may be automatically associated with a existing collection of conversational content and the entire collection of conversational content may be organized systematically. A less ambitious goal is to provide an visualizer and editor that may allow the user to browse the collection of conversational content, organize them into topic clusters, and create new conversational content from existing collection.

In addition to the basic elements mentioned above, high-level functions may be introduced to allow the users to utilize the collection of conversation content in collaboration, discussion, and decision making.

Figure 1 shows a simplified view of how the conversational knowledge circulation might be applied to the industrial environment where communication among customers and engineers are critical. Emphasis is placed on enhancing the lower layers of community knowledge process. It illustrates how conversations at the design, presentation and deployment stages might be supported by conversational knowledge circulation.

At the design stage, the product is designed and possible usage scenarios are developed by discussions among engineers and sales managers. The discussions contain valuable pieces of knowledge, such as intended usage or tips, that may also be useful to the users. Conversational content about the product and service can be composed as a result of the design phase. Conversational content may also used as an additional information source at the fabrication phase to help developers understand the intention of the design.

At the presentation stage, the product and service are displayed to the potential customers in an interactive fashion. In order to make the interactive presentation widely available on the net, embodied conversational agents may be used as a virtual presenter. Embodied conversational agents will be able to cope with frequently asked questions using a collection of conversational content prepared in advance. When questions cannot be answered based on the prepared conversational content, the

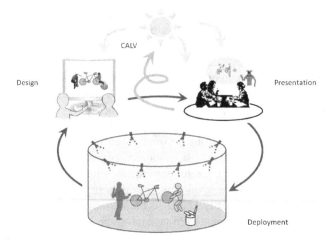

Fig. 1. Conversational knowledge circulation applied to industrial environment

engineer may control the presenter agent as an avatar to create a proper reply. Such communication logs can be saved so that the service division may extend the "FAQ" conversational content for future questions. Embodied conversational agents may be used as a surrogate of the customer to ensure the anonymous communication from the user. The presentation stage can also be employed to train novices when the product and service is introduced to the user.

At the deployment stage, conversations may contain various pieces of knowledge sources, such as the real usage scenario, evaluation from the user, complaints about the current service, demands for new services, etc. The conversation between the user and system engineer may be captured by an intelligent sensing devices. Service robots may be deployed to help the user as well as collect usage data. The collected conversational content may be fed back to the design phase for improvement and further product and service development.

It should be noted that the collection of (potential) customers, salespersons, and engineers forms a community that shares a common product and service. CALV (Community-maintained Artifacts of Lasting Value) [75] is expected to be created as a result of the conversational knowledge circulation. The more information and knowledge is circulated, the richer CALV may be obtained.

3.2 Conversation Quantization

Conversation quantization is a computational framework of circulating conversation quanta that encapsulates discourse units into annotated audio-visual video segments. Conversation quantization is based on the idea of approximating a continuous flow of conversation by a series of minimally coherent segments of discourse called conversation quanta [76].

Augmented conversational environment can be implemented as a smart meeting room or augmented environment that can provide conversation quanta with the participants according to the conversational state and produce conversation quanta by sensing conversational interaction among the participants. The role of the conversation quanta capture is to (semi-)automatically produce a sequence of conversation quanta for a given conversation session. Fully automated conversation quanta capture is considered to be out of the scope of the current technology, for significant knowledge and technological development is required to segment conversations into small pieces and produce semantic annotation for conversational situations. Saito et al [77] discussed human-assisted production of conversation quanta. Vickey [78] is an augmented conversational environment for a driving simulator. It can ground the conversation on the events observed through the simulated window of the vehicle, by analyzing pointing gestures of the participants. IMADE (the realworld Interaction Measurement, Analysis and Design Environment) [79] allows one to capture conversational behavior of a group of people with an optical motion capture device, wearable eye mark recorders, etc. A tool called iCorpusStudio was developed for browsing, analyzing, or annotating an interaction corpus consisting of multimedia data streams obtained from sensing conversation sessions.

Conversational agents have been implemented which will use conversation quanta to make speech acts in conversations. Conversational agents may be virtual or physical. Speech acts contain a full spectrum ranging from linguistic, paralinguistic, and nonlinguistic. EgoChat agents [80], SPOC and IPOC agents [81, 82]. GECA [83] provides a platform on which virtual agents are developed on an open platform using a markup language. We have also developed listener and presenter robots [84, 85], though they still exhibit only basic nonverbal behaviors.

The role of the conversational content server is to circulate conversation quanta in a team/group/community/society. It should be able to deliver conversation quanta to situations on demand or proactively. POC (Public Opinion Channel) [86, 22] implements part of the idea. The role of the conversation quanta manager is to accumulate a collection of conversation quanta. SKG (Sustainable Knowledge Globe) [87] allows the user to visually accumulate a large amount of conversational content on the CG sphere surface so that s/he can establish and maintain a sustainable external memory coevolving with the internal memory. A media converter may be used to translate conversation quanta from/to other information media such as videos or documents. Kurohashi et al [88] developed a method for automatically creating a spoken-language script from a knowledge card consisting of a short text and a reference image.

3.3 Technical Challenges

There are many technical challenges to overcome to put conversation quantization in effect for conversational knowledge circulation.

Challenges on the conversational environment are semantic and contextual processing. Although it is highly desirable to annotate data with semantic information, difficulties may arise from the size and complexity of the semantic domain. The dynamic nature of semantic information need to be addressed. Meaning cannot be predefined, for it arises dynamically in the interaction. Contextual information should be handled properly. When the discourse is well-shared, contextual information should be kept to the minimal, while maximal contextual information should be added when conversation quanta are transported to communities with different background. Thus, contextual information should be properly added or removed depending on the conversational situation.

Challenges on the conversational agents are concerned with the naturalness of interaction and presence of the agent. In order to conduct natural interactions with the user, subtle nonverbal cues in interaction should be captured and reflected on the reactive behaviors of agent. Large varieties of behaviors should be generated efficiently. Situatedness appears to be a key to naturalness. Agents should be situated in the sense that they can allow the human to share information about objects and events in the environment. The agents should at least partly share the way the humans may perceive the world. When the agent lives in the virtual environment, the entire system should permit the user to feel the environment as if s/he is involved. When the agent lives in the physical environment, the system should be able to recognize the objects/event the user is referring to and share the perception. Social awareness must be supported by capturing and forwarding member's status without violating privacy of the sender or disturbing the recipient.

Furthermore, agents should be able to attract and sustain the attention of the user during the conversation session, by recognizing the user's conversational status and managing the utterances based on the user's status. This requirement becomes more evident when the agent is interacting with more than one user. The ultimate goal is to realize empathetic interaction between the human users and the agent. Affective computing need to be introduced to have conversational agents behave naturally based on internal emotion model. Although the agent must be able to recognize the user's subtle change in emotional state, the emotional state of the user must be sensed without distracting her/him.

Challenges on the conversational content server include a self-organizational incremental memory and high-level social functions. Incoming conversation quanta should be able to structurally organized into a collection of conversational content into coherent stories. Automated digesting or summary is needed to navigate the user to a potentially large collection of conversation quanta. The dynamic nature of the memory should be considered since conversation quanta may continuously come in. Automatic visualization might be needed to have the user intuitively grasp the accumulated information. The landscape of the collection of conversation quanta should be transformed gradually so that the users can track the change of the collection. High-level functions might be introduced to account for social awareness and wellness. Social mechanisms such as trust, incentive, reciprocity, fairness, or atmosphere should be properly designed and assessed.

In general, it is challenging to make sure that social intelligence is in fact incorporated in the design of community support system. Green pointed out five challenges for this [89], namely supporting user-centered design for social intelligence; evaluating social intelligence; understanding the effect of social characteristics; ethical considerations for social intelligence; and establishing & maintaining social intelligence.

4   Towards the Breakthrough

In this section, I overview five projects aiming at overcoming the technical challenges in conversational knowledge circulation.

4.1   Attentive Agents

The goal of the attentive agent project is to build a conversational agent that can behave attentively to the dynamics of the interaction with multiple concurrent participants [90]. An agent can be said attentive if it can properly control its conversational behaviors according to the status and behaviors of other participants. For example, the agent should keep quiet when other participants are discussing with each other for a while, whilst it can urge them to come back to the conversation if they have been off the discourse for a long while. In case of talking with multiple concurrent participants, the agent need also to speak to a person who is considered to be a proper addressee. In order to have the conversational agent behave according to such attentive utterance policies, a couple of indices have been introduced: AT (interaction activity) that indicates whether the users are active in their interactions or not, and CLP (conversation leading person) that denotes the participant who is the most likely leading the group during a certain period of the conversation session. In order to overcome the limitation of two-dimensional agent coming from so-called the Mona Lisa Effect, three-dimensional physical pointer have been introduced to point to an intended addressee. The ideas have been implemented into a quiz game agent that can host a quiz session with multiple users and evaluated.

Ohmoto et al [91] addressed visual measurement of involvement of participants. Social atmosphere or extrinsic involvement attributed to the state of the group of participants as a whole is distinguished from intrinsic involvement attributed to that of an individual. Although physiological indices can be used to identify the weak involvement of a person affected by extrinsic involvement, participants of conversation are often reluctant to attach physiological sensing devices. Ohmoto et al investigated the correlation between the physiological indices and visual indices measuring the moving distances and the speed of user's motions, and have found that both intrinsic and extrinsic states of involvement can be detected with the accuracy of around 70% by changing the threshold level.

4.2   From Observation to Interaction

The goal of the autonomous interaction learner project is to build a robot that can autonomously develop natural behavior at three stages [92]. On the discovery stage, the robot attempts to discover the action and command space by watching the interaction. On the association stage, the robot attempts to associate discovered actions with commands. The result of association will be represented as a probabilistic model that can be used both for behavior understanding and generation. On the controller generation phase, the robot converts the behavioral model into an action controller so that it can act in similar situations. A number of novel algorithms have been developed. RSST (Robust Singular Spectrum Transform) is an algorithm that calculates likelihood of change of dynamics in continuous time series without prior knowledge. DGCMD (Distance-Graph Constrained Motif Discovery) uses the result of RSST to discover motifs (recurring temporal patterns) from the given time series. The association algorithm estimates the natural delay between commands and actions so that it can properly associate commands (cause) with subsequent actions (effect).

4.3 Situated Knowledge Management

The goal of the situated knowledge management project is to develop a suite of algorithms so that the system can recognize how knowledge is associated with real world objects and events. The key algorithms are the real-time, light-weight object pose recognition algorithm that takes the CAD model (Piecewise Linear Complex) and the camera-image of the target object to estimate the pose of the object with respect to the camera [93, 94]; the interface for correcting the estimated pose; and a low-overhead three-dimensional items drawing engine [95]. The suite works both in the augmented reality and augmented virtuality environments. In the augmented reality environment, it enables to overlay annotations on the camera-image of the target object. In the augmented virtuality environment, it allows for creating three-dimensional virtualized target object by automatically pasting surface texture. The three-dimensional items drawing engine consists of a hand-held Augmented Reality (AR) system. It allows the user to directly draw free three-dimensional lines in the context of the subject instruments.

4.4 Self-Organizing Incremental Memory

The goal of the incremental self-organizational memory project is to develop a self-organizing incremental neural network that can make incremental unsupervised clustering of given segments of time series. We have developed HB-SOINN (HMM-Based Self Organizing Incremental Neural Network) that uses HMM (Hidden Markov Model) as a preprocessor of SOINN so that the resulting system can handle the variable length patterns into fixed length patterns [96]. The role of HMM is to reduce dimensions of sequence data and to map variable length sequences into vectors of fixed dimension. HMM contributes to robust feature extraction from sequence patterns, which allows for similar statistical features to be extracted from sequence patterns of the same category. As a result of empirical experiments, it has turned out that HB-SOINN can generate a fewer number of clusters than a few competitive batch clustering algorithms.

4.5 Immersive Conversation Environment

The goal of the immersive conversation environment is to build an ambient environment that can provide the human operator with a feeling as if s/he stayed "inside" a conversation robot or embodied conversational agent to receive incoming visual and auditory signals and to create conversational behaviors in a natural fashion [97]. The immersive conversation environment will be used to pursue Wizard of Oz experiments with the embodiment of a conversational robot or an embodied conversational agent. A 360-degree visual display can reproduce an immersive view around a conversational agent. The current display system uses eight 64-inch display panels arranged in a circle with about 2.5 meters diameter. Eight surround speakers are used to reproduce the acoustic environment. It is designed to collect detailed information about how the operator behaves in varying conversational scenes.

5 Conclusion

Knowledge circulation not only decreases uneven distribution of knowledge in a community but also improves the coverage and quality of the shared knowledge. In this paper, I shed light on social aspects of knowledge circulation. First, I have overviewed social intelligence design and discussed how the insights obtained so far might be applied to the design, implementation and evaluation of knowledge circulation. Then, I have presented a generic framework of conversational knowledge circulation in which conversation is used as a primary means for communicating knowledge. Finally, I have presented recent results in conversational quantization for conversational knowledge circulation.

References

[1] Kitsuregawa, M.: Challenge for Info-plosion. In: Corruble, V., Takeda, M., Suzuki, E. (eds.) DS 2007. LNCS (LNAI), vol. 4755, pp. 1–8. Springer, Heidelberg (2007)
[2] Nishida, T.: Social Intelligence Design – An Overview. In: Terano, T., Nishida, T., Namatame, A., Tsumoto, S., Ohsawa, Y., Washio, T. (eds.) JSAI-WS 2001. LNCS (LNAI), vol. 2253, pp. 3–10. Springer, Heidelberg (2001)
[3] Nishida, T.: Social Intelligence Design and Human Computing. In: Huang, T.S., Nijholt, A., Pantic, M., Pentland, A. (eds.) ICMI/IJCAI Workshops 2007. LNCS (LNAI), vol. 4451, pp. 190–214. Springer, Heidelberg (2007)
[4] Fruchter, R., Nishida, T., Rosenberg, D.: Understanding Mediated Communication: the Social Intelligence Design (SID) Approach. AI Soc. 19(1), 1–7 (2005)
[5] Nijholt, A., Nishida, T.: Social Intelligence Design for Mediated Communication. AI Soc. 20(2), 119–124 (2006)
[6] Fruchter, R., Nishida, T., Rosenberg, D.: Mediated Communication in Action: a Social Intelligence Design Approach. AI Soc. 22(2), 93–100 (2007)
[7] Miura, A., Matsumura, N.: Social Intelligence Design: a Junction between Engineering and Social Sciences. AI Soc. 23(2), 139–145 (2009)
[8] Nijholt, A., Stock, O., Nishida, T.: Social Intelligence Design in Ambient Intelligence. AI Soc. 24(1), 1–4 (2009)
[9] http://cdr.uprrp.edu/SID2008/default.htm
[10] http://www.ii.ist.i.kyoto-u.ac.jp/sid/sid2009/
[11] Davenport, T.H., Prusak, L.: Working Knowledge. Harvard Business School Press (2000)
[12] Nonaka, I., Takeuchi, H.: The Knowledge-Creating Company: How Japanese Companies Create the Dynamics of Innovation. Oxford University Press, Oxford (1995)

[13] Azechi, S.: Informational Humidity Model: Explanation of Dual Modes of Community for Social Intelligence Design. AI Soc. 19(1), 110–122 (2005)

[14] Caire, P.: Designing Convivial Digital Cities: a Social Intelligence Design Approach. AI Soc. 24(1), 97–114 (2009)

[15] Katai, O., Minamizono, K., Shiose, T., Kawakami, H.: System Design of "Ba"-like Stages for Improvisational Acts via Leibnizian Space-time and Peirce's Existential Graph Concepts. AI Soc. 22(2), 101–112 (2007)

[16] Ohguro, T., Kuwabara, K., Owada, T., Shirai, Y.: FaintPop: In Touch with the Social Relationships. In: Terano, T., Nishida, T., Namatame, A., Tsumoto, S., Ohsawa, Y., Washio, T. (eds.) JSAI-WS 2001. LNCS (LNAI), vol. 2253, pp. 11–18. Springer, Heidelberg (2001)

[17] Nakata, K.: Enabling Public Discourse. In: Terano, T., Nishida, T., Namatame, A., Tsumoto, S., Ohsawa, Y., Washio, T. (eds.) JSAI-WS 2001. LNCS (LNAI), vol. 2253, pp. 59–66. Springer, Heidelberg (2001)

[18] Nijholt, A.: From Virtual Environment to Virtual Community. In: Terano, T., Nishida, T., Namatame, A., Tsumoto, S., Ohsawa, Y., Washio, T. (eds.) JSAI-WS 2001. LNCS (LNAI), vol. 2253, pp. 19–26. Springer, Heidelberg (2001)

[19] Thomas, J.C.: Collaborative Innovation Tools. In: Terano, T., Nishida, T., Namatame, A., Tsumoto, S., Ohsawa, Y., Washio, T. (eds.) JSAI-WS 2001. LNCS (LNAI), vol. 2253, pp. 27–34. Springer, Heidelberg (2001)

[20] Erickson, T.: 'Social' Systems: Designing Digital Systems that Support Social Intelligence. AI Soc. 23(2), 147–166 (2009)

[21] Luhrs, R., Malsch, T., Voss, K.: Internet, Discourses, and Democracy. In: Terano, T., Nishida, T., Namatame, A., Tsumoto, S., Ohsawa, Y., Washio, T. (eds.) JSAI-WS 2001. LNCS (LNAI), vol. 2253, pp. 67–74. Springer, Heidelberg (2001)

[22] Fukuhara, T., Nishida, T., Uemura, S.: Public Opinion Channel: A System for Augmenting Social Intelligence of a Community. In: Terano, T., Nishida, T., Namatame, A., Tsumoto, S., Ohsawa, Y., Washio, T. (eds.) JSAI-WS 2001. LNCS (LNAI), vol. 2253, pp. 51–58. Springer, Heidelberg (2001)

[23] Fukuhara, T., Murayama, T., Nishida, T.: Analyzing Concerns of People from Weblog Articles. AI Soc. 22(2), 253–263 (2007)

[24] Blake, E.H., Tucker, W.D.: User Interfaces for Communication Bridges across the Digital Divide. AI Soc. 20(2), 232–242 (2006)

[25] Fujihara, N.: How to Evaluate Social Intelligence Design. In: Terano, T., Nishida, T., Namatame, A., Tsumoto, S., Ohsawa, Y., Washio, T. (eds.) JSAI-WS 2001. LNCS (LNAI), vol. 2253, pp. 75–84. Springer, Heidelberg (2001)

[26] Notsu, A., Ichihashi, H., Honda, K., Katai, O.: Visualization of Balancing Systems based on Naïve Psychological Approaches. AI Soc. 23(2), 281–296 (2009)

[27] Miura, A., Shinohara, K.: Social Intelligence Design in Online Chat Communication: a Psychological Study on the Effects of "Congestion". AI Soc. 19(1), 93–109 (2005)

[28] Miura, A., Fujihara, N., Yamashita, K.: Retrieving Information on the World Wide Web: Effects of Domain Specific Knowledge. AI Soc. 20(2), 221–231 (2006)

[29] Matsumura, N., Miura, A., Shibanai, Y., Ohsawa, Y., Nishida, T.: The Dynamism of 2channel. AI Soc. 19(1), 84–92 (2005)

[30] ter Hofte, G.H., Mulder, I., Verwijs, C.: Close Encounters of the Virtual Kind: a Study on Place-based Presence. AI Soc. 20(2), 151–168 (2006)

[31] Morio, H., Buchholz, C.: How Anonymous are You Online? Examining Online Social Behaviors from a Cross-cultural Perspective. AI Soc. 23(2), 297–307 (2009)

[32] Furutani, K., Kobayashi, T., Ura, M.: Effects of Internet Use on Self-Efficacy: Perceived Network-Changing Possibility as a Mediator. AI Soc. 23(2), 251–263 (2009)

[33] Moriyama, J., Kato, Y., Aoki, Y., Kito, A., Behnoodi, M., Miyagawa, Y., Matsuura, M.: Self-Efficacy and Learning Experience of Information Education: in Case of Junior High School. AI Soc. 23(2), 309–325 (2009)

[34] Fruchter, R.: Bricks & Bits & Interaction. In: Terano, T., Nishida, T., Namatame, A., Tsumoto, S., Ohsawa, Y., Washio, T. (eds.) JSAI-WS 2001. LNCS (LNAI), vol. 2253, pp. 35–42. Springer, Heidelberg (2001)

[35] Fruchter, R., Cavallin, H.: Developing Methods to Understand Discourse and Workspace in Distributed Computer-Mediated Interaction. AI Soc. 20(2), 169–188 (2006)

[36] Fruchter, R., Swaminathan, S., Boraiah, M., Upadhyay, C.: Reflection in Interaction. AI Soc. 22(2), 211–226 (2007)

[37] Cornillon, J., Rosenberg, D.: Dialogue Organisation in Argumentative Debates. AI Soc. 19(1), 48–64 (2005)

[38] Cornillon, J., Rosenberg, D.: Experiment in Social Intelligence Design. AI Soc. 22(2), 197–210 (2007)

[39] Rosenberg, D., Foley, S., Lievonen, M., Kammas, S., Crisp, M.J.: Interaction Spaces in Computer-Mediated Communication. AI Soc. 19(1), 22–33 (2005)

[40] González, V.M., Mark, G.: Constant, Constant, Multi-tasking Craziness: Managing Multiple Working Spheres. In: CHI 2004, pp. 113–120 (2004)

[41] Mark, G., Gudith, D., Klocke, U.: The Cost of Interrupted Work: More Speed and Stress. In: CHI 2008, pp. 107–110 (2008)

[42] Su, N.M., Mark, G.: Communication chains and multitasking. In: CHI 2008, pp. 83–92 (2008)

[43] Martin, W.M., Heylighen, A., Cavallin, H.: The Right Story at the Right Time. AI Soc. 19(1), 34–47 (2005)

[44] Fruchter, R.: Degrees of Engagement in Interactive Workspaces. AI Soc. 19(1), 8–21 (2005)

[45] Heylighen, A., Heylighen, F., Bollen, J., Casaer, M.: Distributed (Design) Knowledge Exchange. AI Soc. 22(2), 145–154 (2007)

[46] Stock, O., Zancanaro, M., Rocchi, C., Tomasini, D., Koren, C., Eisikovits, Z., Goren-Bar, D., Weiss, P.L.T.: The Design of a Collaborative Interface for Narration to Support Reconciliation in a Conflict. AI Soc. 24(1), 51–59 (2009)

[47] Merckel, L., Nishida, T.: Enabling Situated Knowledge Management for Complex Instruments by Real-time Reconstruction of Surface Coordinate System on a Mobile Device. AI Soc. 24(1), 85–95 (2009)

[48] Pumareja, D.T., Sikkel, K.: Getting Used with Groupware: a First Class Experience. AI Soc. 20(2), 189–201 (2006)

[49] Cavallin, H., Martin, W.M., Heylighen, A.: How Relative Absolute can be: SUMI and the Impact of the Nature of the Task in Measuring Perceived Software Usability. AI Soc. 22(2), 227–235 (2007)

[50] Suzuki, K., Morimoto, I., Mizukami, E., Otsuka, H., Isahara, H.: An Exploratory Study for Analyzing Interactional Processes of Group Discussion: the Case of a Focus Group Interview. AI Soc. 23(2), 233–249 (2009)

[51] Nijholt, A., op den Akker, R., Heylen, D.: Meetings and Meeting Modeling in Smart Environments. AI Soc. 20(2), 202–220 (2006)

[52] Reidsma, D., op den Akker, R., Rienks, R., Poppe, R., Nijholt, A., Heylen, D., Zwiers, J.: Virtual Meeting Rooms: from Observation to Simulation. AI Soc. 22(2), 133–144 (2007)

[53] Rienks, R., Nijholt, A., Barthelmess, P.: Pro-active Meeting Assistants: Attention Please! AI Soc. 23(2), 213–231 (2009)

[54] Mark, G., DeFlorio, P.: HDTV: a Challenge to Traditional Video Conferences? Publish-only paper, SID-2001 (2001)

[55] Gill, S.P., Borchers, J.: Knowledge in Co-action: Social Intelligence in Collaborative Design Activity. AI Soc. 17(3-4), 322–339 (2003)

[56] Shoji, H., Hori, K.: S-Conart: an Interaction Method that Facilitates Concept Articulation in Shopping Online. AI Soc. 19(1), 65–83 (2005)

[57] Shoji, H., Fujimoto, K., Hori, K.: PLASIU: a System that Facilitates Creative Decision-making in Job-hunting. AI Soc. 23(2), 265–279 (2009)

[58] Gal, E., Bauminger, N., Goren-Bar, D., Pianesi, F., Stock, O., Zancanaro, M., Weiss, P.L.T.: Enhancing Social Communication of Children with High-functioning Autism through a Co-located Interface. AI Soc. 24(1), 75–84 (2009)

[59] Yin, Z., Fruchter, R.: I-Dialogue: Information Extraction from Informal Discourse. AI Soc. 22(2), 169–184 (2007)

[60] Biswas, P., Fruchter, R.: Using Gestures to Convey Internal Mental Models and Index Multimedia Content. AI Soc. 22(2), 155–168 (2007)

[61] Ohmoto, Y., Ueda, K., Ohno, T.: Real-time System for Measuring Gaze Direction and Facial Features: towards Automatic Discrimination of Lies using Diverse Nonverbal Information. AI Soc. 23(2), 187–200 (2009)

[62] Xu, Y., Hiramatsu, T., Tarasenko, K., Nishida, T., Ogasawara, Y., Tajima, T., Hatakeyama, M., Okamoto, M., Nakano, Y.I.: A Two-layered Approach to Communicative Artifacts. AI Soc. 22(2), 185–196 (2007)

[63] Xu, Y., Ueda, K., Komatsu, T., Okadome, T., Hattori, T., Sumi, Y., Nishida, T.: WOZ Experiments for Understanding Mutual Adaptation. AI Soc. 23(2), 201–212 (2009)

[64] Mohammad, Y.F.O., Nishida, T.: Interactive Perception for Amplification of Intended Behavior in Complex Noisy Environments. AI Soc. 23(2), 167–186 (2009)

[65] Mohammad, Y.F.O., Nishida, T.: Toward Combining Autonomy and Interactivity for Social Robots. AI Soc. 24(1), 35–49 (2009)

[66] Yamashita, K., Kubota, H., Nishida, T.: Designing Conversational Agents: Effect of Conversational Form on Our Comprehension. AI Soc. 20(2), 125–137 (2006)

[67] Poel, M., Heylen, D., Nijholt, A., Meulemans, M., van Breemen, A.J.N.: Gaze Behaviour, Believability, Likability and the iCat. AI Soc. 24(1), 61–73 (2009)

[68] Nomura, T., Kanda, T., Suzuki, T.: Experimental Investigation into Influence of Negative Attitudes toward Robots on Human-Robot Interaction. AI Soc. 20(2), 138–150 (2006)

[69] Roest, G.B., Szirbik, N.B.: Escape and Intervention in Multi-agent Systems. AI Soc. 24(1), 25–34 (2009)

[70] Rehm, M., Endraß, B.: Rapid Prototyping of Social Group Dynamics in Multiagent Systems. AI Soc. 24(1), 13–23 (2009)

[71] Mao, W., Gratch, J.: Modeling Social Inference in Virtual Agents. AI Soc. 24(1), 5–11 (2009)

[72] Pan, X., Han, C.S., Dauber, K., Law, K.H.: A Multi-agent based Framework for the Simulation of Human and Social Behaviors during Emergency Evacuations. AI Soc. 22(2), 113–132 (2007)

[73] Cardon, A.: A Distributed Multi-agent System for the Self-Evaluation of Dialogs. In: Terano, T., Nishida, T., Namatame, A., Tsumoto, S., Ohsawa, Y., Washio, T. (eds.) JSAI-WS 2001. LNCS (LNAI), vol. 2253, pp. 43–50. Springer, Heidelberg (2001)

[74] Nishida, T.: Supporting the Conversational Knowledge Process in the Networked Community. In: Bianchi-Berthouze, N. (ed.) DNIS 2003. LNCS, vol. 2822, pp. 138–157. Springer, Heidelberg (2003)

[75] Cosley, D., Frankowski, D., Terveen, L.G., Riedl, J.: Using Intelligent Task Routing and Contribution Review to Help Communities Build Artifacts of Lasting Value. In: CHI 2006, pp. 1037–1046 (2006)

[76] Nishida, T.: Conversation Quantisation for Conversational Knowledge Process. International Journal of Computational Science and Engineering (IJCSE) 3(2), 134–144 (2007)

[77] Saito, K., Kubota, H., Sumi, Y., Nishida, T.: Analysis of Conversation Quanta for Conversational Knowledge Circulation. J. UCS 13(2), 177–185 (2007)

[78] Okamura, G., Kubota, H., Sumi, Y., Nishida, T., Tsukahara, H., Iwasaki, H.: Quantization and Reuse of Driving Conversations. Transactions of Information Processing Society of Japan 48(12), 3893–3906 (2007)

[79] Sumi, Y., Nishida, T., Bono, M., Kijima, H.: IMADE: Research Environment of Real-world Interactions for Structural Understanding and Content Extraction of Conversation. Journal of Information Processing Society of Japan 49(8), 945–949 (2008) (in Japanese)

[80] Kubota, H., Nishida, T.: Channel Design for Strategic Knowledge Interaction. In: Palade, V., Howlett, R.J., Jain, L. (eds.) KES 2003. LNCS, vol. 2773, pp. 1037–1043. Springer, Heidelberg (2003)

[81] Nakano, Y.I., Murayama, T., Nishida, T.: Multimodal Story-based Communication: Integrating a Movie and a Conversational Agent. IEICE Transactions Information and Systems E87-D(6), 1338–1346 (2004)

[82] Okamoto, M., Nakano, Y.I., Okamoto, K., Matsumura, K., Nishida, T.: Producing Effective Shot Transitions in CG Contents based on a Cognitive Model of User Involvement. IEICE Transactions of Information and Systems Special Issue of Life-like Agent and Its Communication, IEICE Trans. Inf. & Syst. E88-D(11), 2623–2532 (2005)

[83] Huang, H.H., Nishida, T., Cerekovic, A., Pandzic, I.S., Nakano, Y.I.: The Design of a Generic Framework for Integrating ECA Components. In: AAMAS, pp. 128–135 (2008)

[84] Nishida, T., Fujihara, N., Azechi, S., Sumi, K., Yano, H., Hirata, T.: Public Opinion Channel for Communities in the Information Age. New Generation Comput. 17(4), 417–427 (1999)

[85] Nishida, T., Terada, K., Tajima, T., Hatakeyama, M., Ogasawara, Y., Sumi, Y., Xu, Y., Mohammad, Y.F.O., Tarasenko, K., Ohya, T., Hiramatsu, T.: Towards Robots as an Embodied Knowledge Medium, Invited Paper, Special Section on Human Communication II. IEICE TRANSACTIONS on Information and Systems E89-D(6), 1768–1780 (2006)

[86] Ohya, T., Hiramatsu, T., Xu, Y., Sumi, Y., Nishida, T.: Robot as an Embodied Knowledge Medium – Having a Robot Talk to Humans using Nonverbal Communication Means. In: Social Intelligence Design 2006 (SID 2006), Osaka (2006)

[87] Kubota, H., Nomura, S., Sumi, Y., Nishida, T.: Sustainable Memory System Using Global and Conical Spaces. J. UCS 13(2), 135–148 (2007)

[88] Kurohashi, S., Kawahara, D., Shibata, T.: Automatic Text Presentation for the Conversational Knowledge Process. In: Nishida, T. (ed.) Conversational Informatics: an Engineering Approach, pp. 201–216. Wiley, Chichester (2007)

[89] Green, W., de Ruyter, B.: The Design & Evaluation of Interactive Systems with Perceived Social Intelligence: Five Challenges. Presented at the Seventh International Workshop on Social Intelligence Design, San Juan, Puerto Rico (2008)

[90] Huang, H.H., Nakano, Y.I., Nishida, T., Furukawa, T., Ohashi, H., Cerekovic, A., Pandzic, I.S.: How multiple concurrent users react to a quiz agent attentive to the dynamics of their game participation. In: AAMAS (2010) (to be presented)

[91] Ohmoto, Y., Miyake, T., Nishida, T.: A Method to Understand an Atmosphere based on Visual Information and Physiological Indices in Multi-user Interaction. Presented at the Seventh International Workshop on Social Intelligence Design, Kyoto, Japan (2009)

[92] Mohammad, Y.F.O., Nishida, T., Okada, S.: Unsupervised Simultaneous Learning of Gestures, Actions and their Associations for Human-Robot Interaction. In: IEEE/RSJ International Conference on Intelligent RObots and Systems, Intelligent Robots and Systems (IROS 2009), pp. 2537–2544 (2009)

[93] Merckel, L., Nishida, T.: Enabling Situated Knowledge Management for Complex Instruments by Real-time Reconstruction of Surface Coordinate System on a Mobile Device. AI Soc. 24(1), 85–95 (2009)

[94] Merckel, L., Nishida, T.: Multi-interfaces Approach to Situated Knowledge Management for Complex Instruments: First Step toward Industrial Deployment. In: AI Soc. (2010), (online first) doi:10.1007/s00146-009-0247-9

[95] Merckel, L., Nishida, T.: Low-Overhead 3D Items Drawing Engine for Communicating Situated Knowledge. In: Liu, J., Wu, J., Yao, Y., Nishida, T. (eds.) AMT 2009. LNCS, vol. 5820, pp. 31–41. Springer, Heidelberg (2009)

[96] Okada, S., Nishida, T.: Incremental Clustering of Gesture Patterns based on a Self Organizing Incremental Neural Network. In: Proceedings of International Joint Conference on Neural Networks, Atlanta, Georgia, USA, June 14-19, pp. 2316–2322 (2009)

[97] Ohmoto, Y., Ohashi, H., Nishida, T.: Proposition of Capture and Express Behavior Environment (CEBE) for Realizing Enculturating Human-agent Interaction (submitted)

Agent-Based Active Information Resource and Its Applications

Tetsuo Kinoshita

Tohoku University, Katahira-2-1-1, Sendai 980-8577, Japan
kino@riec.tohoku.ac.jp

Abstract. A scheme of active information resource (AIR) provides a novel approach to actively support using the distributed information resources over the networked environment. With the AIR, passive information resources are extended into autonomous and active entities. The extended information resources can actively perform tasks to support use of them, so that the burden for users can be reduced. Moreover, multiple extended information resources can organize in a decentralized way to autonomously cooperate with each other, and this will enable a more flexible support for using them in the distributed environment. An agent-oriented design model for AIR is developed, and the agent-based AIR is applied and verified in two tasks of support for using distributed academic information resources and managing the operations of network system. The results of experiments on two prototype systems verified that the proposed approach has better performance than conventional approaches do.

Keywords: Active Information Resource, Information Retrieval, Network Management, Agent, Agent-based System.

1 Active Information Resource: Basic Concept

With the explosion of the amount of information available in digital form, as well as the rapid growth of networked environment, particularly the Internet, numerous electronic materials have been made available from various information sources and these materials are continuously generated, accumulated and updated as valuable information/knowledge sources in the distributed environment. Some characteristics of such information resources include: well organized metadata is available, meaningful relations between information resources exist, researchers' accumulated knowledge about information resources can be reused, etc. Due to the large amount of electronic information resources, the complex relations among them, and the distributed nature of them, it is becoming increasingly difficult for users to efficiently and effectively use these information resources. Therefore, an effective support for using information resources is required. Such a support approach should effectively use the characteristics of various information resources to provide mechanisms and functions such as managing, searching, showing and sharing. Furthermore, it should be adaptive in the distributed environment.

S. Kikuchi, S. Sachdeva, and S. Bhalla (Eds.): DNIS 2010, LNCS 5999, pp. 143–156, 2010.

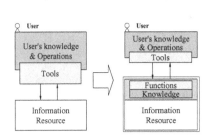

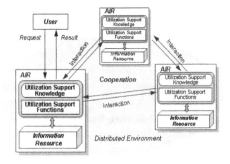

Fig. 1. Extending an Information Resource to Active Information Resource (AIR)

Fig. 2. Cooperation of AIRs

There are many conventional technologies can be applied to support the use of information resources. Typically, information resources in the conventional support system are modeled as passive entities, which can be accessed and manipulated by some functions required by the tasks of support [3]. In such circumstances, there are usually gaps among the characteristics of information resources, the manipulation of these resources, and the support for using them. Due to the gap, the characteristics are not effectively used, and the support is less effective in the distributed environment.

To realize an active support for using various information resources in the distributed environment, this research proposes a novel approach [1,2], as depicted in Fig.1. With this approach, passive information resources are extended into autonomous and active entities. Pertinent knowledge and functions are combined with information resources, and the characteristics of them can be effectively used. The extended information resources can actively perform tasks to support use of them, so that the burden for users can be reduced. Moreover, multiple extended information resources can organize in a decentralized way to autonomously cooperate with each other, and this will enable a more flexible support for using them in the distributed environment as shown in Fig.2. The extended information resource is called Active Information Resource (AIR). In order to design and implement the AIR over the networked environment, a design method for extending information resources into autonomous and active entities and a method how to use the realized AIRs to enable the active support are developed, aiming several application tasks such as searching/sharing useful academic information resources, managing the operation conditions of network system, using web services and so on.

2 Agent-Based Design of Active Information Resource

An AIR is an information resource that is enhanced and extended with information resource specific knowledge and functions for actively and flexibly facilitating use/reuse of it. An agent-oriented design method for AIR is developed and this design method consists of the basic model of AIR, and two agent-based design models of AIR.

Basic Model of AIR: There are 6 main parts in the proposed design model of AIR, which is shown in Fig.3: *Information Resource, Domain Knowledge-base, Knowledge about Information Contents, Information Extraction Unit, Information Processing Unit*, and *Contact & Cooperation Unit*. With the interaction between these 6 parts, an AIR can autonomously perform its tasks [5].

Agent-Based Design Models of AIR: When an AIR is designed with the agent-oriented approach, the knowledge and functions of AIR should be mapped into knowledge and functions of an agent or multiple agents. Thus an AIR can be designed and implemented using one of the following design models:

- *Single-agent-based Design Model*: In this model, the knowledge and functions of an AIR is mapped into knowledge and functions of one agent. Generally, the amount of information is relatively small. Since the structure of such an AIR is relatively simple, its design, implementation and maintenance would be accordingly easy.

- *Multiagent-based Design Model*: For an AIR with fairly large quantity of information, complex functionality of information processing and corresponding knowledge about information use/reuse, it should be more suitable to deploy multiple agents. In this model, each agent is in charge of only a certain part of the AIR's functionality and corresponding knowledge. In such an AIR, the knowledge and ability of a certain agent could be relatively simple, which means the development and maintenance of each agent could be easy.

In the design and implementation of agent-based AIR, we use the Repository-based Multiagent Framework (ADIPS/DASH framework) [13] and the Interactive Design Support Environment for agent system (IDEA) [14,15].

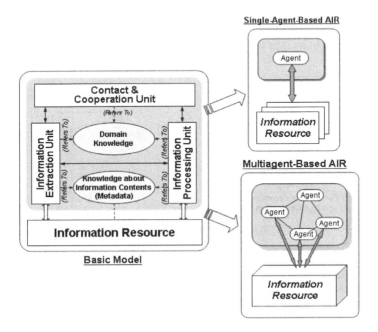

Fig. 3. Design Models of AIR

3   Knowledge Enhanced Search for Academic Information

3.1   AIR for Academic Information Resource

When a personal academic information collection is searched for a researcher's personal use, not only the literal contents, but also the characteristics of them should be considered [4]. However, when a conventional search method is used to find out such a collection, usually the relations among the academic information resources are not properly considered, and the researcher's personal knowledge about the academic information resources are often neglected. Consequently, the recall and efficiency of search could be low [6,7].

To deal with such problem, two types of knowledge about these information resources are introduced and utilized to enhance the search process based on the AIR:

- K_R : Knowledge on Relations among the personally collected academic information resources
- K_U : User's Knowledge about the personally collected academic information resources

The single-agent-based design model of AIR is used to realize the proposed knowledge enhanced search. Each piece of personally collected academic information resource is extended into an AIR called PerC-AIR (Personally Collected academic information resource AIR) [7].

3.2   Design of PerC-AIR

The PerC-AIRs are designed to automatically discover and update their K_R through cooperation. At the meantime, K_U can be automatically maintained by each PerC-AIR which keeps track of its user's operation on a piece of academic information resource.

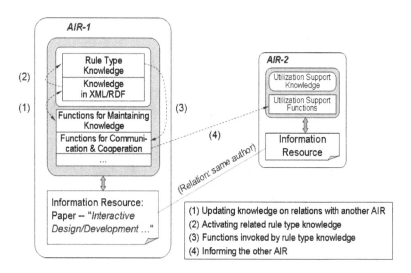

Fig. 4. Architecture of PerC-AIR

Moreover, a knowledge enhanced search in distributed PerC-AIRs is performed through autonomous cooperation between them. The knowledge and functions of PerC-AIR are designed and implemented by using the XML/RDF format representation and the rule-based agent behavior knowledge representation of ADIPS/DASH framework.

```
<!-- AIR-K_B -->                                    Basic knowledge about the
<dc:type>paper</dc:type>                            information contents
<dc:title>Interactive Design/Development Methodology of Agent System</dc:title>
<dc:subject>
  interactive, repository-based agent framework, design/development methodology
</dc:subject>
...

<!-- AIR-K_R -->                                    Knowledge on relations
<air:relatedAIRs>                                   with other AIRs
  <rdf:Seq>
    <rdf:_1 dc:identifier="AIR-Paper0005"
          air:relationType="same-author"
          air:relationDescp="T. Kinoshita"/>
    ...

<!-- AIR-K_U -->                                    User's knowledge about
<air:userComments>                                  the information contents
  <rdf:Seq>
    <rdf:_1 air:commentType="user-marked-relevant"
          air:commentDescp="IDEA"/>
    ...
```

```
// Inform related PerC-AIR about the newly found relation.
(rule inform-relation
  (Relation :airID ?airID :informed NO) = ?relation
  (RelatedAIR :airID ?airID) = ?reltdAIR
  (SelfInfo) = ?selfInfo
  (Status) = ?status                                This rule enables an AIR to
  -->                                               inform another AIR when
  (send :performative information-about-relation    some relations between
        :to      ?reltdAIR:airAgentID               them are found.
        :arrival ?reltdAIR:airWP
        :content (Relation
                    :airID           ?selfInfo:airID
                    :airAgentID      ?status:name
                    :airWP           ?status:environment
                    :relationType    ?relation:relationType
                    :relationDescp   ?relation:relationDescp))
  (modify ?relation:informed YES)
)
```

Fig. 5. Example of knowledge description of PerC-AIR

3.3 Knowledge Enhanced Search by PerC-AIRs

In a K_R enhanced search, relevance spreads from relevant PerC-AIRs to non-relevant ones, and relevance is accumulated in non-relevant PerC-AIRs. The influence of a relevant PerC-AIR on a related one is determined by the relevance value of the relevant PerC-AIR, the strength of their relation, and the depth of iteration. On the other hand, in a K_U enhanced search, the researcher's evaluation or opinion on each item in the collection is reused. Relevant or non-relevant items can be efficiently determined when there is proper K_U of PerC-AIR indicating so.

Fig.6 depicts a process of the combination of K_R and K_U enhanced search, in which basically the protocol and algorithm for the K_R enhanced search are used, except that K_U is used in each step of the K_R enhanced search. The proposed method finds more relevant PerC-AIRs than the conventional search method.

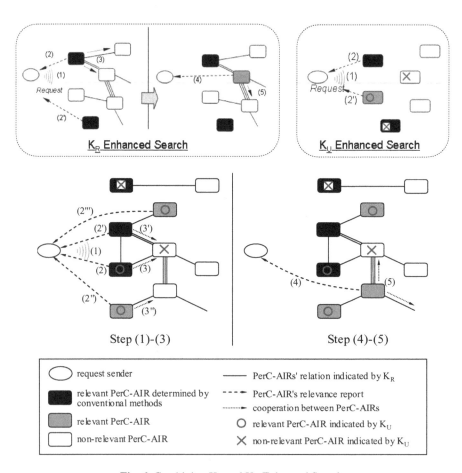

Fig. 6. Combining K_R and K_U Enhanced Search

To verify the proposed search method, some experiments were performed on a prototype system which consists of 110 distributed PerC-AIRs. As demonstrated by the average results of multiple searches shown in Fig.7, the proposed search method has better performance compared to the conventional method. Particularly, the use of a combination of both K_R and K_U can lead to a better performance and a higher efficiency. It can be concluded that the proposed knowledge enhanced search method can be used to search personal academic information collections effectively.

Since the information resources in a researcher's personal academic information collection are potentially valuable to some other researchers, making use of shared personal academic information collections could be an efficient way for researchers to obtain academic information resources [4]. However, in a conventional mechanism for sharing such collections, there is usually a lack of efficient functions for discovery of valuable collections, as well as an effective search method for them.

Using the AIR-based academic information collection, the Active Mechanism for Sharing Distributed Personal Academic Information Collections can also be realized based on the following two main functions: (f1) Autonomous Discovery of Valuable Collections, and (f2) Knowledge Enhanced Search in Shared Collections, in which the Col-AIRs (Collection AIRs) are introduced to automatically gather information about the collection from all PerC-AIRs in the collection, and the proposed mechanism can support academic researchers to efficiently share the academic information resources in the distributed environment.

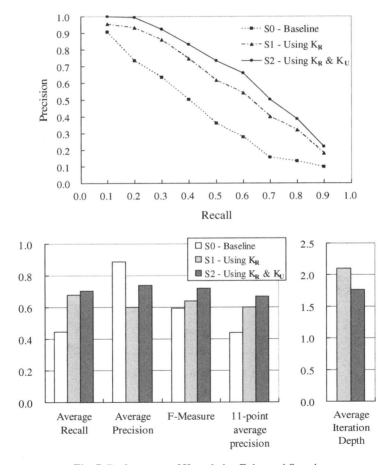

Fig. 7. Performance of Knowledge Enhanced Search

4 Autonomous Monitoring of Network Management Information

4.1 AIR-Based Network Management System (AIR-NMS)

The management activities performed by the network managers/administrators are becoming more demanding and data-intensive because of the rapid growth of modern networks, and automation of network management activities has become necessary. A typical approach to network management is centralized, static, polling-based management that involves high-capacity computing resources at the centralized platform including commercially available management tools [8]. However, in view of the dynamic nature of evolving networks, future network management solutions need to be flexible, adaptable, and intelligent without increasing the burden on network resources. The rapid of network systems has posed the issues of flexibility, scalability, and interoperability for the centralized paradigm. Even though failures in large communication networks are unavoidable, quick detection and identification of the causes of failure can fortify these systems, making them more robust, with more reliable operations, thereby ultimately increasing the level of confidence in the services they provide [9].

Motivated by these considerations, the AIR-based network management system (AIR-NMS) is intended to provide an intelligent, adaptive and autonomous network management support paradigm for various network systems [10,11].

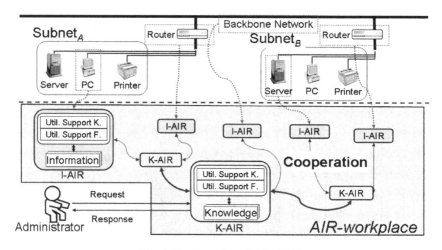

Fig. 8. Configuration of AIR-NMS

The AIR-NMS consists of two types of AIRs; I-AIR (AIR with status information of network) and K-AIR (AIR with knowledge of network management task). I-AIRs manage the status information, which is classifiable into two types: static information and dynamic information. For instance, the relationship between IP addresses and Mac addresses, host names, domain names, IP-routing, etc., are included as static network information, and the dynamic information includes number of packet traffic,

RMON-MIB, SNMPv2-MIB, logs of network services, and so on. I-AIRs are responsible to monitor the operational conditions of network, detect the important conditions to be alarmed, inspect/notify the conditions in response to requests of administrator.

On the other hand, K-AIRs manage heuristics or expertise of expert administrators which can utilized as the generic knowledge of network management tasks. K-AIRs and I-AIRs interact with each other to deal with the given/detected problem of the management task. This paper focus on the network monitoring task of AIR-NMS based on capabilities of I-AIRs [12].

4.2 Design of Agent-Based I-AIR

Conventionally, administrators collect status information through periodical polling, aggregate them, and decide the operational conditions of the network system using his expertise/heuristics. The administrator's task can be disaggregated into three sub-tasks, such as detection, recognition, and specification of the failure (abnormal status). In each sub-task, much experience as a network manager are required, therefore, a beginner cannot be employed as an administrator. The I-AIR is introduced to partially support such empirical tasks of administrator; the distributed and effective monitoring of network system, detection of network failure, processing of collected information according to failure, improvement of reliability of detection, recognition, and specification of failure through cooperation among AIRs.

In I-AIR, two information resource types, plain-text format and RDF/XML format, are utilized to represent and manage the status information. For instance, the log-information is acquired through the Syslog (a standard logging solution on UNIX and Linux systems) in plain-text format and the I-AIR extracts a diverse type of log-information and converts it to RDF/XML format specifications.

On the other hand, I-AIRs hold knowledge about information resources together with the functionality to handle collected information. Essential components of knowledge represented in an I-AIR are as follows:

- *I-AIR Identification Knowledge (ID)*: The ID includes an identification number, task number of I-AIR, etc.
- *Knowledge about Information Resource (IR)*: The IR includes a type, an update-time, a format type, etc.
- *Knowledge about Failure Inspection (FI)*: The FI includes two types of knowledge to inspect the failure: text information to be detected in logs, and a threshold of packets, etc.
- *Knowledge about Periodic Investigation Process Control Method (CM)*: The CM includes the polling time and other conditions for updating of the information resource.
- *Knowledge about Cooperation Protocol (CP)*: The CP includes protocol sequences for cooperation with other AIRs.

The knowledge contained in an I-AIR as ID, IR, and CP is required mainly to operate on the information resource and facilitate communication and cooperation among

I-AIRs. The preeminent characteristic of I-AIR is its autonomous monitoring mechanism, which is supported via FI and CM for the inspection and investigation of obstacles that hinder the normal network operation. Table 1 illustrates the I-AIRs developed in the prototype system and Fig.5 shows an example of describing the knowledge of I-AIR (No.15) based on Object-Attribute-Value description format of ADIPS/DASH agent.

Table 1. Example of implemented I-AIRs

I-AIR No.	Function	I-AIR No.	Function
1	Network Disconnection detector	11	DNS server process checker
2	NIC configuration failure detector	12	SMTP server process checker
3	SPAM mail detector	13	POP server process checker
4	MSBlaster attack detector	14	DNS connection checker
5	Mail send/receive error detector	15	Network route to host checker
6	TCP/IP stack failure checker	16	Kernel information checker
7	NIC configuration failure checker	17	Lease IP address checker
8	HUB failure checker	18	Mail server error checker
9	Router failure checker	19	Number of SPAM mail
10	Communication failure checker		

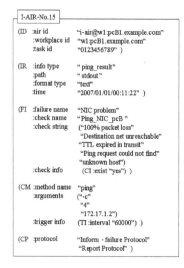

Fig. 9. Example of I-AIR (No.15)

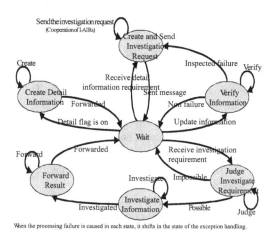

Fig. 10. State transition diagram of I-AIR

Moreover, applying the Single-Agent-based Design Model of AIR, the functionality of I-AIR is designed and implemented based on the state-transition depicted in Fig.10.

4.3   Experiment of Failure Detection by I-AIR

To evaluate the capabilities of I-AIR, an experimental AIR-NMS is set up as shown in Fig.11. The network system comprises a 100BASE-TX Ethernet with a firewall configured as a Network Address Translation (NAT) firewall, a router, and various personal computers (PCs) arranged in four subnetworks. Subnetwork A is configured as a Demilitarized Zone range 172.16.0.0/24. The server (sevA1) DNS and Mail application settings are configured. The other three subnetworks (B, C, D) have IP-addresses in the order given as 172.17.1.0/24, 172.17.2.0/24, and 172.17.3.0/24.

Moreover, the network management console for managing the whole setup resides in pcB1 of subnetwork B. In subnetwork C, there is a desktop-type PC system (pcC1) with a fixed IP address from the DNS server, and a notebook computer (pine) which acquires the IP-addresses through the DHCP. Each node (PC, router, firewall etc.) shows the corresponding AIR workplace where the I-AIRs operate actively. For each node, about 15 AIRs were implemented. This implies that nearly 140 I-AIRs were incorporated within the experimental setup. A Linux operating system was used in each PC.

The network administrator performs the management task according to the conventional manual method, as well as with the I-AIRs based proposed system.

He also measures the performance of the proposed approach adopted for the automation of network functions. In the experiment, the time and the number of procedures executed to correct the obstacle were measured after a network obstacle was reported to a subject. In this paper, the results of detecting specific-failure for multiple causes are demonstrated in below.

In the experiment, two kind of experimental methods have been designed, and for each method, five persons having expertise of managing computer communication systems have been employed:

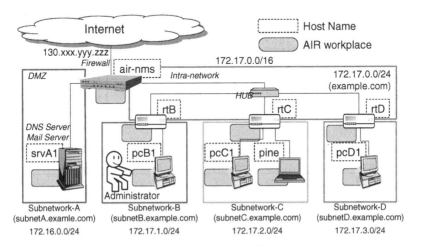

Fig. 11. Construction of experimental AIR-NMS

(i) Monitoring the network with the OS-default network management tools: Several failures obstructing the normal operation of network system are generated and accordingly it is required to restore the network services manually with the client management tools. Also, the time elapsed between the notification of failure to its remedy is measured.

(ii) Monitoring the network utilizing the I-AIRs: The obstacles are detected by the communication / cooperation mechanism of I-AIRs which are then reported to the I-AIR interface, then it is required to rectify the occurring failures. In this case also the time is measured from the point when the obstacle information is presented on the interface to the absolute restoration.

Hence, after some network obstacle has been reported and corrected, the time is measured as well as the number of procedures executed to restore the network to its normal operation.

Table 2. Assumed failure causes: Mail Sending / Receiving Error

Problem	Causes
Cable problem	a. Cable was disconnected.
Port problem	b. The 25th port was closed.
	c. The 110th port was closed.
DNS Server problem	d. DNS Server process was downed.
	e. Configration was not available.
Mail Server problem	f. Mail Server process was downed.

Table 3. Experimental results among individual administrators

		F Time	F Step		G Time	G Step		H Time	H Step		I Time	I Step		J Time	J Step
no I-AIR	d	158	9	b	566	8	e	929	23	f	235	5	a	655	19
	e	743	24	d	871	12	b	339	9	c	615	9	f	182	5
I-AIR	a	51	1	f	104	2	c	82	3	a	40	1	b	86	2
	f	85	4	c	106	2	d	52	3	e	74	2	e	128	6
$\frac{\text{I-AIR}}{\text{no I-AIR}}$ (%)		15.1	15.2		14.6	20.0		10.6	18.8		13.4	21.4		25.6	33.3

Management experience: F. 1year, G. 2year, H. 2year, I. 3year, J. 7year

Table 4. Experimental results among individual failures

	a Time	a Step	b Time	b Step	c Time	c Step	d Time	d Step	e Time	e Step	f Time	f Step
no I-AIR	655	19	566	8	615	9	158	9	743	24	235	5
	-	-	339	9	-	-	871	12	929	23	182	5
I-AIR	51	1	86	2	106	2	52	3	74	2	85	4
	40	1	-	-	82	3	-	-	128	6	104	2
$\frac{\text{I-AIR}}{\text{no I-AIR}}$ (%)	6.9	5.3	19.0	23.5	15.3	27.8	10.1	28.6	12.1	17.0	45.3	60.0

Table 2 depicts the failure situation "Mail Sending / Receiving Error" with some possible causes underlying the occurrence of this anomaly. The task of the subject is to determine the cause of this error. These causes do not occur necessarily in any fixed pattern. The checks to detect these causes are performed randomly. However, using I-AIRs is advantageous because every check is done only once during the course of the fault-localizing process. The failure cause is detected and the main cause behind the failure is reported to the network operator actively.

Experimental results computed by each manager while resolving the mail sending / receiving anomaly were compiled into Table 3. Additionally, the results corresponding to each failure cause were accumulated into Table 4. The results demonstrate that the network management overhead regarding the time taken to resolve a certain fault, along with the number of steps necessary to locate the cause of failure, were reduced to 20% on average.

The foundation of autonomous network monitoring is the use of I-AIRs, which, through active mutual interaction and with the functional network system, can resolve various network-failure situations quite efficiently. A part of I-AIR knowledge is modified dynamically on frequent basis, according to the operational characteristics of the network. The experimental results demonstrated a marked reduction in the administrator workload, through the use of the network monitoring and fault detection functions of I-AIR, as compared to the conventional network management methods.

5 Summary

The concept of active information resource (AIR) is presented and two applications, (i) knowledge enhanced search of academic information collection using PerC-AIRs and (ii) autonomous monitoring of network management information using I-AIRs, are also demonstrated in this paper. The agent-based computing technologies such as the repository-based multiagent framework and the interactive design environment for agent system can successfully be applied in the design and implementation of various AIRs and their agent-based applications.

References

1. Kinoshita, T.: A Method for Utilizing Distributed Information Resources Effectively: Design of Active Information Resource. IEICE Technical Report, AI99-54, pp.13–19 (1999) (in Japanese)
2. Li, B., Abe, T., Sugawara, K., Kinoshita, T.: Active Information Resource: Design Concept and Example. In: Proc. 17th Int. Conf. Advanced Information Networking and Applications (AINA 2003), pp. 274–277 (2003)
3. Chervenak, A., Foster, I., Kesselman, C., Salisbury, C., Tuecke, S.: The data grid: Towards an architecture for the distributed management and analysis of large scientific datasets. Journal of Network and Computer Applications 23(3), 187–200 (2000)
4. Bruce, H.: Personal, Anticipated Information Need. Information Research 10(3) (2005)
5. Bailey, C., Clarke, M.: Managing Knowledge for Personal and Organizational Benefit. Journal of Knowledge Management 5(1), 58–67 (2001)

6. Li, B., Abe, T., Kinoshita, T.: Design of Agent-based Active Information Resource. In: Proc. 1st Int. Conf. Agent-Based Technologies and Systems (ATS 2003), Univ. Calgary, pp. 233–244 (2003)
7. Li, B., Kinoshita, T.: Active Support for Using Academic Information Resource in Distributed Environment. Int. J. Computer Science and Network Security 7(6), 69–73 (2007)
8. Consens, M., Hasan, M.: Supporting network management through declaratively specified data visualizations. In: IEEE/IFIP 3rd International Symposium on Integrated Network Management, pp. 725–738 (1993)
9. Bouloutas, A., Calo, S., Finkel, A.: Alarm correlation and fault identification in communication networks. IEEE Transactions on Communications 42(2,3,4), 523–534 (1994)
10. Konno, S., Iwaya, Y., Abe, T., Kinoshita, T.: Design of Network Management Support System based on Active Information Resource. In: Proc. 18th Int. Conf. Advanced Information Networking and Applications (AINA 2004), vol. 1, pp. 102–106. IEEE, Los Alamitos (2004)
11. Konno, S., Sameera, A., Iwaya, Y., Kinoshita, T.: Effectiveness of Autonomous Network Monitoring Based on Intelligent-Agent-Mediated Status Information. In: Okuno, H.G., Ali, M. (eds.) IEA/AIE 2007. LNCS (LNAI), vol. 4570, pp. 1078–1087. Springer, Heidelberg (2007)
12. Abar, S., Konno, S., Kinoshita, T.: Autonomous Network Monitoring System based on Agent-mediated Network Information. International Journal of Computer Science and Network Security 8(2), 326–333 (2008)
13. Kinoshita, T., Sugawara, K.: ADIPS Framework for Flexible Distributed Systems. In: Ishida, T. (ed.) PRIMA 1998. LNCS (LNAI), vol. 1599, pp. 18–32. Springer, Heidelberg (1999)
14. Uchiya, T., Maemura, T., Xiaolu, L., Kinoshita, T.: Design and Implementation of Interactive Design Environment of Agent System. In: Okuno, H.G., Ali, M. (eds.) IEA/AIE 2007. LNCS (LNAI), vol. 4570, pp. 1088–1097. Springer, Heidelberg (2007)
15. IDEA/DASH Online,
    http://www.ka.riec.tohoku.ac.jp/idea/index.html

Semantic Interoperability in Healthcare Information for EHR Databases

Shelly Sachdeva and Subhash Bhalla

Graduate Department of Computer and Information Systems
University of Aizu, Fukushima, Japan
{d8111107,bhalla}@u-aizu.ac.jp

Abstract. Healthcare information is complex, distributed and non-structured in nature. Integration of information is important to retrieve patient history, for knowledge sharing and to formulate queries. Large scale adoption of electronic healthcare applications requires semantic interoperability. Interoperability of Electronic Health Records (EHRs) is important because patients have become mobile, treatment and health care providers have increased, and also, have become more specialized. The paper analyses the role of semantic interoperability in healthcare. The system modeling approach has been analyzed with a view of supporting system-to-system and user-system interactions. In addition, query interfaces have been considered at varying levels of user and system activities.

Keywords: Electronic Health Records, Semantic Interoperability, openEHR, Healthcare, Archetype Based EHR.

1 Introduction

Digitized form of individual patient's medical record is referred as electronic health record (EHR). These records can be stored, retrieved and shared over a network through enhancement in information technology. An Integrated Care EHR is defined as:" a repository of information regarding the health of a subject of care in computer processable form, stored and transmitted securely, and accessible by multiple authorized users" [1]. The Institute of Electrical and Electronics Engineers (IEEE) defines interoperability as the "ability of two or more components to exchange information and to use the information that has been exchanged" [2]. Semantic Interoperability is a big challenge in healthcare industry. Semantic interoperability states that the meaning of information must be preserved from a user level through logical level to physical level. Users enter information. It should safely reach the designated part of the system, and allow it to be sharable with other users and systems. As we know, there are different vendors for different systems. Thus, Semantic interoperability should be taken into account during exchange of data, information and knowledge. Figure1 indicates that, the meaning of information will be preserved across various applications, systems and enterprises.

Health care domain is complex. It is evolving at a fast rate. Health knowledge is becoming broad, deep and rich with time. Often, different clinics and hospitals have

S. Kikuchi, S. Sachdeva, and S. Bhalla (Eds.): DNIS 2010, LNCS 5999, pp. 157–173, 2010.

their own information systems to maintain patient data. There may be redundancy in data because of distributed and heterogeneous data resources. This may hinder the exchange of data among systems and organizations. There is a need for legacy migration of data to a standard form for the purposes of exchanges. The EHRs should be standardized and should incorporate semantic interoperability. World Health Organization (WHO) has strong desire to develop and implement semantically interoperable health information systems and EHRs [25]. The rest of paper is organized as follows. Section 2 describes the role of dual level modeling approach in achieving semantic interoperability. Section 3 describes the openEHR architecture and its comparison to database management system architecture. Section 4 explains the details of semantic interoperability in EHRs systems. Section 5 describes querying EHR data with emphasis on high-level query interfaces for health professionals. Section 6 describes the discussions. Finally, section 7 presents the summary and conclusions.

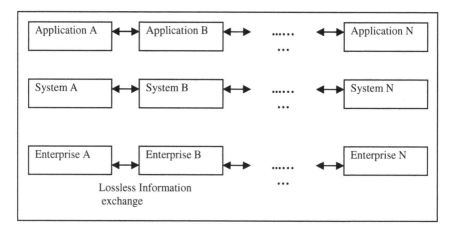

Fig. 1. Semantic Interoperability

2 Dual Level Modelling Approach

In essence, the proposed Electronic Health Records (EHRs) have a complex structure that may include data of about 100-200 parameters, such as temperature, blood-pressure and body mass index. Individual parameters have their own contents (Figure 2). In order to serve as an information interchange platform, EHRs use archetypes to accommodate various forms of contents [6]. The EHR data has multitude of representations. The contents can be structured, semi-structured or unstructured, or a mixture of all three. These can be plain text, coded text, paragraphs, measured quantities with values and units, date, time, date-time, and partial date/time, encapsulated data (multimedia, parsable content), basic types (such as boolean, state variable), container types (list, set) and uniform resource identifiers (URI).

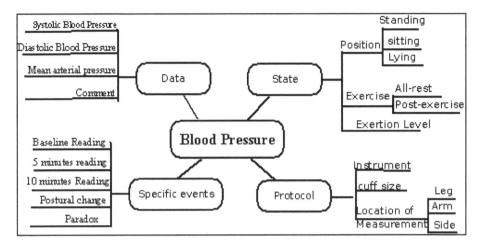

Fig. 2. Blood Pressure Archetype

The classic specifications require that the details of clinical knowledge be simultaneously coded into the software. The disadvantage of the approach has been that with the expansion of clinical knowledge the software becomes unsuitable and outdated. To overcome this, two level modeling approach has been proposed. Dual model approach for EHR architecture is defined by ISO 13606[12] for a single subject of care (patient). The emphasis is on achieving interoperability of systems and components during communication of a part or complete EHR. Examples of Dual Model EHR architecture are CEN/TC251 EN13606 [3] standard (developed by European committee for standardization) and openEHR standard. CEN/TC251 is a regional Standards Development Organization, which is addressing the needs of the stakeholders to have interoperable and implementable standards. It will allow for safe and secure information exchange. OpenEHR [4] foundation was established by University College London and Ocean informatics. It is an international foundation working towards semantic interoperability of EHR and improvement of health care.

In two-level modeling approach, the lower level consists of reference model and the upper level consists of domain level definitions in the form of archetypes and templates. Reference Model (RM) [12] is an object-oriented model that contains the basic entities for representing any entry in an EHR. The software and data can be built from RM. Concepts in openEHR RM are invariant. It comprises a small set of classes that define the generic building blocks to construct EHRs. This information model ensures syntactic or data interoperability. The second level is based on archetypes [5] [10]. These are formal definitions of clinical concepts in the form of structured and constrained combinations of the entities of a RM. A definition of data as archetypes can be developed in terms of constraints on structure, types, values, and behaviors of RM classes for each concept, and in the domain in which we want to use. Archetypes are flexible. They are general-purpose, reusable and composable. These provide knowledge level interoperability, i.e., the ability of systems to reliably communicate with each other at the level of knowledge concepts. Thus, the meaning of clinical data will be preserved in archetype based systems. Standardization can be achieved, so that, whenever there is a change in the clinical knowledge (or requirements), the software need not be changed and only the archetypes need to be modified.

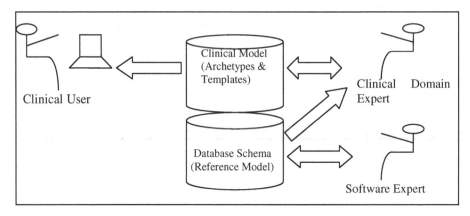

Fig. 3. Two Level Modelling Approach

The clinical user can enter and access information through clinical application. The clinical domain expert can record and maintain the clinical model through modeller. The modeller is software needed to manage the archetypes. Patients can have complete control over access and distribution of their health records.

3   The OpenEHR Architecture

The openEHR is pioneering the field for maintaining semantic interoperable EHRs. It has launched the implementation of the specification project. It aims at a new business model for electronic medical records. The latest edition of Microsoft's Connected Health Framework, includes openEHR (ISO 13606-2) archetypes as part of its domain knowledge architecture. The openEHR Reference Model is based on ISO and CEN EHR standards, and is interoperable with HL7 (Health Level Seven) and EDIFACT (Electronic data interchange for administration, commerce and transport) message standards [12]. This enables openEHR-based software to be integrated with other software and systems. Figure 4 shows how the DBMS architecture [7] can be compared to the openEHR architecture [6].

i) Physical level: The lowest level of abstraction describes the details of reference model. These include identification, access to knowledge resources, data types and structures, versioning semantics, support for archetyping and semantics of enterprise level health information types.

ii) Logical Level: The next higher level of abstraction describes the clinical concepts that are to be stored in the system. They can be represented in the form of archetypes and templates. The implementation of clinical concepts may involve physical-level structures. Users of logical level do not need to be aware of this complexity. Clinical domain experts use logical level.

iii) View Level: The highest level of abstraction describes only part of the entire EHR architecture. This corresponds to the service model. Several views are defined and the users see these views. In addition to hiding details of logical level, the views also provide a security mechanism to prevent users from accessing certain parts within EHR contents.

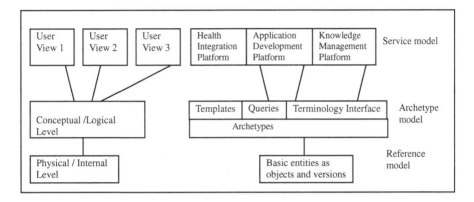

Fig. 4. DBMS architecture compared to openEHR Architecture

4   Semantic Interoperability in EHR: An Overview

In this section, we describe different levels of interoperability, the relationship of archetypes to semantic interoperability and Archetype Description language as a type of language for system level interactions (Figure 1). At the system level, the AQL language is supported for development of initial support infrastructure. In the following sections, this report presents how the application level can benefit from XML conversions and support of query language at application level, in the form of XQuery. Higher-level of support is an active area of research. Many research efforts aim to improve user interaction facilities [18] [22].

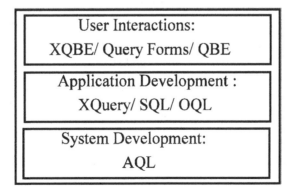

Fig. 5. Query Support at different Levels

4.1 Levels of Interoperability

There are three levels of interoperability [11]:

*   Syntactic (data) interoperability
*   Structural interoperability/Semantic interpretability.
*   Semantic interoperability.

These are described in table 1.

Table 1. Levels of interoperability [11]

Levels of interoperability	Main mechanisms for interoperability	Description
Syntactic interoperability	openEHR Reference Model (RM)	The openEHR reference model alone ensures syntactic interoperability independent of any defined archetypes. The openEHR reference model does not define clinical knowledge. It is defined and communicated by archetypes, separately from the reference model. Hence, data items are communicated between systems only in terms of clearly defined, generic reference model instances. As the reference model is stable, achieving syntactic interoperability between systems is a simple task.
Structural interoperability	Archetypes	Structural interoperability is achieved by the definition and use of archetypes. As agreed models of clinical or other domain specific concepts, archetypes are clinically meaningful entities. An EHR entry (or a part) which has been archetyped will have the same meaning no matter where it appears. Thus, archetypes can be shared by multiple health systems and authorities, enabling information to be shared between different systems and types of healthcare professionals. Clinical knowledge can be shared and clinical information can be safely interpreted by exchanging archetypes.
Semantic interoperability	Domain Knowledge Governance	The use of archetypes and the reference model alone do not guarantee that different EHR systems and vendors will construct equivalent EHR extracts, and use the record hierarchy and terminology in consistent ways. For semantically interoperable systems, archetype development must be coordinated through systematic "Domain Knowledge Governance" tool. For example, it succeeds to avoid incompatible, overlapping archetypes for essentially the same concept.

4.2 Archetypes and Semantic Interoperability

Archetypes specify the design of the clinical data that a Health Care Professional needs to store in a computer system. Archetypes enable the formal definition of clinical content by domain experts without the need for technical understanding. These conserve the meaning of data by maintaining explicitly specified and well-structured clinical content for semantic interpretability. These can safely evolve and thus deal with ever-changing health knowledge using a two-level approach.

In simpler terms, an archetype is an agreed formal and interoperable specification of a re-usable clinical data set which underpins an electronic health record (EHR). It captures as much information about a particular and discrete clinical concept as possible. An example of a simple archetype is WEIGHT, which can be used in multiple places, wherever is required within an EHR. Once the format of an archetype is agreed and published, then it will be held in a 'library'(such as, clinical knowledge manager) and made available for use in any part of a given application, by multiple vendor systems, multiple institutions, and multiple geographical regions. Each group or entity using the same archetype will be able to understand and compute data captured by the same archetype in another clinical environment.

Archetypes are the basis for knowledge based systems (KBS) as these are means to define clinical knowledge (Figure 2). These are language neutral. These should be governed with an international scope, and should be developed by clinical experts in interdisciplinary cooperative way [13]. The developed archetypes need to be reviewed by other clinical experts (e.g., clinical review board) to ensure their completeness and relevance to evidence-based clinical practice. The archetype repository is a place of development, governance and primary source of archetypes. High quality archetypes with high quality clinical content are the key to semantic interoperability of clinical systems [13]. According to archetype editorial group (and clinical knowledge manager (CKM) [9]), the information should be sharable and computable.

Terminologies also help in achieving semantic interoperability. Terms within archetypes are linked (bound) to external terminologies like SNOMED-CT [11]. With the use of reference model, archetypes and a companion domain knowledge governance tool, the semantic interoperability of EHR systems becomes a reachable goal.

The openEHR foundation is developing archetypes which will ensure semantic interoperability. The openEHR archetypes are developed systematically through domain knowledge governance tools. According to the statistics provided by CKM, which is a domain knowledge governance tool, there are 227 numbers of archetypes [9]. Domain knowledge governance will ensure that archetypes will meet the information needs of the various areas.

With CKM, the users interested in modeling clinical content can participate in the creation and/or enhancement of an international set of archetypes. These provide the foundation for interoperable Electronic Health Records. CKM is a framework for managing archetypes. It helps in identifying which archetypes need to be standardized, and which are domain specific. It establishes a frame of reference and helps to train multidisciplinary teams for archetype development. The coordination effort team will inform and support domain knowledge governance. To support this, openEHR has employed the Web Ontology Language (OWL) and the Protégé OWL Plug-In to develop and maintain an Archetype Ontology which provides the necessary meta-information on archetypes for Domain Knowledge Governance. The Archetype Ontology captures the meta-information about archetypes needed to support Domain Knowledge Governance [11].

4.3 Archetype Definition Language (ADL)

Archetypes for any domain are described using a formal language known as Archetype deption language (ADL) [14]. ADL is path addressable like XML. The openEHR Archetype Object Model (AOM) describes the definitive semantic model of archetypes, in the form of an object model [15]. The AOM defines relationships which must hold true between the parts of an archetype for it to be valid as a whole. In simpler terms, all archetypes should conform to AOM. Since EHR has a hierarchical structure, ADL syntax is one possible serialisation of an archetype. ADL uses three other syntaxes, cADL (constraint form of ADL), dADL (data definition form of ADL), and a version of first-order predicate logic (FOPL), to describe constraints on data which are instances of RM [14].

The ADL archetype structure consists of archetype definition (expressed using cADL syntax), language, description, ontology, and revision_history (expressed using dADL syntax), invariant section (expressed using FOPL). The invariant section introduces assertions which relate to the entire archetype. These are used to make statements which are not possible within the block structure of the definition section. Similarly, the dADL syntax provides a formal means of expressing instance data based on an underlying information Model [14]. The cADL is a syntax which enables constraints on data defined by object-oriented information models to be expressed in archetypes or other knowledge definition formalisms [14].

Every ADL archetype is written with respect to a reference model. Archetypes are applied to data via the use of templates, which are defined at a local level. Templates [10] generally correspond closely to screen forms, and may be re-usable at a local or regional level. Templates do not introduce any new semantics to archetypes, they simply specify the use of particular archetypes, further compatible constraints, and default data values.

There are many parameters, such as weight, body temperature and heart rate in an EHR. The ADL for a parameter 'Blood Pressure' (BP) is shown in appendix A (also see Figure 2). ADL for other parameters are available at common repository [16]. ADL has a number of advantages over XML. It is both machine and human processable, and approximately, takes half space of XML. The leaf data type is more comprehensive set (including interval of numerics and date/time types). ADL adheres to object-oriented semantics that do not confuse between notions of attributes and elements. In ADL, there are two types of identifiers (from reference model) - the type names and attributes. Formally, it is possible to convert ADL into XML format and other formats [14]. Table 2 gives the comparison of ADL and XML.

In the near future, there is an important research issue regarding EHR systems, that is, whether all the archetype-based EHR systems will be created as ADL based database systems or ADL-enabled database systems (i.e., traditional database systems with enhanced ADL storage capabilities).

Table 2. Comparison between ADL and XML

Properties	ADL	XML
Machine Processable	Yes	Yes
Human Readable	Yes	Sometimes unreadable (e.g., XML-schema instance, OWL-RDF ontologies)
Leaf data Types	More comprehensive set, including interval of numerics and date/time types	String data; with XML Schema option- more comprehensive set
Adhering to object-oriented semantics	Yes, particularly for container types	XML schema languages do not follow object-oriented semantics
Representation of object properties	Uses attributes	Uses attributes and Sub-elements
Space (for storage)	Uses nearly half of space in XML	May have data redundancy

4.4 Interoperability and Different Levels of Interfaces

In Figure 1, the systems will use a XML/ADL type of language for system-to-system interactions. The healthcare worker will need an additional support layer. The existing support can be compared to (Figure 5)–

A) System Programmers level for development of EHR system- using ADL.

B) Application Programmer level for development of system applications, using XQuery, OQL (object query language) and SQL - type of interfaces (assuming the RDBMSs may support ADL in future).

C) Healthcare worker level interfaces: This is an active research area and no easy-to-use interfaces exist till date. In section 5.2, an attempt to provide one such interface has been outlined. It aims to demonstrate – how interoperability at application programmer level, can be made to support a user interface at healthcare worker level.

5 Querying Archetype Based EHR

EHRs allow multiple representations [17]. In principle, EHRs can be represented as relational structures (governed by an object/relational mapping layer), and in various XML storage representations. There are many properties and classes in the reference model, but the archetypes will constrain only those parts of a model which are meaningful to constrain. These constraints cannot be stronger than those in reference model. For example, if an attribute is mandatory in RM, it is not valid to express a constraint allowing the attribute to be optional in the archetype (ADL). So, the single ADL file is not sufficient enough for querying. The user may want to query some properties or attributes from RM, along with the querying from properties in archetypes. In order to create a data instance of a parameter of EHR, we need different archetypes in ADL, and also these archetypes may belong to different categories of archetypes.

For example, to create a data instance for Blood Pressure, we need two different archetypes-namely, encounter and blood pressure. These archetypes belong to different categories viz., COMPOSITION and OBSERVATION.

The different categories have different structure. At the time of query, a user faces this problem- which archetypes must be included in querying? For example, querying on BP requires the use of two archetypes viz., Encounter archetype (belonging to COMPOSITION category of RM) and Blood Pressure archetype (belonging to OBSERVATION category of RM). This problem can be addressed by the use of templates. Archetypes are encapsulated by Templates for the purpose of intelligent querying [10]. The templates are used for archetype composition or chaining. Archetypes provide the pattern for data rather than an exact template. The result of the use of archetypes to create data in the EHR is that the structure of data in any top-level object conforms to the constraints defined in a composition of archetypes chosen by a template.

At the user level, querying data regarding BP must be very simple. The user only knows BP as a parameter and will query that parameter only.

The EHR system must have an appealing and responsive query interface that provides a rich overview of data and an effective query mechanism for patient data. The overall solution should be designed with an end-to-end perspective in mind. A query interface is required that will support users at varying levels of query skills. These include semi-skilled users at clinics or hospitals.

5.1 Archetype Query Language (AQL)

To query upon EHRs, a query language, Archetype Query Language (AQL) has been developed [8]. It is neutral to EHRs, programming languages and system environments. It depends on the openEHR archetype model, semantics and its syntax. AQL is able to express queries from an archetype-based EHR system. The use of AQL is confined to a skilled programmers' level. It was first named as EQL (EHR Query Language) which has been enhanced with the following two innovations [17]:

i) utilizing the openEHR path mechanism to represent the query criteria and the response or results; and

ii) using a 'containment' mechanism to indicate the data hierarchy and to constrain the source data to which the query is applied.

The syntax of AQL is illustrated by the help of example.

Query: Find all blood pressure values, where systolic value is greater than (or equal to) 140, or diastolic value is greater than (or equals to) 90, within a specified EHR.

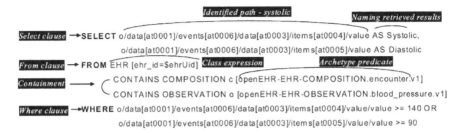

Fig. 6. Syntax of AQL

5.2 High-Level Database Query Interfaces

AQL is difficult for semi-skilled users. It requires the knowledge of archetypes and knowledge of languages such as ADL, XML and SQL. At the present moment, there is no easy-to-use query language interface available for EHRs database. We propose to study for the convenience of healthcare professionals a high-level interface for querying archetype based EHR systems based on the proposed query interface XQBE [18]. An alternative approach proposed by Ocean informatics [19] suggests using a query builder tool, to construct AQL query. It requires form related inputs and more skills on the part of the user. Our goal is similar and it is easy to achieve with the help of XQBE.

XQBE [18] is a user-friendly, visual query language for expressing a large subset of XQuery in a visual form. Its simplicity, expressive power and direct mapping to XQuery are some of the highlighting features for its use. Like XQuery, XQBE relies on the underlying expressions in XML. It requires all data to be converted to XML form. It presents a user with XML sub-tree expressions for the items of user interests. XQBE's main graphical elements are trees. There are two parts, the source part which describes the XML data to be matched against the set of input documents, and the construct part, which specifies which parts will be retained in the result, together with (optional) newly generated XML items.

In order to adopt a XQBE like interface at user level, we propose to convert ADL into XML. ADL can be mapped to an equivalent XML instance. ADL is hierarchical in nature and has a unique identification to each node. Thus, paths are directly convertible to XPath expressions. These can be created. According to Beale and Heard [6], the particular mapping chosen may be designed to be a faithful reflection of the semantics of object-oriented data. There may be need for some additional tags for making the mapping of nested container attributes since XML does not have a systematic object-oriented semantics. Thus, single attribute nodes can be mapped to tagged nodes of the same name. Container attribute nodes map to a series of tagged nodes of the same name, each with the XML attribute 'id' set to the node identifier. Type names map to XML 'type' attributes.

In the present proposal, the patient data description is converted to XML form [21]. It is suitably reformed for adoption of XQBE interface. Thus users can directly use XQBE query interface to access patient data. This process eliminates the need to learn and use the AQL language on the part of the users [21]. The XQBE skills can be learnt with ease [18].

5.3 Mapping ADL to XQBE for EHR Data

Database queries are usually dependent on local database schemas but archetype systems being proposed aim to have portable queries. The queries play a crucial role in decision support and epidemiological situations. The XQBE approach for archetype-based EHRs is being proposed for semi-skilled users (such as doctors, physicians, nurses). The mapping process to create XQBE is shown in following steps (Figure 7).

i)   The conversion of ADL file into XML file.
ii)   Generation of DTD for the XML file.
iii) Generation of XQBE interface structure.

Subsequently, for the semi-skilled user, this three step process will facilitate in querying archetype based EHRs. The step (ii) in above process will aid in the guided construction of query provided by XQBE [18]. However, for some users (skilled in use of XML) the step (ii) may not be needed and XQBE can be used directly for the XML file.

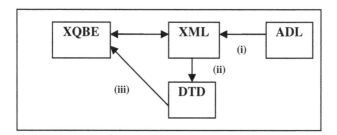

Fig. 7. Mapping process to present XQBE for EHRs [21]

Query Scenario1. Find all Blood Pressure values, having the systolic_BP and dia-stolic_BP, (where systolic_BP >= 140 and diastolic_BP >=90).
    The AQL syntax for the above query is shown in appendix B. By using XQBE approach for querying, we perform step (i) to step (iii) as explained, on BP parame-ter. For each case of query, and for querying different parameters of EHR, we need to convert each parameter (in form of adl) to a corresponding xml for the demon-stration. We propose to develop an automated tool in the subsequent phase. The clinical user will be provided with a substituted XQBE interface (Figure 9) in place of AQL.
    XQBE is a visual interface. A user is presented with graphical image of EHR components, for example, blood Pressure (BP) in this case. In Figure 9, based on the selected source data, the user defines a target sub-tree (in XML form) to express the query (and its outcome). The query is expressed by using the graphical elements of XQBE [18]. The source part of the query is built using the DTD. A guided construc-tion is provided to the user to add predicates for the query. The construct (or result) part of the query is built by the user using the graphical elements of XQBE by drag-ging and dropping them.

```
<!ELEMENT adl_version ( #PCDATA ) >
<!ELEMENT archetype ( original_language, description,
archetype_id, adl_version, concept, definition, ontology ) >
<!ELEMENT archetype_id ( value ) >
<! ELEMENT assumed_value (terminology_id,
code_string, magnitude?, units?, precision? ) >
<! ELEMENT attributes (rm_attribute_name, existence,
children+, cardinality? ) >
<! ATTLIST attributes xsi: type
(C_MULTIPLE_ATTRIBUTE I C_SINGLE_ATTRIBUTE)
#REQUIRED >
<! ELEMENT cardinality (is_ordered, is_unique, interval)
>
<! ELEMENT children (assumed_value I attributes I
code_list I includes I item I list I node_id I occurrences I
property I rm_type_name I target_path I terminology_id)* >
<! ATTLIST children xsi: type
(ARCHETYPE_INTERNAL_REF I ARCHETYPE_SLOT I
C_CODE_PHRASE I C_COMPLEX_OBJECT I
```

Fig. 8. A sample of BP.dtd

Figure 9 shows the element nodes and subelement nodes in the source part. The subelements (systolic and diastolic) of the BP element, one systolic and one diastolic satisfy condition1 (systolic>=140) AND condition2 (diastolic>=90) are described with the help of XQBE convention. As per the convention, an arc containing '+' indicates that 'children' element node may exist at any level of nesting (as in Xpath we use '//'). The construct part consists of element node for BP (set tag T-node), and also element nodes for systolic and diastolic, which relates the projected BP element nodes to its systolic and diastolic subelements. The fragment node (shown by filled triangle) indicates that the entire fragment of systolic and diastolic must be retained in the result.

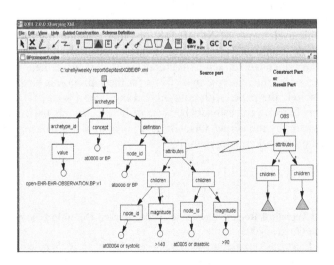

Fig. 9. BP.XQBE- an XQBE template for query

6 Discussions

There are several concerns regarding semantic interoperability which are listed below:

- How will the legacy migration of electronic health record systems to dual model approach based EHR systems be achieved?
- With evolution of domain clinical knowledge concepts, is it possible to achieve semantic interoperability in next 5 years or 10 years?
- At present, few people are trained to work at the intersection of biomedicine and IT. Will clinical domain experts really be interested in the development of archetypes, which are keystones in achieving semantic interoperability?
- Archetype Query Language [8] has been developed by openEHR for querying EHR data. It helps in semantic interoperability. Will it really be a powerful query language?
- How much cost will be involved for maintaining EHR systems?
- In near future, will archetype based EHR systems be semantically interoperable?

Sharing and exchange of knowledge, resources and information pertaining to the care of patients is needed by healthcare professionals. Archetypes and templates are a paradigm for building semantically-enabled software systems. Thus, archetype based EHR system ensures semantic interoperability. Database queries, usually dependent on local database schemas but archetype systems being proposed aim to have portable queries.

7 Summary and Conclusions

Existing query language interfaces are not suitable for healthcare applications. A proposal for developing high level interface suitable for healthcare users has been presented. Further, high-level user facilities, such as QBE sort of interface for ADL using the graphical elements of XQBE [18],or other existing high-level interfaces, such as XGI (A graphical interface for XQuery creation) [23], and GLASS (A Graphical Query Language for semi-structured data) [24] are aimed to be analyzed. Due to the time constraints, the research efforts for these proposals have been omitted. Finally, we conclude the paper with pointing out the strong need of high-level query interfaces for healthcare professionals because of heterogeneous nature and fragmentation of healthcare organizations. Our aim is to support a healthcare user at varying levels to enhance semantic interoperability.

References

1. ISO/TC 215 Technical Report: Electronic Health Record Definition, Scope, and Context. Second Draft (August 2003)
2. Institute of Electrical and Electronics Engineers (IEEE). IEEE Standard Computer Dictionary: A Compilation of IEEE Standard Computer Glossaries, New York, NY (1990)

3. European committee for Standardization, Technical committee on Health informatics, Standard for EHR communication, `http://www.cen.eu`

4. `http://www.openehr.org`

5. ISO 13606-2, Health informatics - Electronic health record communication - Part 2: Archetype interchange specification (2008)

6. Beale, T., Heard, S.: openEHR Architecture: Architecture Overview in The openEHR release 1.0.2. In: Beale, T., Heard, S. (eds.) openEHR Foundation (2008)

7. Silberschatz, K., Sudarshan, S.: Database Systems Concepts, 4th edn. McGraw-Hill, New York (2002) international edition

8. ArchetypeQueryLanguage, `http://www.openehr.org/wiki/display/spec/Archetype+Query+Language+AQL+Ocean`

9. Clinical Knowledge Manager version 1.0.5, http://www.openehr.org/knowledge

10. Beale, T., Heard, S.: Archetype Definitions and Principles in The openEHR foundation release 1.0.2. In: Beale, T., Heard, S. (eds.) The openEHR foundation (2005)

11. Garde, S., Knaup, P., Hovenga, E.J.S., Heard, S.: Towards Semantic Interoperability for Electronic Health Records: Domain Knowledge Governance for openEHR Archetypes. Methods of Information in Medicine (2007)

12. ISO 13606-1, Health informatics - Electronic health record communication - Part 1: Reference Model (2008)

13. Garde, S., Hovenga, E.J.S., Gränz, J., Foozonkhah, S., Heard, S.: Managing Archetypes for Sustainable and Semantically Interoperable Electronic Health Records. Electronic Journal of Health informatics (2007)

14. Beale, T., Heard, S.: The openEHR Archetype Model-Archetype Definition Language ADL 1.4. openEHR release 1.0.2, Issue date December 12 (2008)

15. Beale, T.: The openEHR Archetype Model-Archetype Object Model. openEHR release 1.0.2, Issue date November 20 (2008)

16. `http://www.openehr.org/svn/knowledge/archetypes/dev/html/index_en.html` (ADL for archetypes) (accessed 04/12/2009)

17. Chunlan, M., Heath, F., Thomas, B., Sam, H.: EHR Query Language (EQL)-A Query Language for Archetype-Based Health Records. In: MEDINFO 2007 (2007)

18. Braga, D., Campi, A., Ceri, S.: XQBE (XQueryBy Example): A Visual Interface to the Standard XML Query Language. ACM Transactions on Database Systems 30(2), 398–443 (2005)

19. `http://www.oceaninformatics.com/Solutions/ocean-products/Clinical-Modelling/Ocean-Query-Builder.html` (Query builder, Ocean Informatics) (accessed 10/12/2009)

20. `http://dbgroup.elet.polimi.it/xquery/XQBEDownload.html` (XQBE 2.0.0)

21. Sachdeva, S., Bhalla, S.: Implementing High-Level Query Language Interfaces for Archetype-Based Electronic Health Records Database. In: International Conference on Management of Data (COMAD), December 9-12 (2009)

22. Jayapandian, M., Jagadish, H.V.: Automating the Design and Construction of Query Forms. IEEE Transactions on Knowledge and Data Engineering 21(10), 1389–1402 (2009)

23. Li, X., Gennari, J.H., Brinkley, J.F.: XGI: A Graphical Interface for XQuery Creation. In: AMIA 2007 Symposium Proceedings, pp. 453–457 (2007)

24. Ni, W., Ling, T.W.: GLASS: A Graphical Query Language for semi-structured data. In: Proceedings of the Eighth International Conference on Database Systems for Advanced Applications (DASFAA 2003). IEEE, Los Alamitos (2003)

25. WHO. WHA58.28 - Resolution on eHealth. 58th World Health Assembly (2005), `http://www.who.int/gb/ebwha/pdf_files/WHA58/WHA58_28-en.pdf`

Appendix A

A brief sample of ADL for Blood Pressure Archetype
(BP.adl downloaded from [9])

```
definition
OBSERVATION [at0000] matches {        -- Blood pressure
data matches {
  HISTORY[at0001] matches {  -- history
    events cardinality matches {1..*; unordered} matches {
      EVENT[at0006] occurrences matches {0..*} matches {   -- any event
        data matches {
          ITEM_TREE[at0003] matches {         -- blood pressure
          items cardinality matches {0..*; unordered} matches {
      ELEMENT[at0004] occurrences matches {0..1} matches     -- Systolic
          value matches {
            C_DV_QUANTITY <
                property = <[openehr::125]>
              list = <  ["1"] = <
              units = <"mm[Hg]">
              magnitude = <|0.0..<1000.0|>
              precision = <|0|>
                    > > >
                    }}
      ELEMENT[at0005] occurrences matches {0..1} matches {            -- Diastolic
          value matches {
            C_DV_QUANTITY <
              property = <[openehr::125]>
              list = <          ["1"] = <
                units = <"mm[Hg]">
              magnitude = <|0.0..<1000.0|>
              precision = <|0|>
              > > >        } }
    ELEMENT[at1006] occurrences matches {0..1} matches {-- Mean Arterial Pres-
sure
          value matches {
            C_DV_QUANTITY <
                property = <[openehr::125]>
                list = < ["1"] = <
                units = <"mm[Hg]">
                magnitude = <|0.0..<1000.0|>
                precision = <|0|>
                > > >          } }
    ELEMENT[at1007] occurrences matches {0..1} matches { -- Pulse Pressure
        value matches {
            C_DV_QUANTITY <
                property = <[openehr::125]>
```

```
                    list = < ["1"] = <
                units = <"mm[Hg]">
                magnitude = <|0.0..<1000.0|>
             precision = <|0|>
                > > >        } }
ELEMENT[at0033] occurrences matches {0..1} matches {   -- Comment
        value matches {
          DV_TEXT matches {*}
                       }}}}}
 state matches {
        ITEM_TREE[at0007] matches {  -- state structure
          items cardinality matches {0..*; unordered} matches {
  ELEMENT[at0008] occurrences matches {0..1} matches {   --  Position
             value matches {
                DV_CODED_TEXT matches {
              defining_code matches {
                [local::
                     at1000,         -- Standing
                at1001,        -- Sitting
                at1002,        -- Reclining
                at1003,        -- Lying
                at1013,        -- Trendelenburg
                at1014;        -- Left Lateral
                at1001]        -- assumed value
                       } } } }
```

Appendix B

AQL Syntax for BP Query scenario1

```
   SELECT obs/data[at0001]/events[at0006]/data[at0003]/
   items[at0004]/value/magnitude, obs/data[at0001]/events
   [at0006]/data[at0003]/items[at0005]/value/magnitude
   FROM      EHR      [ehr_id/value=$ehrUid]      CONTAINS
COMPOSITION   [openEHR-EHR-COMPOSITION   .encounter.v1]
CONTAINS    OBSERVATION    obs    openEHR-EHR-OBSER
VATION.blood_pressure.v1]
   WHERE obs/data[at0001]/events[at0006]/
   data[at0003]/items[at0004]/value/magnitude>= 140 AND
   obs/data[at0001]/events[at0006]/data[at0003]/items[at0005]/value/
magnitude>=90
```

Adaptive Integration of Distributed Semantic Web Data

Steven Lynden, Isao Kojima, Akiyoshi Matono, and Yusuke Tanimura

Information Technology Research Institute, National Institute of Advanced Industrial
Science and Technology
(AIST) Tsukuba, Japan
{steven.lynden,a.matono,yusuke.tanimura}@aist.go.jp,
kojima@ni.aist.go.jp

Abstract. The use of RDF (Resource Description Framework) data is
a cornerstone of the Semantic Web. RDF data embedded in Web pages
may be indexed using semantic search engines, however, RDF data is often stored in databases, accessible via Web Services using the SPARQL
query language for RDF, which form part of the Deep Web which is not
accessible using search engines. This paper addresses the problem of effectively integrating RDF data stored in separate Web-accessible databases.
An approach based on distributed query processing is described, where
data from multiple repositories are used to construct partitioned tables
that are integrated using an adaptive query processing technique supporting join reordering, which limits any reliance on statistics and metadata about SPARQL endpoints, as such information is often inaccurate
or unavailable, but is required by existing systems supporting federated
SPARQL queries. The approach presented extends existing approaches
in this area by allowing tables to be added to the query plan while it is
executing, and shows how an approach currently used within relational
query processing can be applied to distributed SPARQL query processing. The approach is evaluated using a prototype implementation and
potential applications are discussed.

1 Introduction

The Resource Description Framework (RDF) [1], published by the World Wide
Web Consortium (W3C), is used to model information in order to support the
exchange of knowledge on the Web. The Semantic Web [2] effort is expected
to lead to an increasing amount of data being published using RDF. In some
cases RDF is embedded in Web pages and sometimes it may be held privately,
but often, RDF data is published in Web-accessible databases which clients can
access using a query interface accepting SPARQL [3], a W3C Recommendation
RDF query language, currently the most widely used method of querying RDF
data. The term "Deep Web" [4] was coined to refer to the part of the web
hidden from search engine indexes behind dynamically generated pages or query
interfaces. Given the abundance of RDF repositories with querying interfaces,

S. Kikuchi, S. Sachdeva, and S. Bhalla (Eds.): DNIS 2010, LNCS 5999, pp. 174–193, 2010.

the Semantic Web also constitutes a Deep Web and there is a need to develop tools for accessing and integrating such data, especially in dynamic application domains such as news and weather reporting, social networks etc. As the number of RDF repositories and the volume of data they contain is increasing, it is anticipated that various applications can benefit from the integration of RDF data from multiple distributed RDF repositories.

The SPARQL language allows sets of triple patterns to be matched against RDF graphs, supported by various features such as conjunctions, disjunctions, filter expressions, optional triple patterns and multiple ways of representing query results. The SPARQL query language has an associated protocol, also a W3C Recommendation, the SPARQL Protocol for RDF [5], which defines an interface by which queries may be executed on a SPARQL data resource along with bindings for HTTP and SOAP. The purpose of the SPARQL protocol is to promote interoperability, where clients can interact with SPARQL data resources in a consistent way. For query results, another W3C Recommendation, SPARQL Query Results for XML [6], provides a way of encoding results from SPARQL queries. Many RDF data repositories use the SPARQL query language, SPARQL Protocol and XML results format in conjunction to provide an interface for clients. Examples include DBPedia [7], an extraction of structured information from Wikipedia, and the RDF-based instantiation of the DBLP computer science publication bibliography database [8]. The use of the SPARQL query language, protocol and result format in conjunction eliminates syntactic heterogeneity between different data sources allowing them to be accessed consistently regardless of the underlying RDF database implementation. Alternatives to the SPARQL protocol exist in the form of the Open Grid Forum (OGF) Data Access and Integration Service (DAIS) Working Group's specifications for accessing RDF data resources [9], which may eventually provide an alternative to the SPARQL protocol, and middleware support such as OGSA-DAI-RDF [10], however at present the SPARQL protocol is far more widely used.

Federated queries across multiple SPARQL endpoints allow data from such endpoints to be integrated, and is also of potential benefit in heterogeneous information systems where individual components may use SPARQL wrappers to expose data. Data integration in this context is made particularly appealing by the Linked Open Data Project [11], which aims to promote widespread usage of URI-based representations to allow RDF terms to be consistently defined. The use of consistent URIs to represent the same terms in different databases means that joining data across two data sources is possible, therefore federated queries over multiple repositories can provide results that are not obtainable from any one individual repository.

Existing systems for integrating RDF data from distributed SPARQL endpoints generally rely on the availability of statistics about the data which are then used by an optimiser to compute a join order. This works well when the optimiser has accurate statistics about the data present in each of the repositories, allowing it to minimise the size of the data retrieved from each endpoint and effectively optimise joins and other operations. In contrast this paper focuses on an adaptive approach that can respond to the characteristics of the data and SPARQL endpoints from which the data is retrieved (for example join predicate selectivity, rate at which the data is produced by the service etc.) while the data is being integrated.

The framework described compiles a federated SPARQL query into a number of source queries which are sent to individual SPARQL endpoints to retrieve the data required to answer the query. The results from source queries are used to construct a set of vertically partitioned RDF tables, which act as a temporary buffer to hold data while it is integrated. The vertical partitioning of RDF triples into relational predicate tables (one table for each predicate; subject and object values as the two columns of each table) has been shown to be effective in [12]. The vertically partitioned tables are processed by an extension of the adaptive query processing techniques presented in [13], which allows join order to change during query processing. This approach is adopted for optimising queries that perform time-consuming joins between multiple RDF data sources by dynamically building a query plan that can adapt to the characteristics of the data as the query is being processed. Although the technique presented uses relational database processing techniques to join the predicate tables, the approach in general aims to address challenges specific to integrating data from RDF repositories, where SPARQL endpoints are autonomous and managed by individuals resulting in unpredictability and a lack of accurate statistics about the data, meaning that adaptive query processing techniques can provide a significant benefit. In addition to this, adaptive approaches also have the potential to perform better in unpredictable environments, which is the case with integrating data on a large scale. For example, data sources may be busy or temporarily unavailable and therefore process queries more slowly than anticipated, and some endpoints may vary with respect to their support for specific features of the SPARQL query language and their efficiency in supporting those features.

As the basic data model for knowledge representation on the Semantic Web, RDF is currently becoming widely adopted. RDF is being used by various different individuals and organisations to publish information and a number of publicly available RDF databases contain millions of triples [1]. RDF forms the basis of other Semantic Web components such as RDF Schema (RDFS) and The Web Ontology Language (OWL). Together these components have the potential to provide an interoperable description of information that can be interpreted unambiguously and processed by automated reasoning and inference systems. Community efforts, such as the the Linked Data project aim to promote the sharing of data using RDF, and furthermore, RDF is being used by some publishers to offer dynamic content, for example BBC Backstage [2], which publishes various media in RDF, for example information about TV and radio programmes. All this means that publicly available RDF data is constantly being updated and

---

[1] The Uniprot protein database [http://www.uniprot.org], DBPedia and the Linked Open Data project are examples.

[2] BBC Backstage [http://ideas.welcomebackstage.com/data]

growing rapidly, presenting a significant challenge to the developers of applications that need to access this data. Furthermore, one of the key challenges related to the proliferation of RDF data is that it is widely distributed. A distributed RDF query processor addresses this issue by providing a technique for querying multiple RDF repositories as if querying a single repository using SPARQL. In particular, the use of adaptive query processing techniques is useful in scenarios where the statistics about the contents of the repositories are incomplete or unavailable, the data is constantly updated, and joins need to be performed between data in different repositories.

The technique described in this paper has been used to develop a federated SPARQL query processing interface, the Adaptive Distributed Endpoint RDF Integration System (ADERIS) [14], which allows users to compose SPARQL queries and execute them over multiple SPARQL services. This application is suitable for users with a knowledge of SPARQL allowing them to compose queries, however the approach is also useful for developing other applications where queries are composed automatically and hidden from the user behind the scenes.

2   Related Work

As SPARQL is relatively new, having become a W3C recommendation in 2008, work focusing on distributed data integration over SPARQL endpoints is at an early stage. Work in the wider area of distributed RDF query processing can be roughly divided into the following categories:

1. Search engine based approaches which aim to index a large number of individual documents containing Semantic Web data, for example Swoogle [15] and YARS2 [16].
2. Top-down approaches where an RDF data set is partitioned into smaller subsets which are processed in parallel. Examples of such approaches include decomposition using RDF molecules [17], data partitioning and placement strategies for parallel RDF query processing [18] and the use of peer-to-peer/distributed hash table based approaches for distributing query processing over a set of peers [19].
3. Mediator-based approaches where data sets from multiple autonomous endpoints are combined together by the mediator, which optimises a federated SPARQL query, providing transparent access to the individual endpoints as if they were a single RDF graph.

Work related to (3) is discussed here, where SPARQL is used as the query language and protocol via which data sources are accessed.

Firstly, although not a comprehensive distributed query processing solution, it is worth mentioning that ARQ [20] (the SPARQL query processor for the Jena [21] framework for developing semantic web applications) provides query language extensions for executing remote queries, extending SPARQL with a "SERVICE" construct which forwards triple patterns to a remote endpoint.

ARQ is extended by DARQ [22] (Distributed ARQ), a comprehensive RDF distributed query processor capable of parsing, planning, optimising and executing queries over multiple SPARQL endpoints. When using DARQ, it is not necessary to refer to named graphs or specify anything other than a standard declarative SPARQL query - DARQ parses the query, determines which data sources should be queried and optimises the process of retrieving and integrating data from individual data sources. The system optimises queries based on information about individual data sources provided by *service descriptions*, a metadata format introduced in order to describe an RDF data source. Service descriptions provide the DARQ optimiser with information such as predicate selectivity values and other statistics, and are utilised during the generation of source queries and join ordering, which is combined with physical optimisation implemented using iterative dynamic programming.

FeDeRate [23] is a system for executing distributed queries over multiple data sources supporting a variety of different interfaces, focusing on applications in the domain of bioinformatics. Queries are submitted to FeDeRate in SPARQL, mapped to a set of source queries which are sent to the individual data sources, the results of which are combined and returned as the result of the federated query. FeDeRate provides support for distributed queries with named graph patterns using SPARQL's syntactic support for queries over multiple named graphs. Multiple named graphs may be referred to in a query, each of which is accessed in the order that they appear in the query (as is the case with ARQ). FeDeRate aims to minimise query result sizes in order to more efficiently execute queries by using results from previously executed source queries as constant values in subsequent queries. Optimisation such as query re-writing and join ordering are not performed.

SemWIQ [24] is another system implemented using ARQ that offers a similar approach to DARQ, supporting RDF distributed queries with an optimiser that uses statistics about endpoints obtained by a monitoring component that issues SPARQL queries in order to generate statistics. In contrast to DARQ, SemWIQ is able to support queries over endpoints for which DARQ's statistics and metadata can not be obtained due to restrictions of autonomy or privacy etc., which is also the aim of the work presented in this paper. The monitoring component, RDF-Stats [25], available as a separate component to the distributed query processor, uses an extended SPARQL syntax supported by many SPARQL endpoints which allows the aggregate construct COUNT. The authors state that the monitoring component pulls a large amount of data from data sources and should therefore be installed close (or ideally on the same cluster/node) as the data source it is monitoring. SemWIQ implements various query optimisation strategies such as push-down of filter expressions, push down of optional group patterns, push-down of joins and join and union reordering. However, as is the case with DARQ, optimisation is done statically and requires statistics about data sources.

The work presented in this paper differs from the approaches discussed in this section in the sense that adaptive, rather than static optimisation, is used.

Furthermore, the approach presented in this paper focuses on efficiently executing a specific kind of query, that of ordering multiple joins with large input sizes. In practice, the adaptive approach could be used in conjunction with the optimisation techniques implemented by systems such as DARQ and SemWIQ.

3   Federated SPARQL Query Processing Framework

The framework is based on a mediator that accepts a federated SPARQL query, decomposes the query into a set of source queries and adaptively processes the results. The system's behavior is divided into two phases:

- Setup phase: the mediator is initialised with a list of SPARQL endpoints over which queries are to be executed in the next phase.
- Query processing: federated queries are accepted and executed by the mediator.

3.1   Setup Phase

To efficiently generate source queries, some metadata is required about each RDF repository. As one of the key assumptions made in this work is that repositories are autonomous and it may be difficult to retrieve detailed metadata and statistics, the metadata used is restricted to information obtainable via a straightforward SPARQL query. As a minimum requirement, the mediator requires requires knowledge of whether a particular predicate is known to be present or absent in a given data source. During the initialisation of the system, a list of SPARQL endpoint URIs is passed to the mediator, which submits the following SPARQL query to each endpoint:

```
SELECT DISTINCT ?p WHERE { ?s ?p ?o }
```

Regarding RDF data and SPARQL queries, in general the following observations made in [18] hold true:

- The number of distinct predicate values is much less than the number of distinct subjects or objects.
- In SPARQL queries, predicates are usually specified as query constraints, whereas subjects and objects are more likely to be variables.

These properties are exploited later during querying, and also here during initialisation, as the number of distinct predicate terms tends to be much smaller than subjects and objects, the above query can usually be executed in a reasonable amount of time by most endpoints. In cases where the query cannot be executed due to limits on query processing time or result sizes, an alternative method such as the RDF-Stats tool for generating statistics about RDF data can be used.

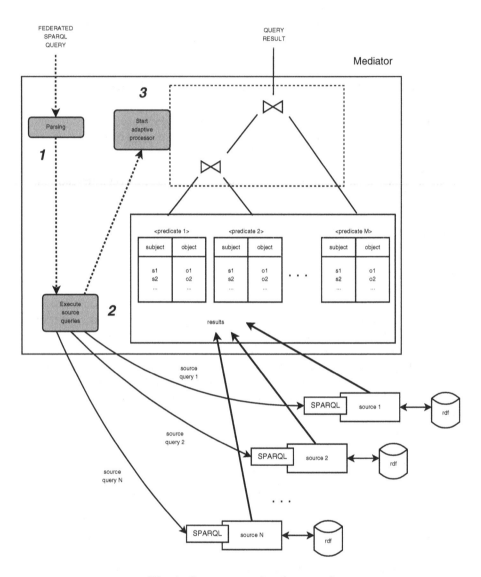

Fig. 1. Query processing framework

This figure illustrates the steps performed during query processing. In (1) the federated SPARQL query is parsed and a set of source queries over N data sources are compiled which are then executed in (2). An adaptive optimiser/evaluator is then started (3) which joins the M predicate tables, constructed using source query results, to answer the query.

3.2   Query Processing

Query processing, illustrated in Figure 1, consists of the following three steps:

1. Source query generation: the query is parsed and the data sources that need to be utilised in order to answer the query are determined. Source queries are constructed for each of these data sources, with the aim of retrieving all the data needed to answer the query while minimising the execution time of each query.
2. Execution of source queries and construction of predicate tables: source queries are sent to the SPARQL endpoints and the results are used to construct the predicate tables. Each triple returned from a query is placed, according to its predicate value, in the appropriate table.
3. Adaptive join processing: the predicate tables are joined dynamically, additionally any other operations in the federated SPARQL query, such as FILTER predicates that could not be pushed down to source queries, are applied here.

Steps 2 and 3 may overlap and take place concurrently, for instance, source queries may still be executing while some predicate tables have been constructed and are being joined. Each of the above steps is described in more detail below. The following example federated query will be referred in order to illustrate the process.

3.3   Example Federated SPARQL Query

To illustrate the approach, an example federated query over four data sources is used. The data sources contain triples from the Friend of a Friend (FOAF) [26] and DBPedia ontologies, distributed among data sources as follows.
Data source S1: Contains triples predicates `foaf:name` or `foaf:homepage`, i.e. triples of the following form exist in the data source:

```
subject <http://xmlns.com/foaf/0.1/name> object
subject <http://xmlns.com/foaf/0.1/homepage> object
```

Data source S2: like S1, also contains only triples with predicate values `foaf:name` and `foaf:homepage` and no triples where the predicate is `foaf:depiction` or `dbpedia:occupation`. Data source S3: contains triples where the predicate is `foaf:depiction` only. Data source S4: contains triples where the predicate `dbpedia:occupation` only.
Each of the above data sources exposes its data via a separate SPARQL endpoint. The following query, which fetches the name, URL and image properties associated with all entities classified as musicians, is executed over the data sources:

```
PREFIX foaf: <http://xmlns.com/foaf/0.1>
PREFIX dbpedia: <http://dbpedia.org/property>
SELECT ?name ?url ?img WHERE {
   ?p foaf:name ?name .
```

```
    ?p foaf:homepage ?url .
    ?p foaf:depiction ?img .
    ?p dbpedia:occupation ?occupation
    filter (?occupation=<dbpedia:Musician>)
}
```

3.4   Source Query Generation

The generation of source queries, sent to remote endpoints to retrieve data needed to execute the federated query, is the only point at which the system requires metadata, collected in the setup phase, about the SPARQL endpoints being queried. For each data source, a source query is generated which contains all of the triple patterns from the federated query that could possibly match triples in the given data source, pushing down any filter expression predicates and joins where possible. For the example query, the source queries generated for data sources S1 & S2 are:

```
SELECT ?p ?o1 ?o2 WHERE {
  ?p foaf:homepage ?o1
  ?p foaf:name ?o2
}
```

The above query pushes down a part of the join between the triple patterns involving foaf:name and foaf:homepage to data sources S1 and S2 because these data sources contain triples with both of these predicate values. The source query for S3 retrieves all triples with the predicate value foaf:depiction as no filter predicate or join can be pushed down to this data source:

```
SELECT ?p ?o WHERE {
  ?p foaf:depiction ?o
}
```

Finally, the source query for S4 pushes down one filter predicate as follows:

```
SELECT ?p ?o WHERE {
  ?p dbpedia:occupation ?o
  FILTER (?o=<dbpedia:Musician>)
}
```

The metadata provides a mapping from predicates to data sources allowing source queries such as those exemplified above to be constructed, retrieving from each data source only triples with predicate values that are needed to evaluate the query.

3.5   Construction of Predicate Tables

Each source query produces a stream of result triples used to construct a set of predicate tables, where a table exists for each unique predicate value contained in the results produced by the source queries. Each predicate table possess two columns (subject and object) and maintains an index on any column that can be potentially used as a join predicate during the subsequent phase when the tables are joined to answer the query. A predicate table is complete when all source queries that can possibly produce triples to be inserted into the given table have finished. In the example introduced in the previous section, the four predicate tables in Figure 2 are created.

In the work presented in this paper, predicate tables are stored in main memory, but they could potentially be written to a file system if necessary. To provide true scalability, a distributed file system could be used to store the tables over multiple nodes and parallelise the process of sending source queries and generating predicate tables from the results. Further performance enhancements are also possible, for example, caching predicate tables used in previous queries in order to speed up subsequent queries may be possible in some scenarios.

dbpedia:Occupation		Source query:
Subject	Object	S4
dbpedia:Beck	dbpedia:Musician	(sample row)
...	...	

foaf:Depiction		Source query:
Subject	Object	S3
dbpedia:Batman	http://../batman.png	
...	...	

foaf:Homepage		Source query:
Subject	Object	S1 and S2
dbpedia:Beck	www.beck.com	
...	...	

foaf:Name		Source query:
Subject	Object	S1 and S2
dbpedia:Ian_Rankin	foaf:Ian_Rankin	
...	...	

Fig. 2. Example predicate tables created by the source queries in Section 3.4

3.6   Adaptive Join Processing

Producing query results involves computing a set of joins between the constructed predicate tables. Here, all joins are index nested loop joins that consume each tuple from their left input, use an index on join attributes to lookup matching tuples from the right input and output joined tuples to the next operator in the plan. [13] presents a technique for reordering pipelined index nested loop join-based query plans where the notion of *depleted state* is introduced to encapsulate the moment in which a sequence of joins is in a state whereby the join order can be changed without throwing away results that have already been produced. Here, this technique is applied when joining the predicate tables as it allows the joins to be reordered based on run-time selectivity statistics (which

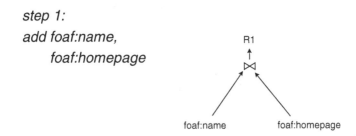

Fig. 3. Adaptive join processing is initialised when two tables become complete. This plan can be executed while the other predicate tables are being generated.

initially have to be roughly estimated). As some predicate tables become complete before others, it is necessary to extend the reordering approach in a way that allows the processing of joins between tables as soon as they become complete rather than being idle until the entire set of query queries has finished. When joins are executed, the selectivity of each join predicate is monitored so that subsequent optimisation can use accurate selectivity values to find optimal join orders. Monitoring is implemented within each join operator by recording the number of input and output tuples processed by the operator. This whole process takes place dynamically as the predicate tables constructed from source query results become available.

The approach is explained using the ongoing example, the query processing of which is illustrated in figures 3, 4 and 5. Initially, consider a state in which source queries S1 and S2 are the first ones to finish. This means that the predicate tables foaf:name and foaf:homepage are complete and may be joined, as illustrated by Step 1 in Figure 3. The optimiser chooses to join with foaf:name as the left input and use the index on foaf:homepage to execute the join. Following this, source query S4 finishes and the dbpedia:occupation table is complete. At this point foaf:homepage has only been used as the probed table (this is the the right input table to the join from which tuples are retrieved using the index); some rows from foaf:name have been consumed by the join and some joined tuples have been output (R1). In order to avoid losing the joined result, the optimiser creates two sub-plans as illustrated by Step 2 in Figure 4; the first sub-plan has the result produced in Step 1, R1, as the left input table to a join with dbpedia:occupation; the second sub-plan uses the unprocessed part of foaf:name (the part of this table that wasn't consumed by the join in Step 1), referred to as foaf:name(*), and the optimiser is able to fully reorder this plan based on the selectivity statistics gathered as the joins are executed. Both sub-plans access the dbpedia:occupation table simultaneously; the first sub-plan using the index to probe the table, the second plan either using it as a probed input or possibly a left input to the first join, as is the case in Step 3 in Figure 5. This can be achieved by keeping a separate set of pointers for each sub-plan to determine which rows to read.

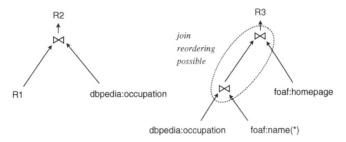

Fig. 4. A third table is added during adaptive join processing. Two independent plans are formed that will be integrated when all predicate tables are complete. The join order of the plan generating R3 may be changed (at any time, independently of the other plan) in response to monitored join selectivities. The other plan's join order cannot be changed because there is no index on the intermediate results R1 that was generated in the previous step.

When the remaining source query, S3, has finished, all predicate tables are now complete and the plan can be finalised. As two independent query sub-plans have been created in Step 2, this must take place when both plans are in states where reordering is possible (a 'depleted state' as defined in [13]), which can be achieved by pausing the first sub-plan to reach this state after the completion of S3 and waiting for the next sub-plan to enter this state. At this point the final plan may be constructed, which consists of the results from the two sub-plans created in Step 2 joined with the now-available foaf:depication table, and additionally a sub-plan processing the non-consumed parts of each of the four tables. It should be noted that the plan in Step 2 that produces R2 needs to be executed until is has completely consumed its input from R1. The plans that produce R1 (in Step 1) and R3 (in Step 2) are discarded once the next step is reached as they do not read tuples from a cache operator, and in these cases only the cache needs to be preserved. Each of the sub-plans in Step 3 are combined by union operators to produce the query result as no new predicate tables need to be added. Processing joins as described here has the following advantages:

- Nothing is known about the selectivity of join predicates before the query is executed so it is difficult to produce a statically optimised query plan. Using selectivity monitoring and join reordering alleviates the need to produce a statically optimised plan.
- The predicate tables are processed as they become available so there is no need to wait until all source queries have finished before executing the query. This has the potential to reduce query response time in certain cases.

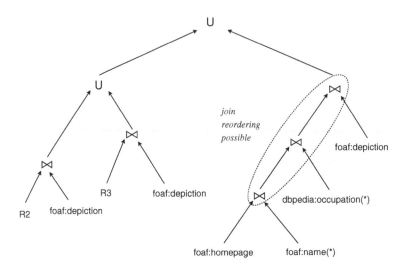

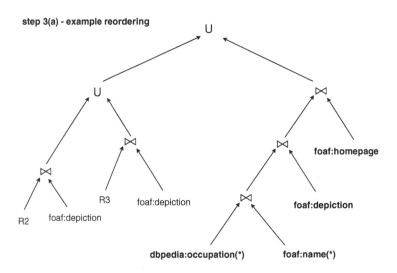

Fig. 5. The final table is added during adaptive join processing. Step 3 (a) shows an example reordering which can take place if join predicate selectivity monitoring shows that estimated selectivity is substantially different from the monitored selectivity. The reordered elements of the plan are highlighted in bold. Note that sub-plans can be reordered independently of other sub-plans.

– Predicate tables that have been joined can potentially be discarded. Compared to waiting until all predicate tables become complete, this can be beneficial in terms of memory usage when processing large amounts of data.

4   Performance Analysis

A prototype query processor has been implemented in Java, currently supporting only SPARQL SELECT queries with subject/object variables (predicates must be constrained) and without various language features such as support for the OPTIONAL construct. The prototype does however allow for an evaluation of the approach based on the execution of a simple federated query over multiple SPARQL endpoints.

4.1   Data Sources and Queries

An evaluation is made using QUERY-1:

```
PREFIX dbpedia: <http://dbpedia.org/property>
PREFIX rdfs: <http://www.w3.org/2000/01/rdf-schema>
PREFIX foaf: <http://xmlns.com/foaf/0.1>
PREFIX skos: <http://www.w3.org/2004/02/skos/core>
SELECT ?ref ?comment ?page ?subj WHERE {
   ?obj dbpedia:reference ?ref .
   ?obj rdfs:comment ?comment .
   ?obj foaf:page ?page .
   ?obj skos:subject ?subj
}
```

This federated SPARQL query combines data from four separate RDF data sources, each of which contains data from one of the following four DBPedia data sets which are available for download individually from the DBPedia site:

1. The "External Links" data set containing triples with the predicate term `dbpedia:reference`.
2. The "Short Abstracts" data set containing triples with the predicate term `rdfs:comment`.
3. The "Links to Wikipedia Article" data set containing triples with the predicate term `foaf:page`.
4. The "Article Categories" data set containing triples with the predicate term `skos:subject`.

The SPARQL endpoints are constructed by downloading the files from DBPedia and using Joseki [27] to provide a SPARQL front end. Each endpoint is located on a separate (3GHz Intel Xeon) machine with 1MB available memory for the Java Virtual Machine providing the endpoint. Endpoint machines are connected to the machine on which the mediator is deployed (2GHz AMD Athlon X2, 2GB RAM) via a 100Mbs Ethernet LAN. The experiments compare the following four query processing strategies:

- *wait no-adapt*: The system does not allow join reordering (no-adapt) and the joins are not processed until all source queries have finished and the predicate tables are complete (wait).
- *no-wait no-adapt*: The system does not allow join reordering (no-adapt) but as soon as predicate tables are complete they may be joined if possible (no-wait).
- *wait adapt*: Join reordering is allowed (adapt), but the plan is not executed until all source queries have finished and the predicate tables are complete (wait).
- *no-wait adapt*: The full application of the technique presented in this paper; join reordering is allowed (adapt) and predicate tables may be joined as soon as they become available (no-wait).

Distributed query processing over autonomous data sources can be complicated by the unpredictability of the data sources and the communication channels between them. Data sources may sometimes be slow to respond if demand from multiple clients is high or maintenance/updates to the data are taking place. In some cases, queries may fail completely if data sources are temporarily unavailable or a network problem occurs when communicating with the data source. There may also be differences between the underlying implementation used by different data sources, for example, indexes used and support or lack of support for different aspects of the SPARQL query language. Adaptive query processing can help to mitigate the effects of such unpredictability by responding to different data source response times while a federated query is being processed. To model such an environment, source query failures are incorporated into the experiments. Source query failures are modeled as follows: for each federated query processed by the system, one of the source queries (randomly selected) encounters a fatal problem necessitating the source query to be executed again. This behaviour is introduced with the aim of simulating real-life situations where a SPARQL service is busy and cannot respond (temporarily unavailable) or a service is down and data must instead be retrieved from an alternative (replicated) service. Two experiments are performed, as described below.

Experiment 1. The first experiment investigates the performance with respect to scalability by varying the size of the data sets. Subsets of the complete data sets are used to execute the experiments with each data source containing the same number of triples, starting with each data source containing 100,000 triples. The experiment is repeated for larger data set sizes in increments of 100,000 triples until all data sources contain 600,000 triples.

Experiment 2. The second experiment introduces FILTER expressions into QUERY-1 to evaluate performance where predicates are evaluated by each of the data sources - this results in variance of the size of the data set returned by each data source and the amount of time taken by each data source to answer source queries. Randomly selected letters or common string patterns such as 'the', 'as', 'in' etc. are inserted into REGEX (regular expression) functions associated with each triple pattern. In any randomly generated query, for each triple pattern there is a 0.25 probability of inserting a randomly selected word. For example, a randomly generated query could be:

```
PREFIX dbpedia: <http://dbpedia.org/property>
PREFIX rdfs: <http://www.w3.org/2000/01/rdf-schema>
PREFIX foaf: <http://xmlns.com/foaf/0.1>
PREFIX skos: <http://www.w3.org/2004/02/skos/core>
SELECT ?obj ?ref ?comment ?page ?subj WHERE {
   ?obj dbpedia:reference ?ref .
   ?obj rdfs:comment ?comment .
   ?obj foaf:page ?page . FILTER regex(str(?page),'in')
   ?obj skos:subject ?subj
}
```

Inserting predicates into the query in this manner introduces increases the variance of the processing time per result triple for source queries (since REGEX functions may need to be implemented by data sources), the size of the result set returned from each data source, and hence, the size of the federated query result set and overall query processing time. 20 queries were generated randomly using the approach described, and for each generated query, the response times of the four different query processing strategies were compared.

4.2   Results

Response times for Experiment 1 are shown in Figure 6. It can be seen that the no-wait adapt strategy provides the fastest response times, in particular with large data sizes. Response times for Experiment 2 are shown in Figure 7, where each item on the x-axis corresponds to a randomly generated query (the order in which the queries appear is not significant and comparisons should be made between the four different response times of the different strategies within the same query only). Again, the no-wait adapt strategy generally performs better when compared to the other strategies, in particular when overall query processing times are relatively high. As with Experiment 1, results show some advantage of the two adaptive techniques (wait adapt and no-wait adapt) over the non-adaptive techniques, in particular when query response times are high. As the no-wait adapt strategy appears to perform best when the number of triples is large, e.g. when each data source contains 600k triples in in Figure 6 where it has a clear advantage over the non-adaptive strategies, future work will involve performing experiments with larger data sets in order to confirm that this observed trend continues as join input sizes grow. For low query response times, corresponding to queries that involve relatively small volumes of data being returned from source queries, there is little difference between the four strategies. Where the adaptive strategies perform worse than the non-adaptive strategies, this can be accounted for by the overhead in adapting, including monitoring for states at which reordering is possible and starting/stopping the executing plan.

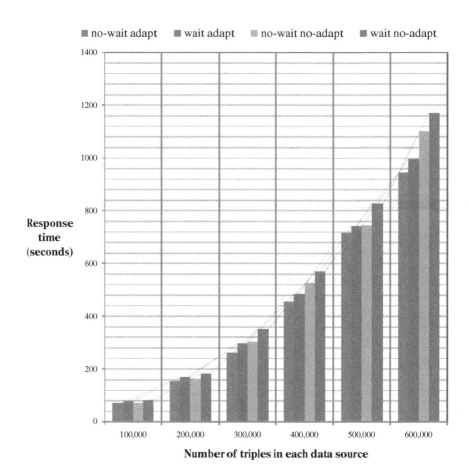

Fig. 6. Experiment 1 response times

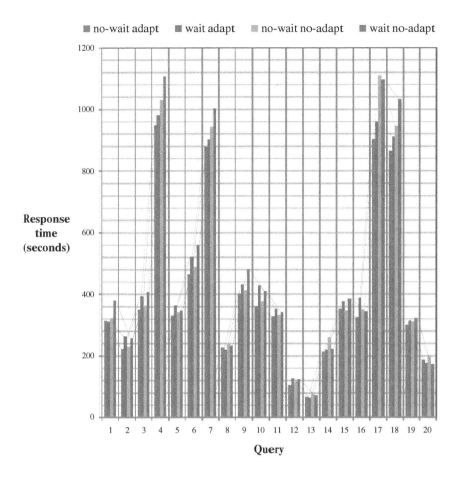

Fig. 7. Experiment 2 response times

5   Conclusions

An adaptive framework has been presented for executing queries over multiple
SPARQL endpoints that differs from existing approaches which use static query
optimisation techniques. Many SPARQL web services are currently available and
the number of them is growing. The work presented in this paper is a framework
for executing queries over federations of such services. The framework proposed
in this paper, which allows adaptive query processing over dynamically constructed predicate tables to be performed in conjunction with the construction
of the predicate tables, was shown to perform relatively well in unpredictable environments where source query failures may occur. The prototype implemented
was evaluated using real data, showing some advantage in terms of response
times of adaptive over non-adaptive methods using a subset of DBPedia. Future
work will aim to investigate other data sets with different characteristics and

larger data sets. As the approach presented in this paper focuses on efficiently executing a specific kind of query, that of adaptively ordering multiple joins, further work will focus on optimising other kinds of queries and implementing support for more SPARQL query language features. Future work will also concentrate on investigating how the work can be applied in various domains.

Acknowledgments

This work was supported by the Strategic Information and Communications R&D Promotion Programme (SCOPE) of the Ministry of Internal Affairs and Communications (MIC), Japan.

References

1. Klyne, G., Carroll, J.J.: Resource description framework (rdf): Concepts and abstract syntax. Technical report, W3C (2004)
2. Berners-Lee, T.H., Lassila, O.J.: The Semantic Web. Scientific American 284(5), 28–37 (2001)
3. Eric Prudhommeaux and Andy Seaborne. SPARQL Query Language for RDF. Technical report, W3C (2008)
4. Bergman, M.K.: The Deep Web: Surfacing Hidden Value. The Journal of Electronic Publishing 7 (2001)
5. Clark, K.G., Feigenbaum, L., Torres, E.: SPARQL Protocol for RDF. Technical report, W3C (2008)
6. Beckett, D., Broekstra, J.: SPARQL Query Rresults XML Format. Technical report, W3C (2008)
7. DBPedia, http://dbpedia.org/
8. D2R Server publishing the DBLP Bibliography Database, http://www4.wiwiss.fu-berlin.de/dblp/
9. Gutiérrez, M.E., Kojima, I., Pahlevi, S.M., Corcho, Ó., Gómez-Pérez, A.: Accessing RDF(S) data resources in service-based grid infrastructures. Concurrency and Compututation: Practice and Experience 21(8), 1029–1051 (2009)
10. Kojima, I., Kimoto, M.: Implementation of a Service-based Grid Middleware for Accessing RDF Databases. In: Proceedings of Workshop on Semantic Extensions to Middleware: Enabling Large Scale Knowledge Applications (SEMELS 2009) (November 2009)
11. Linked Data - Connect Distributed Data across the Web, http://linkeddata.org/
12. Abadi, D.J., Marcus, A., Madden, S., Hollenbach, K.: SW-Store: a vertically partitioned DBMS for Semantic Web data management. VLDB J 18(2), 385–406 (2009)
13. Li, Q., Sha, M., Markl, V., Beyer, K., Colby, L., Lohman, G.: Adaptively Reordering Joins during Query Execution. In: Proc. ICDE, pp. 26–35. IEEE Computer Society, Los Alamitos (2007)
14. Lynden, S., Kojima, I., Matono, A., Tanimura, Y.: ADERIS: Adaptively integrating RDF data from SPARQL endpoints (Demo Paper). In: Proceedings of the Database Systems for Advanced Applications (DASFAA) Conference 2010 (2010) (to appear)
15. Ding, L., Finin, T., Joshi, A., Peng, Y., Cost, R.S., Sachs, J., Pang, R., Reddivari, P., Doshi, V.: Swoogle: A Semantic Web Search And Metadata Engine. In: 13th ACM Conference on Information and Knowledge Management (2004)

16. Harth, A., Umbrich, J., Hogan, A., Decker, S.: YARS2: A Federated Repository for Querying Graph Structured Data from the Web. In: Aberer, K., et al. (eds.) ASWC 2007 and ISWC 2007. LNCS, vol. 4825, pp. 211–224. Springer, Heidelberg (2007)
17. Newman, A., Li, Y.-F., Hunter, J.: Scalable Semantics, The Silver Lining of Cloud Computing. In: 4th IEEE International Conference on e-Science (e-Science 2008) (2008)
18. Tanimura, Y., Matono, A., Kojima, I., Sekiguchi, S.: Storage Scheme for Parallel RDF Database Processing Using Distributed File System and MapReduce. In: International Conference on High Performance Computing in the Asia Pacific Region (2009)
19. Liarou, E., Idreos, S., Koubarakis, M.: Continuous RDF Query Processing over DHTs. In: International Conference Semantic Web Computing (2007), http://iswc2007.semanticweb.org/papers/323.pdf
20. ARQ SPARQL query processing framework, http://jena.sourceforge.net/ARQ/
21. Carroll, J.J., Dickinson, I., Dollin, C., Seaborne, A., Wilkinson, K., Reynolds, D., Reynolds, D.: Jena: Implementing the semantic web recommendations. Technical Report HPL-2003-146, Hewlett Packard Laboratories (2004)
22. Quilitz, B., Leser, U.: Querying Distributed RDF Data Sources with SPARQL. In: Bechhofer, S., Hauswirth, M., Hoffmann, J., Koubarakis, M. (eds.) ESWC 2008. LNCS, vol. 5021. Springer, Heidelberg (2008)
23. Prudhommeaux, E.: Optimal RDF access to relational databases. Technical report, W3C (2005), http://www.w3.org/2004/04/30-RDF-RDB-access/
24. Langegger, A., Woss, A., Bloch, W.: A Semantic Web Middleware for Virtual Data Integration on the Web. In: Bechhofer, S., Hauswirth, M., Hoffmann, J., Koubarakis, M. (eds.) ESWC 2008. LNCS, vol. 5021. Springer, Heidelberg (2008)
25. RDFStats home (subproject of Semantic Web Integrator and Query Engine), http://semwiq.faw.uni-linz.ac.at/rdfstats/
26. The Friend of a Friend (FOAF) Project, http://www.foaf-project.org/
27. JOSEKI - A SPARQL Server for Jena, http://www.joseki.org/

Managing Groups and Group Annotations in
MADCOW

Danilo Avola, Paolo Bottoni, Marco Laureti,
Stefano Levialdi, and Emanuele Panizzi

Department of Computer Science
University of Rome "La Sapienza"
Via Salaria 113, 00198, Rome. Italy
{bottoni,levialdi,panizzi}@di.uniroma1.it,
{danilo.avola,marco.laureti}@gmail.com

Abstract. An important feature for annotation systems of Web pages is the possibility to create and manage users groups focused on specific topics. This allows the users belonging to a given group to access both meta information (annotations) and information (related Web pages) in order to establish a common shared knowledge. The increasingly complex activities involved in the life cycle of a group (e.g. e-learning groups, students groups, scientific groups) have highlighted the need for advanced tools supporting heterogeneous operations, such as: research of authoritative contributors, trust evaluation of Web page contents, research of relevant additional digital resources, research of related groups, and so on. While most annotations systems have ordinary group functionalities (e.g. searching, invitation, sharing), MADCOW 2.0 offers a proof of concept for advanced group management, based on trust and content's research criteria, aimed at supporting group formation and enlargement policies.

1 Introduction

Annotations are becoming a common feature of current user experience with social networking sites, as well as an important feature in the cooperative construction of documents. However, several limitations hinder current annotation systems, both Web-based and document-based ones.

In the first case, social networking sites offer limited forms of annotation: addition of tags or meta-information on specific documents (e.g. user generated content for Google Maps or Flickr), insertion of textual comments on a video in YouTube, addition of comments or corrections in PDF or Word Documents. One can observe that in these cases notes - intended here as any addition to the original document - are presented in an all-or-nothing fashion: one can disable or filter them, if interested only in notes from some author, for example, or not interested at all. However, in principle, every user of the document (map, video, image, etc.) is a potential viewer of the associated notes and every user with some level of ownership on the document is entitled to add, modify or delete notes. As a consequence, restrictions to the circulation of annotated documents have

S. Kikuchi, S. Sachdeva, and S. Bhalla (Eds.): DNIS 2010, LNCS 5999, pp. 194–209, 2010.

to be managed, if their content is sensitive. Alternatively as is often the case, the content of notes remains generic or understandable only to few individuals, without this having been devised explicitly.

Hence, it is becoming important for modern Web annotation systems to offer the possibility to create and manage user groups focused on specific topics. Users belonging to a given group would thus access both meta information (annotations) and information (related Web pages) in order to establish a common shared knowledge. The increasingly complex activities in a group life cycle have highlighted the need for advanced tools to support heterogeneous operations, such as: trust evaluation of Web page contents, search for authoritative contributors, relevant additional digital resources, related groups, etc.

MADCOW (Multimedia Annotation of Digital Content Over the Web) is a system for the annotation of textual, image and video material contained in Web pages, allowing the generation of Web notes which in turn can comprise text or any other digital content and which are associated with meta-information favoring their retrieval and allowing the creator of the note to define their visibility to other users [10,11,12]. Notes can in fact be *public* or *private*. In the first case, every user with a MADCOW-enabled browser will be able to notice the presence of a note on a page and interact with it. In the second case, the notes will be visible only in a session where the user of the browser has been identified as the author of the note, through a username-password combination.

While most annotation systems offer ordinary group functionalities (e.g. searching, invitation, sharing), MADCOW2.0 implements a proof of concept of an advanced group management, based on trust and content's research criteria. This fosters MADCOW's use as a collaborative system. By stating that a note is a group note, the author makes it public to a restricted set of MADCOW users, namely those registered to the group for which the note is published. Of course, the author must be a member of such a group. All the other visitors of the page will instead remain unaware of the presence of such a note.

This feature achieves two effects. On the one hand, it makes navigation easier for users who might not be interested in all possible comments on a certain page. Rather than disabling the presentation of notes, they can decide at any time which notes they are interested in. Second, it allows the realization of cooperative tasks through MADCOW services, so that users can communicate by following threads of discussion on single topics, without recurring to tools other than their browsers, and without setting up elaborate schemes for securing the communication content.

The paper illustrates the realization of the notion of group in MADCOW and some scenarios for its use. Moreover, it presents several mechanisms which have been devised on the server side to support group creation and management. In particular, we have defined a set of services which help group administrators to find new people potentially interested in joining a group, to define credibility scores for group participants, and to identify possible disturbing elements.

In the rest of the paper, Section 2 illustrates related work on Web annotation and other approaches to the definition of groups. Section 3 gives a brief overview of MADCOW's architecture and use. Section 4 introduces the notion of group and the attributes by which they are specified, and presents three scenarios of use. Section 5 illustrates the management tools offered to group administrators and gives an account of the *lookup* function for identifying potential members of a group. Finally, Section 6 draws conclusions and points to future work.

2   Related Work

Several studies have been conducted to define the nature and the objectives of annotations, both digital and traditional ones. In particular, Pei-Luen and others, considering uses in e-learning, have indicated a number of reasons for taking notes, from highlighting interesting points, to improving understanding and memorization of content, to facilitating further writing [13]. In this sense, the original observation of O'Hara and Sellen seeing annotation as an intermediate process between reading and writing [7] is confirmed and strengthened. A study on functionalities of digital annotations is reported on in [14].

The theme of ownership and visibility of annotations has been investigated by Marshall and Brush [4,15] and dealt with as a formal property in [6].

The classical categorization of annotations objectives in [5] (*remember, think, clarify, share*) refers mainly to the personal goals of a single in making an annotation. However, as annotations see their usage expanded in cooperative settings, the need arises to focus on annotation-centered collaborative *processes*. As an example, in the field of e-learning, a new category of objectives, *correct*, appears to be relevant. In this sense two of the authors have proposed to use categories from the Rhetorical Structure Theory [8] to identify the role of an annotation, and have integrated the specification of such a role, together with the accessibility of the note, as one of the properties specifying an annotation [9].

Among recent systems offering ways to share annotations, we quote Web Orchestration, which allows users to customize and share web information, enhancing collaboration via shared metadata-based web annotations, bookmarks, and their combinations [16].

Digital Libraries (DLs) offer several opportunities for text-focused collaboration. DLNotes [17] can be embedded into a DL to enable users to perform free-text and ontology-based annotations. It supports supervised annotation activities and allows the association of discussion threads with annotations.

The functionalities supporting a group can be classified as *passive* or *active*. A passive function enriches the resources of a group by simply following users' instructions (e.g sending invitations). Conversely, an active function autonomously tries to reach the same purpose by using methodologies based on heuristics, content or reasoning. Regardless the annotation target (e.g. text, video, image), current systems increasingly implement collaborative oriented features. A pioneering system introducing advanced collaborative oriented features is Vannotea, presented in [18]. It enables the real-time collaborative indexing, browsing, description, annotation and discussion of high quality digital film or video content. The system supports activities in large-scale group-to-group collaboration such as: discussion threads, supervised working, resources sharing, etc. The framework in [19] covers, in a more modern style, several collaborative aspects faced by Vannotea. In particular, it allows collaborative annotation of generic data streams. SpaTag.us [20] is a simple and powerful framework providing easy keyword tagging while browsing Web content. It also lets users highlight text snippets in place and automatically collects tagged or highlighted paragraphs into a personal notebook. This system provides a set of Web 2.0 oriented functionalities by which notebooks, tags, and highlights can be shared among organized groups of users. Similarly, the Diigo system [1] provides advanced functionalities in text and image annotation including the identification of sets of overlapped notes (of different authors) related to a specific part of text. Diigo allows users to define group content features (e.g. description, category, visibility) in order to favor both search-related groups and interaction with linked communities. Stickis [3] and Fleck [2] offer creation and management of working groups in order to personalize the annotation experience, supporting collaborative work and sharing resources. The LEMO Annotation Framework [21] is a complete tool for annotation of digital resources, providing a uniform model for various types of annotation of multimedia content. The framework enables users to define working groups by which to perform interoperable activities on the stored Web resources. However, all these works exploit advanced passive functionalities during the management of group set environment. Active functionalities (such as those offered by MADCOW) are present in non-annotation systems (e.g. see [22,23]) exploiting information content and trust-oriented evaluation methods to perform information retrieval activities on digital corpora of documents.

3   A Brief Overview of MADCOW

MADCOW is an annotation system, introducing innovative and advanced features in order to simplify the user annotation experience. During the progressive releases of the MADCOW project [10,24,11,12] different matters related to the annotation concept of objects (e.g. text, image, video) have been successfully faced, such as: *location* of notes inside the related document, typically expressed as its position, *identification* of note presence via highlighting or iconic placeholders, or watermarking techniques, semantic *roles* of annotations and their *accessibility*. As far roles are concerned, they are defined in terms of categories of annotation purposes. MADCOW identifies nine main roles: *Announcement, Comment, Example, Explanation, Integration, Memorandum, Question, Solution, Summary* [9]. The accessibility defines which operations (e.g. creation, modification, deletion) can be performed on an annotation by users of the system. This characteristic is closely linked to the annotation *visibility* which can be: *private, public,* or *group*. Different types of annotation can in fact coexist in the same document: a private level of annotation that is only accessible by the authors of the annotations, a collective level of annotations shared by a user group, and finally a public level of annotation accessible by all users [AGF07, MRB02, MRB04]. Others minor attributes that has to be considered during an annotation are title, author and date that further characterize the digital process. MADCOW faces also common negative aspects of the digital annotations, such as: the decontextualization of the annotations with respect to the digital document and the impossibility to perform the annotation activity in a natural way. The Web 2.0 oriented features of MADCOW overcome the mentioned problems. Moreover, within specific contexts (e.g. e-learning), it is important to have mechanisms able to trace accesses, user behavior, authorizations, for the annotation activity. With the aim to accomplish a suitable group management, the MADCOW digital annotation is logically structured into two components: metadata and content. More specifically, the first one is a set of attributes, such as: author, title, date, location (in our context an XPointer), URL (of the original document), and type of the annotation (i.e. an attribute that classifies the annotation content). The second one regards an extension of the Range interface (a W3C standard) by which to represent the annotation content. In this way, MADCOW uses a single data structure to represent annotations on text, video, and image elements.

4   Group Annotations

In general, a group is defined as a set of people, whose status and roles are interrelated, interacting with each other in an orderly manner on the basis of shared expectations. In particular, we consider groups with common interests, where notes become a tool for the construction of shared knowledge. A MADCOW group provides an innovative way for team learning and research, allowing a user community to pool resources and lead discussions for specific purposes or interests. In the context of a group, annotations are not just a way to explain and enrich a resource with information or personal observations, but also a method for transmitting and sharing ideas to improve collaborative work practices. In this section we consider the attributes that characterize user groups, describing the domain of values they can take and the related applications. It is assumed that for each user group the following attributes are specified:

- creator
- title
- visibility
- goalTags
- usersList
- profilesList
- domainsList
- whiteUserList
- blackUsersList

The *creator* is a member who, designated within the group or as a personal initiative, begins the process of creating the user group. A creator is a MADCOW user holding the relevant qualifications, who will carry on the execution procedure of creating the user group, who will send the invitations and perform the maintenance activities devoted to group management. The *title*, as specified by the creator, defines the name of the group, intended as a brief description of the group activity. The *visibility* attribute indicates the status visibility of the group, assumed here to be either private or public. A group with public visibility is included in the process of looking for groups and is therefore detectable by any user of the system. On the other hand, private groups are excluded from the logic of research, so they are not visible. The *userlist* is the attribute that, in addition to outlining the list of users participating in the group, specifies their roles and profiles. A user may participate in multiple groups with different user profiles. The system provides some predefined user profiles; it also allows the generation of additional and more specific definitions for particular needs. The *profilesList* is the list of profiles of group owners. The remaining attributes express additional information on the group of users, useful to parameterize the searches for annotations carried out with the *lookup* procedure analyzed specifically in Section 5.2. *goalTags* is the set of tags specifying the interests of the user group. The *domainsList* provides the list of the group's favorite web domains, for topics and reliability of published information. The *whiteUserList* identifies users whose annotations were evaluated positively by the group. The *blackUsersList* identifies users who obtained negative evaluations.

4.1   Scenarios

We present three scenarios to illustrate the utility of user groups for a web annotation system in different significant contexts. For each scenario we assume users to be logged in the system; moreover in the first two scenarios we assume that the creator of the group has direct knowledge of the users to whom to forward invitations, while in the last scenario a specific function called *Look-Up* is used to search for potential subscribers of the group. Figure 1 illustrates the modes available in MADCOW to send invitations.

Scenario 1. Three professors teaching the same course to different classes have decided to set up a new user group in order to annotate the web page where the program of the course is published. Not only do they want to annotate the program, but also to share some reference material with students. The faculty secretariat thus creates a new group, sends invitations to the teachers with profile *super user group* and send invitations to students using profile *reader group*. Hence, only teachers can create annotations within the group, while students can only read the generated annotations.

Scenario 2. A teacher wants to manage student exams via annotations. Hence, she creates a group, invites students of her course to join, and associates the students with a profile allowing them to read only teacher's annotations besides their own. Then she prepares a web page containing questions for the students and asks them to answer by annotating this page. Each student will create annotations with group visibility on this page and the teacher will read them. However, each student will not be able to read other student annotations.

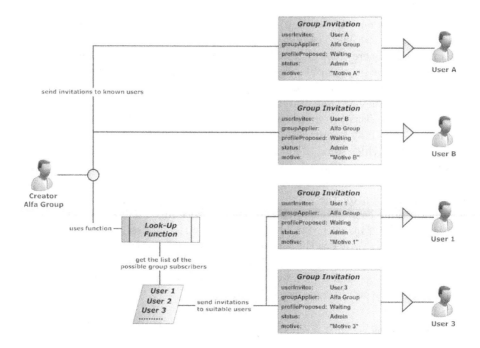

Fig. 1. Different modes to send invitations

Scenario 3: who should I invite?. A researcher is interested in studying the
"personality of voters". He intends to create a team to work with, but he does
not know personally anybody who could help gather useful material for the purpose of her study. Therefore he decides to look for team members among people
who have annotated some web sites about related topics. He needs a list of users
together with the public annotations they made relative to this research. He
sends invitations to those he deems appropriate for the intended research. Thus
he creates a new group and indicates "psychology", "personality" and "voters"
as goalTags; then, he defines the domainsList sources "www.psychoinfo.it" and
"w3.uniroma1.it/bibvalentini". Finally, he performs the lookup procedure, getting the list of users with the annotations they made, from which he selects
receivers of invitations and uses the whiteUserList and the blackUsersList to
include or exclude users from the following search operations.

4.2   Properties of Groups

If, during the creation of an annotation, the user selects the "Group" visibility,
the system provides a list of the groups to which the user participates. By selecting a group, the list of its members is shown, so that the user can make the
annotation visible to selected users only; the system default "All users" makes
the annotation visible to all group members. In Figure 2, the user is creating an
annotation with group visibility.

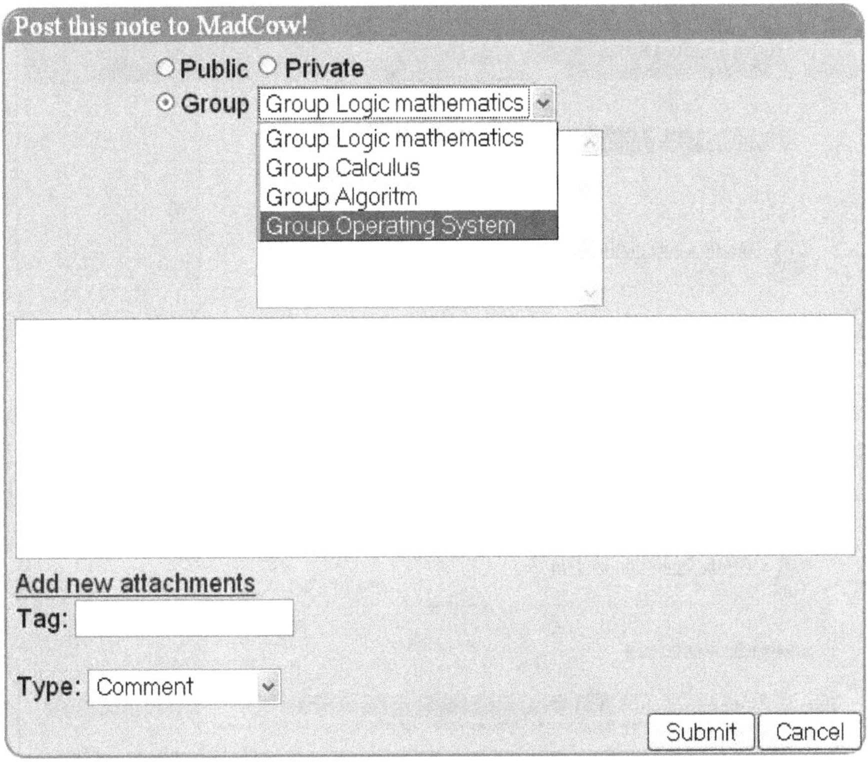

Fig. 2. Creation of a note with group visibility

In the `Group admin` section (see Figure 3), users may access general information on the groups they belong to. Editing and deletion of a group are accessible
only to authorized users. The *delete* operation is performed by clicking the trash
icon to the right of the group name. Clicking on the group title, it is possible
to change the group attributes: *title, visibility, goalTags, usersList, domainsList,
whiteUserList, blackUsersList*. Changing attribute visibility is via a combo box,
simply selecting the new value, while other attributes are managed through list
boxes. After selecting an element of the list box, the command to the side of
the component can be executed. The commands allowed are: `add`, `mod`, `del`. By
clicking on the `Apply` button, the user updates the group database.

5   Managing Groups

MADCOW has a form-based user authentication process, by username and pass-
word. If the process is successful, the MADCOW portal displays to the user, among
others functionalities, the personal `HomePageUser`; this page enables user to
use the following functions related to the groups management: `Group Admin`,

Fig. 3. Group Admin Section of the MADCOW website

Invitation Admin and **Annotation List**. In the Group Admin section, shown in Figure 3, any user can access the page containing the Member group list (i.e. the list of groups the user is a member of) and the Creator group list (list of groups created by the user). The section also provides the opportunity to create new groups. This is carried out, as shown in Figure 4, by filling a suitable form composed by the following specific fields:

title a text box, to define an expressive group name;

visibility a combo box, to set the visibility status of the group;

goalTags a list box, where each element presents name and relevance-value of a goalTag. New elements are added following the **Insert new group** link. This opens a new interface where the name of the new tag is inserted through a text box and its relevance-value adjusted through a slider. The relevance-value ranges between 1 (low) to 10 (high) according to the relevance of the tags with respect to the group target.

userList a list box, where each element presents email and profile of a group member. New elements are added following the **Insert new user** link. This opens a new interface to specify both the mail of the invited user (through a

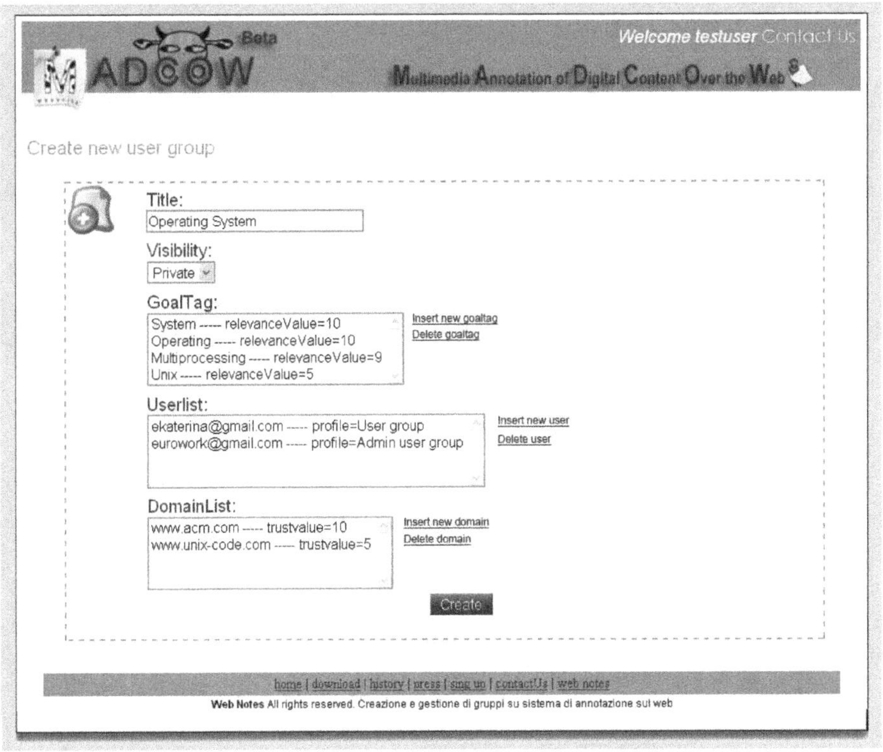

Fig. 4. Form to create a new group

text box) and the name of the profile related to the user. MADCOW provides suitable default user profiles. Moreover, the `Insert new profile` function allows the definition of new profiles;

`domainsList` a list box, where each element presents name and trust-value of a Web domain. New elements are added following the `Insert new group` link. This opens a new interface to specify both the domain name (through a text box) and the trust-value related to the new Web domain (through a slider). The trust-value ranges between 1 (low) and 10 (high) based on the authoritativeness of the information published on the Web domain.

When the application form has been processed, MADCOW will check the completeness of the acquired data. If the check process is passed, a new user group is created and possible invitations sent; otherwise, it will generate an alert message that highlights the error occurred during the form filling.

5.1 Populating Groups

This section explains how to populate a group in MADCOW. A user can participate in a group both by accepting the invitation sent by the group creator or by

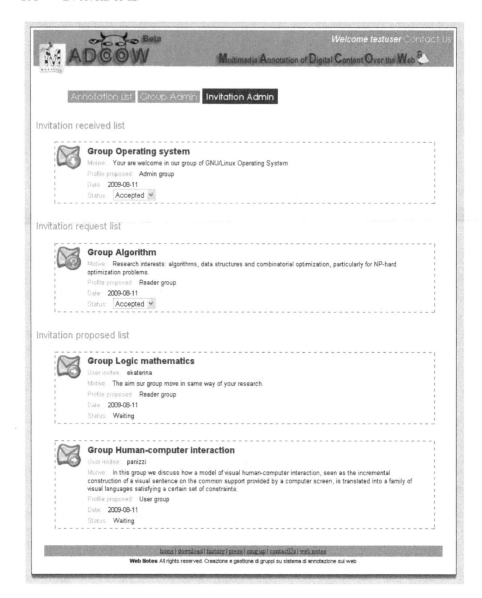

Fig. 5. Invitation admin

the agreement of the group creator to her request of participation. In the latter case, the user can require the participation to the group only if the group has public visibility. Figure 5 shows the MADCOW functionalities available to send invitations to every users group. In order to explain the process by which to populate the different groups it is necessary to introduce some basic concepts about the attributes that characterize the just mentioned two cases. Every invitation (first case) is characterized by the following attributes: groupApplier, userInvitee, profileProposed, status, motive, date. The GroupApplier attribute identifies the group that has sent the invitation; userInvitee defines the recipient user of the invitation; profileProposed specifies the user profile, proposed by the group (later editable), by which the user will participate to the group activity; status describes the current condition related to an invitation, in particular it presents the following states: accepted (if the invitation has been accepted by the userInvitee), refused (if the invitation has been refused by the userInvitee), waiting (if the invitation has to be examined by the userInvitee), *undelivered* (if the system is not able to deliver the invitation to the userInvtee); *motive* shows a brief message, performed by the creator of the groupApplier, where the group targets and the motivation of the invitation are specified; finally, *date* indicates the date in which MADCOW sent the invitation. Every request for participation in a group (second case) is characterized by the following attributes: *userApplier*; *groupRequested*; *status*; *motive*; *dates*. UserApplier identifies the user requiring participation in a group; groupRequested specifies the group to which the user is applying; status describes the current condition related to a request for participation, in particular it presents the following states: accepted (if the request for participation has been accepted by the creator of the groupRequired), refused (if the request for participation has been refused by the creator of the groupRequired), waiting (if the request for participation has to be examined by the creator of the groupRequired); motive shows a brief message, performed by the userApplier, where targets and motivations for the request for participation are specified; finally, date indicates the date in which the MADCOW accepted the request for participation.

The just two mentioned MADCOW functionalities are performed within the section Invitation Admin, accessible from HomePageUser. Besides, as shown in Figure 5, in the same section, it is possible to display: *Invitation proposed list*, *Invitation request list*, and *Invitation received list*. The invitation proposed list contains the invitations sent by the creator of the group. It shows information about: *userInvitee*, *profileProposed*, *motive*, *date*, and *status*. The invitation request list contains the requests that the user has sent to the groups with which she intends to cooperate. It shows information about: *groupRequested*, *motive*, *date* and *status*. Invitation received list contains the invitations sent by the user. It shows information about: *groupRequested*, *profileProposed*, *motive*, *date*, and *status*. For every invitation the user can choose between the following actions: accept (become a member), refuse (not become a member), wait (analyzing the group description). In this context, it is important to highlight the meaning of the general MADCOWuser profile. It can be considered as the data collection linked to a specific user. More specifically, profiling is the process, related to the creation and storing of a profile, performed by extracting a set of data which describe a user's characteristics. Whenever it is possible to recognize a single user, or a user level in order to provide specific contents or services, then MADCOW is using a specific user profile. MADCOW has a pyramidal structure for profiling, supporting the management of rules on every single user in order to provide the more suitable contents and services. A user can participate to groups under different user profiles; for this reason it is necessary to know the group topic in order to enable it to specific contents and services. For example, the reader can only access the annotations of a group of which she is member, while the user group can also access insert new annotations. Within MADCOW, a hierarchy about profile authorizations has been defined, which can be integrated with others personalized rules in order to relate every profile to specific contents, functionalities and services. Further definitions of particular profiles belonging to the related group, are not accessible from others users belonging to different groups. The procedure for the creation of the personalized user profile is available both during the creation of a group and under the Group Admin section.

5.2   The Lookup Function

The lookup function is used, while populating a group, to obtain a set of annotations related to the group topics. The user can analyze the resulting set of annotations to decide to send the invitations to the most suitable users. Every result is linked to a Web page where the related annotation is posted. For this reason, this function can be also used to search new digital documents related to the group topic. The lookup function in the Group Admin section is only usable by users with the suitable `Run lookup` authorizations. The list containing the annotations is produced by considering only public annotations. Every public annotation is examined by the following evaluators: *User-Trust*, *Goal-Tag*, *Domain-Trust*, *Annotation-Type*. The sum of the results produced by these evaluators provides a numerical index (Global evaluation) expressing the connection degree between the examined annotation and the group target. The User-Trust evaluation provides a numerical value for an annotation according to the judgment expressed by the group members on the author of the annotation. MADCOW also provides procedures to manage the whiteUserList and the blackUsersList. The first list contains the set of users whose annotations have been positively judged by the group members, while the second list contains the set of users whose annotations have been negatively evaluated by the group. If the annotator belongs to the whiteUserList, the User-Trust evaluation returns a weighted average of the trust-values for the annotator. Conversely, if the annotator belongs to the blackUsersList the procedure ends excluding the examined annotation. The Goal-Tag evaluator takes into account the linked tags and matches the tags belonging to the annotation to the tags in the Goal-Tag list of the group. The result is computed as the ratio between the sum of the tag relevance-values detected during the matching process and the sum of the tag relevance-values belonging to the Goal-Tag of the group. The Domain-Trust evaluator takes into account the information source. More specifically, the domainList provides a set of Web domains that are positively judged from the users group according to both content reliability and relevance with the group topic. The Domain-Trust procedure checks if the source URL, related to the annotation, belongs to the domainList of the group. The procedure returns a value between 1 to 10 according to the relevance of the Web page content with the group topic. The Annotation-Type evaluation examines the kind of annotations (Announcement, Comment,

Example, Explanation, Integration, Memorandum, Question, Solution, Summary). In this way it is possible to associate a type-value attribute (between 1 to 10) to identify the preference degree.

The results returned from the lookup function are shown in a Web page, containing both the summarization of the parameters for the search process and a list of annotations returned by the lookup function, as shown in Figure 6.

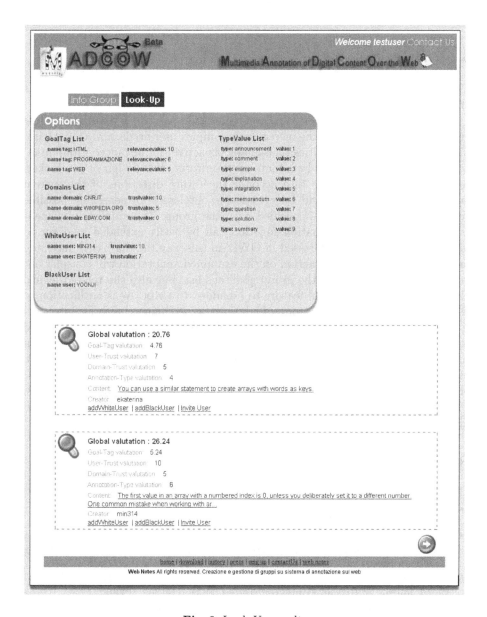

Fig. 6. Look-Up results

Every annotation is shown with the following information: *creator, tags, creation date, annotation content* and *hyperlink* to the related page. The lookup function is configured according to the following attributes: *goalTags, domainsList, whiteUsersList* and *blackUsersList*. The function is very flexible as it can be performed by using different research parameters which can also be combined in order to provide more specific searches. It is necessary to set at least a kind of evaluation in order to start the research. Only MADCOW users with access to the lookup function can add users to the whiteUserList and to the blackUserList.

6   Conclusions

The opportunity to create and manage users groups focused on specific topics is an added value of any digital annotation system.

The current functionalities (usually classifiable as passive) supported by these groups do not seem to meet the increasingly complex user needs. MADCOW2.0 implements a set of functionalities (classifiable as active) for advanced group management, based on trust and content search criteria. The design of the lookup research function has taken into account the future directions of the system aimed to model, manage and query suitable ontologies for searching possible goalTags lexical alternatives specified with the group. The current implementation of the goalTag evaluation function can be easily extended in order to interact with ontologies expressed in OWL; in this way the tag-based matching process performed by the function on the examined annotation will consider not only the tags belonging to the group goalTags list, but also the related lexical alternatives. A new practical feature to be added to MADCOW is notification to users belonging to a specific group of administrative changes, such as: introduction of a new tag within the goalTags list, chances within the user profile, and so on. In this way, the user would not be forced to access the group information summary section in order to check possible updates. Notifications could be also used to advise the user according to the answers related to own annotations, or to apprise the user regarding occurred annotations on their own Web pages. This would allow quick access to comments and questions related to one's own digital documents, and quick reactions to group modifications.

References

1. Diigo (2009), http://www.diigo.com/index
2. Fleck (2009), http://fleck.com/
3. Stickis (2009), http://stickis.com/
4. Marshall, C., Brush, A.: From personal to shared annotations. In: Proc. Human Factors in Computing Systems. Extended Abstracts, pp. 812–813. ACM Press, New York (2002)
5. Ovsiannikov, I.A., Arbib, M.A., Mcneill, T.H.: Annotation technology. International Journal of Human-Computer Studies 50(4), 329–362 (1999)
6. Agosti, M., Ferro, N.: A formal model of annotations of digital content. ACM Transactions on Information Systems 26(1), Article 3 (2007)

7. O'Hara, K., Sellen, A.: A comparison of reading paper and on-line documents. In: Proc. Human Factors in Computing Systems. ACM Press, New York (1997)

8. Mann, W.C., Thompson, S.A.: Rhetorical structure theory: A theory of text organization. Technical Report ISI/RS-87-90, Information Sciences Institute, University of Southern California (1987)

9. Bottoni, P., Levialdi, S., Rizzo, P.: An analysis and case study of digital annotation. In: Bianchi-Berthouze, N. (ed.) DNIS 2003. LNCS, vol. 2822, pp. 216–230. Springer, Heidelberg (2003)

10. Bottoni, P., Civica, R., Levialdi, S., Orso, L., Panizzi, E., Trinchese, R.: Madcow: a multimedia digital annotation system. In: Proc. AVI 2004, pp. 55–62 (2004)

11. Bottoni, P., Levialdi, S., Panizzi, E., Pambuffetti, N., Trinchese, R.: Storing and retrieving multimedia web notes. IJCSE 2(5/6), 341–358 (2006)

12. Bottoni, P., Cuomo, M., Levialdi, S., Panizzi, E., Passavanti, M., Trinchese, R.: Differences and identities in document retrieval in an annotation environment. In: Bhalla, S. (ed.) DNIS 2007. LNCS, vol. 4777, pp. 139–153. Springer, Heidelberg (2007)

13. Pei-Luen, P.R., Sho-Hsen, C., Yun-Ting, C.: Developing web annotation tools for learners and instructors. Interacting with Computers 16, 163–181 (2004)

14. Waller, R.: Functionality in digital annotation: Imitating and supporting real-world annotation. Ariadne 35 (2003)

15. Marshall, C., Brush, A.: Exploring the relationship between personal and public annotations. In: Proc. 4th ACM/IEEE-CS Joint Conference on Digital Libraries, pp. 349–357. ACM Press, New York (2004)

16. Nishino, F., Fu, L., Kume, T.: Web orchestration: Customization and sharing tool for web information. In: Proc. 1st Int. Conf. on Human Centered Design, HCI International 2009. LNCS, pp. 689–696. Springer, Heidelberg (2009)

17. Da Rocha, T.R., Willrich, R., Fileto, R., Tazi, S.: Supporting collaborative learning activities with a digital library and annotations. In: Proc. 9th FIP TC WCCE 2009. LNCS, pp. 349–358. Springer, Heidelberg (2009)

18. Schroeter, R., Hunter, J., Kosovic, D.: Filmed-collaborative video indexing, annotation, and discussion tools over broadband networks. In: Proc. 10th International Multimedia Modeling Conference, pp. 346–353. IEEE Computer Society, Los Alamitos (2004)

19. Huang, T., Pallickara, S., Fox, G.: A distributed framework for collaborative annotation of streams. In: Proc. Collaborative Technologies and Systems, pp. 440–447. IEEE Computer Society, Los Alamitos (2009)

20. Hong, L., Chi, E.H.: Annotate once, appear anywhere: collective foraging for snippets of interest using paragraph fingerprinting. In: Proc. CHI 2009 Human factors in computing systems, pp. 1791–1794. ACM, New York (2009)

21. Haslhofer, B., Jochum, W., King, R., Sadilek, C., Schellner, K.: The lemo annotation framework: weaving multimedia annotations with the web. Int. J. Digit. Libr. 10(1), 15–32 (2009)

22. Ma, Q., Miyamori, H., Kidawara, Y., Tanaka, K.: Content-coverage based trust-oriented evaluation method for information retrieval. In: Proc. SKG 2006, p. 22. IEEE Computer Society, Los Alamitos (2006)

23. Chen, W., Chen, J., Li, Q.: Adaptive community-based multimedia data retrieval in a distributed environment. In: Proc. ICUIMC 2008, pp. 20–24. ACM, New York (2008)

24. Bottoni, P., Civica, R., Levialdi, S., Orso, L., Panizzi, E., Trinchese, R.: Storing and retrieving multimedia web notes. In: Bhalla, S. (ed.) DNIS 2005. LNCS, vol. 3433, pp. 119–137. Springer, Heidelberg (2005)

Semantic Network Closure Structures
in Dual Translation of Stochastic Languages

Lukáš Pichl

International Christian University
Osawa 3-10-2, Mitaka, Tokyo, 181-8585, Japan
lukas@icu.ac.jp
http://www.icu.ac.jp/

Abstract. Iterated semantic translation as reflected in dual space of
two different languages provides - through a normalized reference set
of dictionary expressed vocabulary that is inclusive of double sided link
direction correspondence - a mechanism for assessment of word extent
breadth and meaning delineation. This measure resolves according to
the number of source language passages in round translating steps, until
content exhaustion. Semantic clusters emerge in above procedural format through graph decomposition of the united word set above language
pair to isolated segments of connected semantic vertices. Such a linkage is
also present in single language morphological context of thesaurus format
with enumerated word content proximity, typically farther subjected to
additional support that arises from accompanying clauses, applied level
of grammatical compatibility, and overall linguistic usage purity. Resort-
ing to algorithmic reverberation and numerical case studies on stochastic
data samples, it is argued that resemblance of contained and mirrored
semantic vertex structures follows cultural interaction, although shad-
owed at various degrees of virtualization, in evolving linguistic relation
that ultimately leads to a knowledge cohesion level frame of scientific
depth. Since it has been known that complexity information depth as
content purity grade paraphrases itself in Kolmogorov efficiency sense,
either to stability measure of fixed structures or manifests as a catalytic
growth factor in nurturing substances, this work provides a particular
instantiation in topological context of semantic graphs.

1 Introduction

Content information measured in terms of the shortest available program to en-
code a given dataset is known to stochastically equate the system entropy value
as expressed by means of the compositional probability function in constructive
proof of such statement [1]. Application context of computing machinery, library
science and language standardization provides interesting filter frames in practical instantiation so as to encompass incompleteness and organizational restrains
for optimal content reconstruction.

To assess a vocabulary content in its entirety, a complete composition scheme in
knowledge space is required from alphabet specification via meaning dimensions

S. Kikuchi, S. Sachdeva, and S. Bhalla (Eds.): DNIS 2010, LNCS 5999, pp. 210–224, 2010.
© Springer-Verlag Berlin Heidelberg 2010

and notion typing stencils, both physical and abstracted, towards emergence of knowledge accumulation formalism. Such a process may start from replication of a single element (existence) to enumerated finite set of alphabet (replication and separation) and syllable formation, leaving aside a certain minimal subset with economic functions (determination, preposition), and continues further to morpheme formation, thereby usually subjected to notational combinatorial complexity, and consequently adjusted instead to fundamentals of actual semantic processing architecture, such as typical thinking and categorizing patterns native to a processing brain (action, emotion, subject), in a process gradually forcing outward and inward views to postulated coincidence patterns, which manifests as a reflection of the stabilizing processing organization onto semantic output fixation (function selector catalogues for instance).

In terms of self composition, nature selected and seed encoded complex structuring processes in plant growth mechanisms, while human civilization increasingly accumulates instances of such external knowledge for abstracted utilization elsewhere, a mechanism used in education that itself may alter or produce vocabulary items to above effect. This work concerns with a word level snapshot of language in two frequented semantic tool frames, first a thesaurus, then a dictionary, both reduced to word correspondence grains with logical, integer, and fractional weight scales adopted for numerical analysis. Such approach establishes a transform from construction parameters of probabilistic word data samples to the semantic cluster appearance. A reference numerical text corpus is then composed using derived cluster probability density of words, and subjected to metalevel analysis, namely text abstraction first by means of sentence extraction with local and global similarity coefficients, and then using absolute hierarchical ranking, in order to address representative features that arise from finite and restrained vocabularies, and apply to natural language datasets. Since the field of semantic processing is long established [2,3], this work focuses on purely numerical capture aspects that recently regained attention [4].

The article is organized as follows. After a section dedicated to graph formalism of numerical semantic circuits, construction of derived text documents is given along with the examples of related substance extraction methods. Replication of computational formalism is then recognized as a quantifiable cultural personalization mechanism, which is accompanied with actual natural language processing illustrations to conclude this text with a brief summary.

2   Numerical Semantic Circuits

Since all computational text processing methods reduce content manipulation to number processing at certain depth of representation layer (program, software, compilation, middleware, instruction set) an enumerated semantic dataset is denoted as a graph collection of vertices,

$$V = \{v_i\}_{i=1}^{n_V}, \tag{1}$$

where domain meaning relations expressed in vertex annotation patterns are given in form of multiple connection edges,

$$E = \{e(v_i; \tau_i, v_j; \tau_j) | \forall v_i, v_j \in V \wedge \tau_i \in T\}, \tag{2}$$

where T is a set of tags (synonym domain, antonym domain, sets of cultural links, economic links, object resource links, processing links, or different languages). Edges are naturally represented as logical weights, integer weights, and rational weights,

$$e(;,;) = \{0 \vee 1\}$$
$$e(;,;) = \{\infty \vee p \vee p - 1 \vee \dots \vee 1\}$$
$$e(;,;) = \{\infty \vee 1 \vee 1/2 \vee \dots \vee 1/q\} \tag{3}$$

so as to map elementary arithmetic thinking patterns. Numerical labels in Eq. (3) are sorted from edge absence value to the highest proximity. Connection circuits can be enabled with $\tau_i \neq \tau_j$ thus forming an overlay multigraph structure. In what follows edges are directed ($e(i; , j;)$ is not necessarily symmetric) and appear as a result of serial probabilistic generative process that can be conditioned with current graph snapshot topology.

Derived semantic evaluation criteria of practical interest include graph transitive closure with number and sizes of extracted strongly connected segments. Weighed graphs subresolve transitive closure to the shortest path version characterized with segment wide square pattern of minimal connection cost. Transitive closure is also a suitable aggregation level to derive and utilize complement graphs for semantic closure.

It is noticed here that optimized subgraph vertex skeleton layout, from which each word locates at most l-edges away, can provide a core vocabulary set for self-learner extension, whereas determination of all complete subgraphs (prior to transitive closure, and furthermore in comparison) shows aggregation depth in original dataset. Both approaches are nondeterministic polynomial complete problems even if tractable with low size of content nucleation typical in natural languages, and as such are not pursued here. Similarly, sparse nature of data content allows us to leave aside graph theoretic applications of vertex and edge set reconstruction paths as frame transforms of access compression and topological sort. Instead, multigraph structure of vertex colorization is applied using tags $\tau_i \in T$ for self-consistent frame-compatible content enhancement.

2.1   Thesaurus Format

Enumerated single layer word graph structure as given above (source and reverse coincide, $\tau_1 = \tau_2 = \tau$, is initialized by expanding list of source vertices one by one while assigning a n end vertex thus forming a thesaurus edge. Structure formation then continues according to three generative procedures,

1. uniform method selects a pair of vertices at no bias,
2. shaped method is binomially modulated (parameter p),
3. cluster method adopts current vertex degree statistics,

with vertex selection probability functions given as

$$\pi(v_k) = \begin{cases} \frac{1}{n_v} & \text{uniform} \\ {}_nC_k p^k (1-p)^{n-k} & \text{shaped} \\ \frac{\deg(v_k)}{\sum_{l=1}^{n_V} \deg(v_L)} & \text{cluster} \end{cases} \quad k = 1, \ldots, n_V, \tag{4}$$

where indegree is selected for reverse end, outdegree is selected for source end, and denominator of the last fraction stands for the current number of graph edges. Weighted graphs, in addition, adopt in either case label value of $d+1$ or $1/(d+1)$, where d is the outdegree preceding current edge addition.

Symmetric version of the above algorithm automatically includes opposite directed edge to the graph at every generation step. In either case, data structure is represented in adjacency matrix $A[; \tau_i][; \tau_j]$ (to be resolved into τ pair instances in what follows).

The structure generation process grows above semantic circuits to edge concentration levels at fixed inverse proportion to graph size (constant average outdegree as common for natural language dataset) or higher, up to a numerical fabric with connection density $c \in (0,1)$ that mirrors parallel destruction of semantics on behalf of intense vertex feature generalization that turns to bypass specialization importance.

2.2 Dictionary Format

Vertex side duplication in thesaurus procedure provides dictionary structure as a naturally derived extension, in which both vertex lists are link initialized (category format inclusion operator) and extended (structure builder) separately. Denoting total graph size as $2n_v$, there are following possibilities ($T = \{\tau_1, \tau_2\}$)

1. pure dictionary ($e(; \tau_1, ; \tau_2) = 0 = e(; \tau_2, ; \tau_1)$)
2. semicomplete base ($e(; \tau_i, ; \tau_{3-i}) = 0$ for $i = 1 \vee i = 2$)
3. complete base (full dictionary and thesaurus pair).

To compare thesaurus and pure dictionary structures, the size of the latter doubles graph size of the former. Uniformly structured dictionary (on both sides, $\tau_i \to \tau_{3-i}$) combined with uniformly structured thesauri (source and reverse) scales a single thesaurus format from n_V to $2n_V$ as expected, to which mixed structures provide suitable means for semantic layout comparison.

2.3 Content Separation

Strongly connected graph segments correspond to semantic domain clustering as contained between original graph and its transitive closure version, and are conditioned by the path length of the shortest indirect connection (twice the number of translation rounds in case of pure dictionary closure). Whenever thesaurus and dictionary are combined, transitive closure of thesaurus should precede the phase that allows language translation, at least in classical linguistic way of frame categorization.

The algorithm for transitive closure in compact form reads [5].

```
procedure Floyd_Warshall(A[1:n][1:n])
  extern binop1, binop2
  C[][]:=A
  for k:=1 to n /*get links through 1 to k*/
    for j:=1 to n
      for l:=1 to n
        C[j][l]=binop1(C[j][l],binop2(C[j][k],C[k][l]))
  return C
end ! transitive closure and shortest path
```

Here the two binary operators are$\{ \vee, \wedge\}$ on logical graphs and $\{\min, +\}$ on weighed graphs. To determine strongly connected components (the transform $C[i][j]=((C[i][j]==\infty)?0:1)$ is used for weighted graphs) it suffices (in cubic time complexity) to perform a straightforward postprocessing analysis below.

```
function Strong_Connected_Components(C[1:n][1:n])
  inlabel[1:n]=1, c=0
  for i=1 to n
    if (inlabel[i])
      print(line end, i), c=c+1
      for j=1 to n
        if (C[i][j] .and. C[j][i] .and. inlabel[j])
          print(space,j), inlabel[j]=0
  return c
end ! a posteriori analysis to Floyd_Warshall
```

If only strong connected graph components are needed (interlink statistics between full blocks is omitted), graph search is known to provide the required result in quadratic time complexity [6].

```
! uses stack of graph node labels
visited[1:n]=0

procedure search(i,first) ! recursive
  visited[i]=1;
  for j=1 to n
    if(C[i][j] .and. !visited[j])
      visited[j]=1, search(j,first)
      if(first==1) push(j)
end ! direction conditioned depth first search

procedure Kosaraju(c)
  for i=1 to n
   if(!visited[i]) search(i,1)
   for i,j=1 to n C[i][j]<->C[j][i] ! graph transpose
   visited[1:n]=0
   while(!stack_empty())
```

```
      i=pop(), print(line end,i), c=c+1 ! new component
      search(i,0) ! reversed traverse
      for j=1 to stack_size()
        if(visited[k=pop()]) print(space, k)
        else (push(k), break)
end ! strongly connected components
```

One sided (source or reverse) connected components rooted at a particular source word v_i are obtained with breadth first algorithm that is merely augmented with the tree depth labels along the node link extraction process.

```
! uses a first in first out queue of graph labels
depth[1:n]=-1

procedure translation_set(i,A[1:n][1:n])
! matrix A is (tau_1-> tau_2 and tau_2->tau_1) projected
  put(i), depth[i]=0
  while(queue_empty==0)
    j=get(), level=depth[j]
    for k=1 to n
      if (A[j][k]!=0 .and. A[j][k]<infty .and. depth[k]==-1)
        depth[k]=level+1, put(k)
end ! tree-depth-tracked breadth first search
```

When the above search procedure of translation set is depth conditioned in addition with each vertex inclusion to the strongly connected component of the root node v_i, the term

$$\sigma(i : \tau_i, j; \tau_i) = |\text{depth}[i] - \text{depth}[j]|/2 \qquad (5)$$

gives the minimal number of translation passes via language τ_j to connect words i and j in language τ_i.

The above components provide notational and procedural basis for numerical patterns analyzed below.

2.4 Stochastic Patterns

Figure 1 shows an inset view of two graphs, one composed with uniformly based binomially-augmented fixed vertex selection, and one augmented with a scale free growth procedure emphasizing the vertex indegree when determining endpoint of a new oriented edge. All graphs are initialized in the semantic frame proper sense, meaning each vertex indegree and outdegree is at least one. The extra average vertex valence is set to five in all computations with the exception of dense numerical image illustration of Fig. 1 ($N_m = 50$). Figure 2 gives an account of indegree and outdegree statistics resulting from the above outlined numerical circuit prototype composition, including the indegree power law behavior of the scale free component.

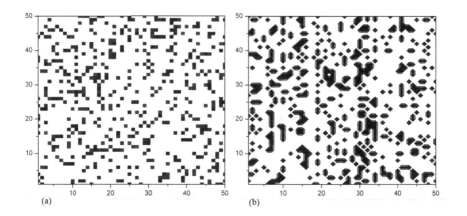

Fig. 1. Connection weight pattern in a semantic subgraph with (a) binomial and (b) clustering vertex selection methods

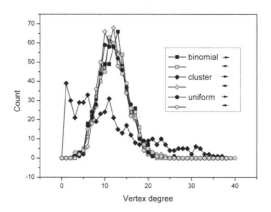

Fig. 2. Layout of vertex indegree and outdegree population for vertex selection methods explained in the text

Figure 3 gives vertex changing frequencies for the shortest paths among all vertex pairs obtained from the breadth first search generated traverse trees. Since in order to utilize the number of strongly connected components a substantial degree of frame incompleteness is needed, such a particular illustration is deferred and can instead be found in statistical work on natural language processing. Thus the current subgraph connection patterns, vertex linkage lengths, and histogram component shapes provide necessary elements for document processing and meaning depth inference.

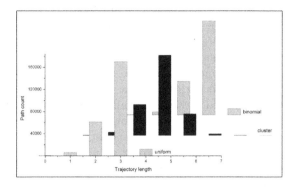

Fig. 3. Histogram of shortest path edge numbers for vertex selection methods explained in the text

3 Numerical Circuit Documents

Prototyped semantic layout of vocabulary circuitry generated above practically links to a document content as the induced meaning range, whether in full or additionally restrained, for instance up to a certain depth or an edge weight cut off.

Naturally languages for human communication are complicated due to personalization of prototypes (incompatible layouts) or an incomplete personal acquisition depth pattern of the same prototype standard (varying education depth). A communication problem in the latter situation is how to assess author depth of meaning intent, while the former circumstance may also include multigraph editing preprocessing to match conventional semantic circuit etalons. Exclusion of author intended meaning depth assessment through emphasis on recipients effect or a personal semantic utility naturally leads to acknowledgement maximization of received document text patterns.

Elementary document composition rules are outlined next and processed for a numerical provision of illustrative examples. The number of clauses is fixed as m_d; each is long from 1 to n_c according to a binomial shape of probability density with $p = 0.5$.

3.1 Markov Composition

Defining a current state as the word at the end of a clause fragment, the probability of the next word can be taken as

$$m_{pq} = \frac{M_{pq}}{\sum_q M_{pq}}. \tag{6}$$

Without introducing further composition parameters, M_{pq} may stand either for graph matrix or its complement (computed for weighted graphs as a reversal $w_i \leftrightarrow w_{|w|-i+1}$ in the set of predefined instantiated values). Since either case is a path traverse on matrix M that is rather sparse in the original semantic domain, complement matrix corresponds to a syntax-adjusted equally randomized text population, whereas the original graph matrix would employ semantic shifts that may cycle (or freeze if isolated vertices were allowed). Therefore the matrix type is altered with probability ratio q set as 0.5 unless mentioned otherwise. Sentence opening words in each clause are taken at random.

Although transitive closure matrices would provide for higher delineated transitions among strongly connected components, out of which instances could then be taken at random or in above manner for more realistic semantic content nucleation, the procedure in Eq. (6) suffices to provide an unbiased probabilistic test document corpus that serves as a suitable comparative reference to the following method.

3.2   Scale Free Assembly

Dynamic collaborative systems such as communication providers or certain types of metabolic networks are known to exhibit hub structures and overal organization that is called scale free when the vertex degree density follows a power law with negative exponent [7]. Such a complexity pattern may appear in text documents and is simulated as follows.

An empty dynamic list to contain separate document included words and their frequencies is initialized. Another static list is made for vocabulary and initialized with vertex outdegrees. New clause members are selected from either list with probability ratio q and $1 - q$. All word inclusion operations are accompanied with document word list update and frequency increment.

3.3   Substance Extraction

Document content compression to core information is usually performed by means of abstract generation. Provided the actual vertex choice pattern in a clause largely includes grammar effects (whether after accurate morpheme annotation with grammatical T tags in a clause, or directly adopting the sinogram system that is complete as a sequence of graphical morpheme characters), the abstract can be generated by means of clause extraction (and a posteriori indirect grammar pattern restoration, if any is required). In such a representation, substance extraction reduces to a clause selection algorithm. To this aim, first the similarity clustering method of [10] is reviewed (as a document relative tool) and then compared to clause ranking based on scores derived from the semantic prototype (as its own knowledge-base-rooted absolute tool that can emphasize any depth of semantic extent structure).

Content Similarity Clustering. Document substance extraction based on similarity of clauses is an effective clustering method that selects new cluster parents as well as candidates for seed merger by using the percentage of shared component vertices. The terminology is as follows [10].

Global similarity coefficient of two clauses C_i and C_j reads

$$\text{GSC}(C_i, C_j) = \frac{2 \times n(\text{common words of } C_i \text{ and } C_j)}{n(C_i) + n(C_j)} \quad \forall i \neq j \leq l \qquad (7)$$

Local similarity coefficient of a clause C_i and a cluster C reads in the same way

$$\text{LSC}(C_i, C) = \frac{2 \times n(\text{common words of } C_i \text{ and } C)}{n(C_i) + n(C)} \quad \forall i \leq l. \qquad (8)$$

The method starts with a set of word lists and computes their pairwise global similarity coefficient at first.

The pair with the highest GSC value, provided it is above a threshold level τ, is replaced with a new global cluster clause that enlists only the common words. Then the local similarity coefficient is computed for all other lists, and if the value is above τ, such a list is deleted at the end of round while the cluster list updates to commonly shared words. The procedure then recurs.

Each cluster is annotated with a label to their single original clause placed high at source document. As the cluster clauses diminish through common word mergers in the length of representation words, similarity coefficients eventually decrease below the parameter τ. Clusters are ranked with the clause count in which order their labels enter the abstract text. The label with a higher position in the original document is retained in symmetric mergers.

Vocabulary Template Scoring. Provided a given semantic prototype, here expressed in its multigraph structure, the absolute vertex meaning relevance can be represented as word outdegree (or negative cumulative weight of outgoing edges in weighted version), and the relevance of entire clause then assessed as the total of such values, rescaled to the average per contained vertex, so as to obtain a scoring method independent of the numerical clause length,

$$S(C) = \frac{1}{l} \sum_{i=1}^{l} w(C_i) \quad C = \{C_i\}_{i=1}^{l}, \quad w = w(V, E, v_i(c_i)). \qquad (9)$$

Natural variations motivated above include scoring with sizes of connected trees or strongly connected components, and, importantly, application of complement graphs.

A word scoring method based on original graph emphasizes commonality of meaning whereas the one derived from complement version accentuates the content specialization degree.

3.4   Stochastic Patterns

Above enumerated semantic composition structure in its procedural closure transforms a particular set of probability density patterns (stochastic operators) forward from the level of prototype assembly to the level of expanded document set construction, and backward at the level of substance extraction for abstract composition. Figure 4 provides numerical coordinates for an illustrative run of one hundred documents that have word vertices either selected in random path attachment with interlaced straight connection matrices of the original and complement graphs (a), or in a hub scaling growth process (b), for each of the three sets of previously outlined semantic prototypes. The number of clauses in a document, abstract, and the maximal word count parameters are selected as 20, 5, and 50, respectively. The abscissa values represent the ratio of clauses nucleated as clusters into the resulting abstract substance, and consequently are rather sparse, which allows to apply a small horizontal shift among all three datasets to avoid their overlap. The ordinate values use vertex outdegree to weight the relevance domain of each word, at the levels of the abstract and the entire document. It is confirmed that the scale free procedure of document content composition makes the similarity ranking based method of abstract extraction more effective. Since furthermore bottom semantic circuit structures imprint onto machine extracted abstracts differently, numerical spaces of extraction method conditioned efficiencies could be used in complexity assessment, genre categorization, and other linguistic applications.

It may be worthwhile noting that representation of Fig. 4, possibly with more superposed evaluation dimensions, naturally brings material science issues of point concentration, geometric separation, and focus location into a dynamic picture of document evolution and message elaboration, for instance as it is driven and represented in the genetic programming formalism for novel applications.

4   Replication Layer Saturation

Since above graph notation is optimally generic, it is worthwhile to review its application domain in linguistics. Thesaurus is a language tool that can be termed self explanatory. As it certainly appears in various editions of what is widely considered the same standard, this fact implies equivalent numerical representation, excepting perhaps context emphasizing depth and content volume effect. Such equivalence relation can be studied for practical purpose using strongly connected component analysis as a tool for instance. Online encyclopedias provide another illustration although at a higher level of topical documents.

Partial instantiation schemes of the same graph semantic structure may correspond to personal learning staging. If errors are admitted, interesting educational tasks emerge on how as to infer them indirectly from documents and what fixation mechanisms are effective for what purpose even in the absence or scarcity of error signaling documents supplied for evaluation.

It is understandable that education and personal experience in a given environmental setting solidify with certain priors, that are learnt to flex in part when a different language experience is met. A saturation problem then arises

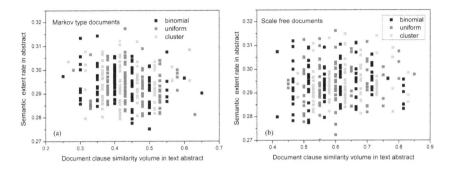

Fig. 4. Numerical documents as positioned by summary efficiency ratio pairs for (a) fixed transition and (b) current state preference text growth methods

for each personal semantic fixation skeleton on as to what other semantic graph structures have to be met before all acquired semantic path obstacles can be overcome. Similarly language separation measures in a knowledge space including innovative potential of new members could be hypothesized using linked vertex structures, which applications are though not in scope of the present static frame based article.

The field of numerical semantics extends to a number of other computer science applications of practical interest [8,9].

5   Natural Language Excurse

This part provides application software examples for above notions in Asian languages. In particular, Korean-Japanese-Chinese language family is known to instantiate semantic patterns in a scientific manner from the processing operator emphasized form to the knowledge system empowered formalism.

On practical footing, unicode's CJK encoding long ago provided the practical standardization platform to numerical text processing that was instantiated in two Java virtual machine programming endeavors to natural language processing that are referred hereafter.

5.1   Korean Syllabi Cycles

The applet environment shown in Fig. 5 implemented a Korean vocabulary set with its Japanese dictionary counterpart, and illustrates translation routes among both languages as a part of a broader study beyond scope of this article [11].

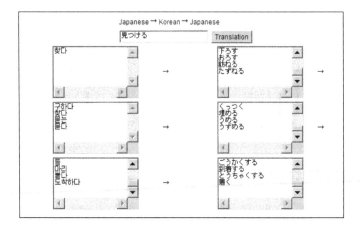

Fig. 5. Java applet for dictionary closure loops between Japanese and Korean languages based on a graph search

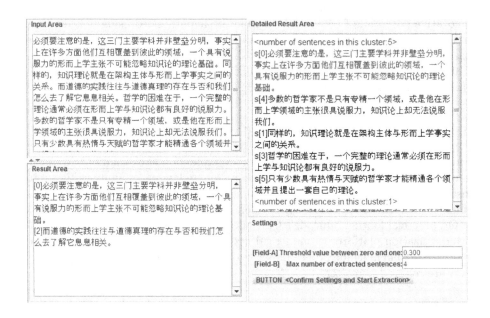

Fig. 6. Java applet for abstract summarization of short sinogram text based on the similarity clustering method

5.2 Sinograph Characters

The applet environment shown in Fig. 6 implemented a Chinese character set with the single language prototype left to readers, and illustrates similarity co-efficient based abstract extraction as a part of a broader study beyond scope of this article [12].

6 Concluding Remarks

Semantic translation closure was formulated using graph theory formalism and instantiated in a metaphoric frame that numerically impersonalizes points of view and thinking patterns as those culturally represented in closed thesaurus and open dictionary knowledge structures. Apart from reinstating enumeration nature of language processing in the machine framework that provides for fundamental stochastic transforms on knowledge probability spaces, significance of direct semantic prototype linkage to document summarization and substance extraction algorithms was found, namely one that can be reinforced as terminological processing coincidence at notation and organization levels. Numerical semantic circuit frame assembled in the present article has been accompanied with recited actual natural language applications to illustrate structural interface in either frame. Completion effort in sense of pure efficiency guides computerized categorization efforts to incorporation of nature structures as imprinted onto material organization in mosaic decomposition of knowledge. Confluent structures of dictionary mediated knowledge culture interface outlined here may numerically illustrate an old proverb the more languages you know, the more times a person you are, perhaps most significantly in view of irreducible knowledge components.

References

1. Shiryaev, A. (ed.): Selected Works of Andrey Kolmogorov: Information Theory and Theory of Algorithms. Springer, Heidelberg (1993)
2. Minsky, M. (ed.): Semantic Information Processing. MIT Press, Cambridge (1968)
3. Sowa, J.F. (ed.): Principles of Semantic Networks: Explorations in the Representation of Knowledge. Morgan Kaufmann Publishers, San Francisco (1991)
4. Theoharis, Y., Tzitzikas, Y., Kotzinos, D., Christophides, V.: On Graph Features of Semantic Web Schemas. IEEE Transactions on Knowledge and Data Engineering 20(5), 692 (2008)
5. Floyd, R.: Algorithm 97: Shortest Path. Communications of the ACM 5(6), 345 (1962); Warshall, S.: A theorem on Boolean matrices. Journal of the ACM 9(1), 11 (1962)
6. Aho, A., Hopcroft, J., Ullman, J.: Data Structures and Algorithms. Addison-Wesley, Reading (1983)
7. Barabasi, A.-L., Albert, R.: Emergence of scaling in random networks. Science 286, 509 (1999)

8. Raghavan, S., Rohana, R., Leon, D., Podgurski, A., Augustine, V.: Dex: A semantic-graph differencing tool for studying changes in large code bases. In: Proceedings of the 20th IEEE International Conference on Software Maintenance, pp. 188–197. IEEE Computer Society, Los Alamitos (2004)
9. Ramamritham, K., Chrysanthis, P.: Semantics-based Concurrency Control. In: Encyclopedia of Database Systems 2009, p. 2591 (2009)
10. Kumar, A., Kumar, P., Rao, S., Reddy, K.: An Improved Approach to Extract Document Summaries Based on Popularity. In: Bhalla, S. (ed.) DNIS 2005. LNCS, vol. 3433, pp. 310–318. Springer, Heidelberg (2005)
11. Takagi, M.: Translation Routes and Implicit Language Structure of Japanese and Korean, thesis work, International Christian University Library (2008)
12. Mikami, N.: Popularity-based Approach to Automatic Extraction of Abstracts from Chinese Text Documents, thesis work, International Christian University Library (2006)

VisTree: Generic Decision Tree Inducer and Visualizer

Vasudha Bhatnagar*, Eman Zaman, Yayati Rajpal, and Manju Bhardwaj

Department of Computer Science, University of Delhi, India
vbhatnagar@cs.du.ac.in

Abstract. Decision tree is one of the most popular and commonly used technique for predictive modeling. Interpretability and understandability makes decision trees an attractive option among various classification induction algorithms. There are several freewares available for decision tree induction which can be used in data mining education and practice. However, these freewares have limited capability to interactively visualize the induced tree and experiment with the induction process.

In this paper, we describe the design of a generic decision tree inducer and visualizer which gives options for multiple splitting criteria (e.g. information gain, gain ratio etc.) and pruning criteria (e.g. minimum error pruning, cost complexity pruning etc.) for decision tree induction. The induced tree can be visualized interactively by the user and even saved for future visualization and comparison with another tree. These options are available through a user friendly GUI. The performance statistics for the induced tree can also be viewed by the user. The package has been designed using open source softwares including JDK 1.6, Netbeans 6.5 and Prefuse (for visualization of the constructed tree).

1 Introduction

Two most commonly performed tasks in data mining and machine learning are predictive modeling and descriptive modeling. The former task is concerned with constructing a model from the available training data and using it to predict a predefined attribute in the data set, while the latter is concerned with discovering an understandable description of the model. Predictive modeling is also known as supervised learning, since the algorithm induces a predictive model from the training instances that are labeled with the correct categorization. If the attribute to be predicted is categorical, the technique is called classification, else if numerical value is to be predicted the task is called regression [2,7].

In classification, often the focal point of the exercise is the performance of classifier in terms of accuracy or loss estimation due to wrong prediction. Artificial neural networks, support vector machines, k-NN are some of the classification techniques known for inducing models with good predictive power, though these models are abysmally low on understanding [2,9]. Inducing decision trees is a

---

* Corresponding author.

S. Kikuchi, S. Sachdeva, and S. Bhalla (Eds.): DNIS 2010, LNCS 5999, pp. 225–243, 2010.

classification technique which is extremely useful in applications like medical diagnosis, insurance and banking applications, where understanding the predictive model plays an important role in the overall success of the application [6].

In this paper we describe a recent (and continuing) project on development of a package for generating, evaluating and visualizing decision trees while varying splitting criterion, pruning criterion etc.. The package called VisTree, is a meta algorithm for decision tree induction integrated with a visualizer, which facilitates visual comparison of decision trees induced by varying splitting criterion and pruning criterion (meta parameters) for VisTree.

The package is intended to meet the requirements of the community using decision tree induction as described below:

1. Student: A platform for decision tree induction, where he or she can play with the training set and visualize the effect of variations in splitting/pruning criteria on the structure of the tree.
2. Instructor: A tool for enhancing student understanding of mathematics of splitting/pruning criteria by visual display of decision trees.
3. Researcher: A laboratory to plugin newly designed splitting/pruning criteria and evaluate it by means of predefined metrics. Visualization of the tree enhances the understanding.
4. Practitioner: A visual tool to understand the reasons underlying the decisions.

The package has been developed using open source technologies and is intended to be a open source available for academic purposes.

The paper is organized as follows. We present a gentle introduction to the classification problem and decision trees in Sections 2 and 3 respectively. Section 4 discusses the freewares available for decision tree induction and presents motivation for this project. Section 5 describes the technologies used and architecture of the package. Finally we conclude the paper in Section 6.

2   Classification

Classification, also known as supervised learning is a process of constructing a model from training data and using this model to predict category or class of the unknown data. The data consist of records that are categorized into mutually exclusive classes. A classification algorithm learns the training data characteristics to construct a model with power to discriminate between the classes.

Formally the classification problem is described as follows. Given a data set $D = \{x_1, x_2, \ldots, x_N\}$ of tuples and a set of classes $C = \{C_1, C_2, \ldots, C_m\}$. Each x_i has p attributes and an attached class label, which is the $(p+1)^{th}$ attribute. Thus, D consist of tuples that have already been classified by well defined process and form the basis of learning. The classification problem is to define a mapping $f : D \longrightarrow C$, such that class for a tuple x with unknown class label can be predicted using the mapping f. Since x_i's are labeled, D is also called a training set and classification is referred to as supervised learning.

In practice, classification is a two step process [1]. First step is the learning step where a classification algorithm builds the classifier by learning from the given training set. In the next step, the model is used to predict the class of records whose class label is not known [2]. Some of the common approaches to induce a classifier are Artificial Neural Network, k-Nearest Neighbor, Support Vector Machines and Naive Bayes classifier, details of which can be found in several well known books on data mining and machine learning [10,2,3,4,5]. We briefly describe them below.

1. Artificial Neural Network (ANN): ANN consists of a set of connected input/output nodes in which each connection has a weight associated with it. During the learning phase, the network learns by adjusting weights so as to be able to predict the correct class label of the input tuples. After learning is complete, the ANN is used as a black-box for prediction.

2. k-Nearest-Neighbor: k-Nearest Neighbor classifiers are lazy classifiers based on the intuition that the unlabeled object is likely to belong to the same class as that of the k closest objects in the n-dimensional feature space defined by the training set. k-NN algorithm uses a similarity measure to discover k-most similar records (neighbors) in the training set, to the record whose class is to be predicted. Once the neighbors have been identified, the majority class is assigned to the unlabeled record.

3. Decision trees: Decision trees are hierarchical predictive models learnt from training tuples. A decision tree is a flowchart like tree structure, where each internal node denotes a test on an attribute, each branch represents an outcome of the test and each leaf node holds a class label. The internal nodes are also called condition nodes and the leaf nodes are decision nodes. To predict the class of a unlabeled record, the tree is traversed starting from the root, following the branch which satisfies the condition nodes and the label of the leaf node is assigned to the unlabeled record.

4. Naive Bayes Classification: Bayesian classifiers are statistical classifiers. They are based on Bayes theorem, and assume independence of the attributes. They predict class membership probabilities, such as the probability that a given tuple belongs to a particular class. The unlabeled record is assigned the class with highest probability.

5. Support Vector Machines(SVM): SVMs perform classification by constructing an n-dimensional hyperplane that optimally separates the data into two categories. The separating hyperplane is optimal and separates data in such a way that cases with one category of the target variable are on one side of the plane and cases with the other category are on the other size of the plane. SVMs are naturally binary classifiers and used as black-box.

Each of these methods have their own strengths and weaknesses. For example, neural networks are robust and accurate but have lengthy training process. They are useful in applications like hand writing recognition, e-spam filtering etc., where accuracy is of prime importance and reasoning for classification is

---

[1] Lazy classifiers like k-nearest neighbor algorithms complete the task in one step.

not very important. Naive Bayes classifier is a probabilistic classifier that is based on Bayes theorem. It is a purely data driven classifier but offers no reasoning of the decisions made. Decision tree is also a purely data driven classifier but has a unique advantage of understandability of the decisions that assign the class label to unlabeled record. However they suffer from the problem of instability [1]. SVMs are also robust predictive models with high accuracy but their construction is computationally expensive and they too are used as black-boxes.

2.1   Performance Evaluation Measures

A classifier learnt on a given training set T is expected to predict the labels of new records correctly. However, since T captures only a small subset of data distribution of the population, the classifier may not be able to generalize well for unlabeled data. Before the classifier is put to use, it is necessary to assess how it is likely to perform on unseen data. This assessment is made by testing the classifier on 'test' data. The test data is labeled data, which may be either a subset of the training data or may be available explicitly. Several measures can be computed for the testing exercise, to assess the expected performance of the classifier.

Confusion matrix, most commonly used performance evaluation measure, is a summarization of the performance of the classifier and helps to assess if the system is confusing between classes (i.e. commonly mislabeling one as another). Each column of the matrix represents the number of instances in a predicted class, while each row represents the instances in an actual class. Entry (i, j) denotes the number of instances that belong to class i but are predicted as of class j. Table 1 shows the confusion matrix for a two class classifier. The entries in the confusion matrix have the following meaning: TN is the number of correct predictions of negative instances, FN is the number of positive instances that are incorrectly predicted as negatives, FP is the number of negative instances that are incorrectly predicted as positives and TP is the number of correct predictions of positive instances.

Table 1. Confusion matrix for a binary classifier

	Classified Positive	Classified Negative
Actual Positive	TP	FN
Actual Negative	FP	TN

Some commonly used performance measures defined for the two class confusion matrix are listed below. More measures can be found in [6].

1. The **accuracy** (A) is the proportion of the total number of predictions that were correct. It is determined using the equation:

$$A = \left(\frac{TN + TP}{TN + FN + FP + TP} \right)$$

2. The **recall** (R) or **true positive rate** is the proportion of positive cases that were correctly identified and is calculated using the equation:

$$R = \left(\frac{TP}{TP + FN} \right)$$

3. The **false positive rate** (*FP Rate*) is the proportion of negatives cases that were incorrectly classified as positive and is calculated using the equation:

$$FP\ Rate = \left(\frac{FP}{TN + FP} \right)$$

4. The **true negative rate** (*TN Rate*) or **specificity** is defined as the proportion of negatives cases that were classified correctly and is calculated using the equation:

$$TN\ Rate = \left(\frac{TN}{FP + TN} \right)$$

5. **Precision** (*P*) is the proportion of the predicted positive cases that were correct and is calculated as follows:

$$P = \left(\frac{TP}{FP + TP} \right)$$

6. **F-measure** (*F*) is the harmonic mean of precision and recall. It can have values between 0 to 1. Specifically, this measure is defined as:

$$F = \frac{2 \times P \times R}{P + R}$$

2.2   Validation Methods for Classifiers

Validation methods are useful for model selection on a particular dataset. These methods help to estimate how well the model is likely to perform on unseen data. Some of the commonly used validation methods described in [2] are:

1. **Holdout method**: It is the simplest validation method. The dataset is randomly partitioned into two sets, called the training set and the test set. The classifier is induced on the training set and tested on test set. The flip side of this method is that the evaluation may depend heavily on which data points end up in the training set and the test set, and hence may be significantly different depending on how the division into two sets is done.
2. **K-fold cross-validation**: In this method original dataset is randomly partitioned into K subsets. Of the K subsets, one is retained for testing the model, and remaining (K-1) subsets are used as training data. The process is then repeated K times (the folds), with each of the K subsets used exactly once as the test data. K results from the folds then are averaged (or otherwise combined) to produce a single estimate of the performance measure.
3. **Leave-one-out cross-validation**: This method involves using a single observation from the original dataset as the test data and the remaining observations as the training data. This is repeated such that each observation in the sample is used once as test data. This is same as a K-fold cross validation with K being equal to number of observations in the original dataset. Leave-one-out cross-validation is usually very expensive from computational point of view because of the large number of times the training process is repeated.

4. **Bootstrap method**: In bootstrapping, instead of repeatedly analyzing subsets of the data, we repeatedly analyze subsamples of the data. Each subsample is a random sample with replacement from the full dataset.

3   Decision Trees

A decision tree is a hierarchical model constructed from the training set and used for prediction of unlabeled data. It is a flowchart like tree structure, where each internal node denotes a test on an attribute and each branch represents an outcome of the test on the basis of which the training set is partitioned. The topmost node in a tree is the root node. Each leaf node holds a class label and denotes a decision. Given a tuple **x**, for which class label is unknown, the attribute values of the tuple are tested against the decision tree. A path is traced from root to a leaf node such that conditions on the path are satisfied by x and label at the leaf is assigned as predicted class [2].

At each level, decision tree induction algorithm tries to find the attribute that best discriminates between the classes. The measure used for identifying this attribute is called the splitting measure and the best attribute selected is called the splitting attribute. The first splitting attribute forms the root of the tree. The training data is partitioned on the splitting attribute and for each partition, the process is applied recursively till the stopping criteria is met. With loose stopping criteria, the decision trees often become so large and unwieldy that they become inaccurate for predictive purposes and also, difficult to understand. Therefore they need to be pruned.

There are several decision tree induction algorithms like ID3 [12], C4.5 [19], CART [8] etc.. They primarily differ in respect of the type of splitting criterion, stopping criterion and pruning method used during the induction process. We briefly describe these aspects in the following subsections. For details, please refer to the text by Rokach and Maimon [6]. Fig. 1 shows the decision tree for one of the most cited datasets used by Quinlan [12].

3.1   Splitting Criterion

Splitting Criterion is used at each condition node to select the attribute that best discriminates between the classes in the training set. The condition node has branches corresponding to the split point (in case of continuous attributes) or splitting subset (in case of cardinal attributes) [2]. Rokach and Maimon [6] list more than a dozen splitting criteria, of which we describe three most commonly used criteria.

Let the given training set D consist of m classes. Then, p_i is the probability that a tuple belongs to class C_i and is estimated as :

$$p_i = \frac{\#\ Tuples\ of\ class\ C_i}{\#\ Tuples\ in\ set\ D}$$

OUTLOOK	TEMPRATURE	HUMIDITY	WINDY	PLAY
sunny	85	85	false	Don't_Play
sunny	80	90	true	Don't_Play
overcast	83	78	false	Play
rain	70	96	false	Play
rain	68	80	false	Play
rain	65	70	true	Don't_Play
overcast	64	65	true	Play
sunny	72	95	false	Don't_Play
sunny	69	70	false	Play
rain	75	80	false	Play
sunny	75	70	true	Play
overcast	72	90	true	Play
overcast	81	75	false	Play
rain	71	80	true	Don't_Play

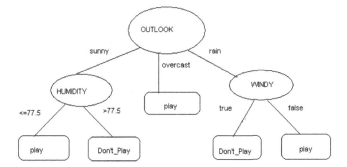

Fig. 1. Example tree for the Golf data set [12]

Let a be the attribute which is to be evaluated using splitting criterion and it's domain consists of v values.

1. **Information gain**: Information gain can be defined in terms of an impurity measure 'entropy' that measures the (im)purity of examples in the training set. Given a set D, consisting of m classes, entropy can be defined as:

$$E(D) = \sum_{i=1}^{m} -p_i \log_2 p_i$$

To measure the effectiveness of an attribute in classifying the training data, information gain is defined as the expected reduction in entropy caused by partitioning the examples according to this attribute. Thus, information gain of an attribute a, relative to the training set D, is defined as:

$$IG(D, a) = E(D) - IG_a(D)$$

where

$$IG_a(D) = \sum_{j=1}^{v} \frac{|D_j|}{|D|} \times IG(D_j)$$

Here D_j is the subset of D for which attribute a takes the value a_j.

2. **Gain Ratio**: Information gain is biased towards attributes that have many values. Gain ratio removes this bias by normalizing information gain using split information value. It is defined as:

$$GainRatio(D, a) = \frac{IG(D, a)}{Splitinfo(D, a)}$$

where $IG(D, a)$ is the information gain when the dataset D is split on attribute a as formulated above and Splitinfo(D,a) is defined as:

$$Splitinfo(D, a) = -\sum_{j=1}^{v} \frac{|D_j|}{|D|} \times log_2 \frac{|D_j|}{|D|}$$

3. **Gini Gain**: Gini gain is similar to information gain, but uses Gini Index instead of entropy. Thus, Gini gain is defined as:

$$Ginigain(D, a) = Gini(D) - Gini_a(D)$$

Here Gini(D) is the impurity measure on dataset D with respect to class labels calculated as :

$$Gini(D) = 1 - \sum_{i=1}^{m} p_i^2$$

$Gini_a(D)$ is defined as:

$$Gini_a(D) = \sum_{j=1}^{v} \frac{|D_j|}{|D|} \times Gini(D_j)$$

3.2   Pruning Methods

One of the questions that arise in a decision tree algorithm is the optimal size of the final tree. This is important not only from the point of view of understanding, but also from the accuracy perspective. A tree that is too large risks over-fitting the training data and poorly generalizing to new and unlabeled data. Pruning methods identify the least reliable branches and remove them.

There are two possible strategies used to reduce the size of the tree (pruning). A conservative approach is to grow the tree in restrictive manner and a liberal approach is to grow the tree fully and then prune to remove nodes that do not add to performance of the tree. Mingers [13] presents an excellent description of the pruning methods along with the intuition and mathematics of each method. Some commonly used methods are: (i) Cost complexity pruning, (ii) Reduced error pruning and (iii) Minimum error pruning.

4   Freewares for Decision Tree Induction

There are several freewares that are available for inducing decision trees. Weka [20] provides J48 (a variation of C4.5), ID3 and several other decision tree algorithms with user friendly interface for varying parameters. Source codes of C4.5 [19] and ID3 [18] are freely available, which can be compiled and used. All these packages provide facilities for validation and evaluation of induced decision tree using more than one methods. We describe functionality, inputs and outputs of the C4.5 and Weka packages in next two subsections, followed by the agenda for VisTree project.

4.1   C4.5 Package

C4.5 is a program for inducing predictive model as a set of classification rules as well as decision trees from a given training set [19].

All files read and written by C4.5 are of the form file-stem.ext, where 'file-stem' is a file name stem that identifies the induction task and 'ext' is an extension that defines the type of file. The program expects to find at least two files: a names file file-stem.names defining class, attribute and attribute value names, and a data file file-stem.data containing a set of objects, each of which is described by its values of each of the attributes and its class.

The program can generate trees in two ways. In batch mode (the default), the program generates a single tree using all the available data. In iterative mode, the program starts with a randomly-selected subset of the data (the window), generates a trial decision tree, adds some misclassified objects, and continues until the trial decision tree correctly classifies all objects not in the window or until it appears that no progress is being made. Since iterative mode starts with a randomly-selected subset, multiple trials with the same data can be used to generate more than one tree.

All trees generated in the process are saved in file-stem.unpruned. After each tree is generated, it is pruned in an attempt to simplify it. The "best" pruned tree (selected by the program if more there is more than one trial) is saved in machine-readable form in file-stem.tree.

All trees produced, both pre- and post-simplification, are evaluated on the training data. If required, they can also be evaluated on unseen data in the file file-stem.test.

Fig. 2 shows the output generated for the golf dataset by C4.5 at the default verbosity level.

4.2   Weka Package

Weka is the most popular data mining package freely available at [20]. It is a comprehensive software that provides a wide variety of pre-processing and visualization options to support commonly used data mining tasks. This open source software is an outcome of a project at the university of Waikato, New Zealand. The project started more than a decade ago with overall goal to build

```
C4.5 [release 8] decision tree generator    Thu Jun 15 09:15:50 2000
----------------------------------------------


    Options:
    File stem <golf>


Read 14 cases (4 attributes) from golf.data

Decision Tree:

outlook = overcast: Play (4.0)
outlook = sunny:
|   humidity <= 75 : Play (2.0)
|   humidity > 75 : Don't Play (3.0)
outlook = rain:
|   windy = true: Don't Play (2.0)
|   windy = false: Play (3.0)



Tree saved



Evaluation on training data (14 items):


    Before Pruning          After Pruning
    ----------------    ----------------------------

    Size    Errors    Size    Errors   Estimate

     8     0( 0.0%)     8    0( 0.0%)   (38.5%)  <<
```

Fig. 2. Output delivered by C4.5 for golf dataset

```
=== Run information ===

Scheme:       weka.classifiers.trees.J48 -U -M 2
Relation:     weather
Instances:    14
Attributes:   5
           outlook
           temperature
           humidity
           windy
           play
Test mode:    evaluate on training data

=== Classifier model (full training set) ===

J48 unpruned tree
------------------

outlook = sunny
|   humidity <= 75: yes (2.0)
|   humidity > 75: no (3.0)
outlook = overcast: yes (4.0)
outlook = rainy
|   windy = TRUE: no (2.0)
|   windy = FALSE: yes (3.0)

Number of Leaves  :      5

Size of the tree :       8


Time taken to build model: 0.03 seconds

=== Evaluation on training set ===
=== Summary ===

Correctly Classified Instances         14          100     %
Incorrectly Classified Instances        0            0     %
Kappa statistic                         1
Mean absolute error                    0
Root mean squared error                 0
Relative absolute error                0     %
Root relative squared error            0     %
Total Number of Instances              14

=== Detailed Accuracy By Class ===

         TP Rate  FP Rate  Precision  Recall  F-Measure  ROC Area  Class
           1        0        1        1        1         1          yes
           1        0        1        1        1         1          no
Weighted Avg.  1      0        1        1        1         1

=== Confusion Matrix ===

 a b   <-- classified as
 9 0 | a = yes
```

Fig. 3. Text output delivered by Weka for golf dataset

a state-of-art facility for machine learning technique which is applicable to real world data mining problems. A recent project update is available as a SIGKDD Explorer paper [14], while the package is publically available at [20].

In the area of classification, it offers most of the well known classification algorithms including several for decision tree induction. The induced tree is displayed in text mode (shown in Fig. 3) as well as graphical model (shown in Fig. 4). The graphical model is easier to understand than the text structure especially if the induced tree is big.

Recently Weka has added Prefuse as plugin for interactive visualization of the tree. As of Weka version 3.5.8 (only developer version, not stable-3.6 branch) one can easily add tree visualization plugins in the Explorer (Classify and Cluster panel) [21]. Though it is claimed that it is easy to implement custom visualizations, despite best efforts the authors could not make it work. The screenshot shown in Fig. 5 was taken from [21].

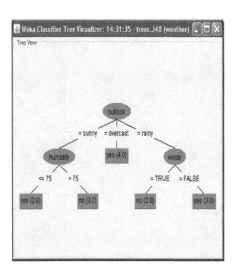

Fig. 4. Graphical output of Weka for golf dataset

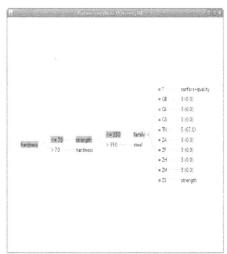

Fig. 5. Prefuse output of Weka using J48 on UCI dataset anneal with default parameters [21]

4.3 Motivation and Agenda for VisTree

VisTree project was conceived when one of the authors experienced the difficulty in explaining to the graduate students, the intuition behind the mathematics of some of the pruning and splitting criteria used in decision tree induction algorithms. It was further felt that visualization of decision tree structure could aid in understanding while experimenting with different combinations of splitting and pruning criteria. The desiderata required

(i) Ease to vary the splitting/pruning criteria for decision tree induction,
(ii) Ease to visualize the tree interactively,
(iii) Ease to store the induced tree and it's performance metrics for visual/ automatic comparison at some later time.

Both C4.5 and Weka provide means to vary the splitting/pruning criteria (item (i)); but in somewhat limited way. Both packages deliver rich set of performance statistics using different validation methods. Weka 3.5.8 provides interactive visualization of the induced tree. Neither of the two facilitate the functionality mentioned in item (iii). The authors consider this functionality important for pedagogical reasons.

VisTree aims to provide a focused environment for exploring and experimenting with the decision tree induction process. It intends to provide a platform where tree induction experiments with varied pruning and splitting criteria can be saved as xml files and can be viewed, explored or compared at a later time without touching the training set. Desired performance statistics are available in a separate text file. We believe this is a useful functionality for students and teachers as visualization and interactiveness helps in better understanding the impact of variations in induction process. Since decision trees are known to be unstable classifiers [1], visual representation of trees aids understanding of the instability for practitioners.

5   VisTree: Generic Decision Tree Inducer and Visualizer

VisTree is a generic decision tree induction and visualization environment written in Java adapted from [18]. It can be established by a single click through jar file, which can be downloaded and installed [22]. It allows the users to induce different decision trees by varying splitting criteria, pruning criteria through a GUI and to visually/automatically compare the trees with respect to their accuracies and structures.

5.1   Technologies Used

The software has been built using open source technologies and is available for download [22]. It is developed using JDK 1.6 [15], NetBeans [16] and Prefuse [17].

The JDK (Java Development Kit) includes tools useful for developing and testing programs written in the Java programming language and running on the Java platform. These tools are designed to be used from the command line. Except for the applet viewer, these tools do not provide a graphical user interface[15].

NetBeans version 6.5 is used to build the Graphical User Interface (GUI) for the package. It refers to both a platform for the development of applications for the network and an integrated development environment (IDE). The NetBeans Platform provides a reliable and flexible application architecture that encourages sustainable development practices. Since the platform architecture is modular, it's easy to create applications that are robust and extensible [16].

Prefuse is a set of software tools for creating rich interactive data visualizations. The toolkit provides a visualization framework for the Java programming language. It also provides optimized data structures for tables, graphs, and trees, a host of layout and visual encoding techniques, and support for animation, dynamic queries, integrated search, and database connectivity. Prefuse is written in Java, using the Java 2D graphics library, and is easily integrated into Java Swing applications or web applets [17]. Prefuse is used for visualization in VisTree.

5.2   The Meta Algorithm

The core of VisTree package is a meta algorithm for decision tree induction. The algorithm takes as input the training set, meta parameters i.e. splitting criterion, pruning criterion and induces a tree. A choice of validation methods is available to the user. The induced tree is displayed in both rule format (as text) (shown in Fig. 10) and as an interactive tree in graphical form (shown in Fig. 11). Both are saved in separate files in addition to the performance statistics. The meta algorithm employed in VisTree is given below.

Algorithm 1. Tree Growing

Require: Training set S, input feature set A, target feature y, SplitCriterion, PruneCriterion, StoppingCriteria

Create a tree T with root node R.
SplitNode(R, S, A, y, SplitCriterion, StoppingCriterion)

return TreePruning(T, S ,y, PruningCriterion)

Interactive display of the induced decision tree is the most useful and attractive feature of VisTree. This is achieved by integrating Prefuse [17] with the core algorithm. In this mode, a click on a node shrinks or expands the tree, giving an opportunity to explore the structure of the tree, particularly if the induced tree is large. Internal node are attribute names (in capital letters) or the values(test) of(on) the attribute. In this way, a user can see entire tree in an easy way and can also move the tree to view a particular path from root to a leaf in one go. Fig. 11 shows induced decision tree for golf dataset. Search option in right most corner of the Prefuse window allows the user to search for a particular node. When user types a name of a node, it is highlighted. This allows for interactive exploration of the tree.

Algorithm 2. SplitNode

Require: Node R, Training set S, input feature set A, target feature y, SplitCriterion, StoppingCriterion

  if StoppingCriterion(S) **then**
    Mark node R as a leaf with the most common value of y in S as a label.
  else
    for all $a_j \in A$ **do**
      Find attribute a_j that obtains the best SplitCriterion(a_j,S).
    end for
    Label node R with a_j
    for all outcome v_i of a_j **do**
      Create a node R_i and connect node R to R_i with an edge that is labelled as v_i.
      SplitNode(R_i, ($\sigma_{a_j=v_i}S$), A, y, SplitCriterion, StoppingCriterion)
    end for
  end if

  return.

Algorithm 3. TreePruning

Require: Training set S, tree to be pruned T, target feature y, PruningCriterion.

  repeat
    Select a node t in T such that pruning it maximally improves PruningCriterion(T,y).
    if t $\neq \emptyset$ **then**
      T = pruned(T,t).
    end if
  until t $= \emptyset$

  return T

5.3　Input Data Format

Currently, the input training data is made available to the package by two files like C4.5 [19]. The ".names" file describes the attributes and the ".dat" file contains the actual data. The attribute description file contains description of the attributes in the following format:
```
Attribute_1 : type
Attribute_2 : type
......
```
where *type* is either continuous or categorical. It is recommended to specify attribute names in large cap. For categorical attributes, domain is specified preceded by a colon(:). For the golf dataset shown in Figure 1, "golf.names" file looks as shown below.
```
OUTLOOK : categorical : sunny overcast rain
TEMPERATURE : continuous
```

```
HUMIDITY : continuous
WINDY : categorical : false true
PLAY : categorical : Don'tPlay Play
```

The ".dat" file has the first line as list of attribute names separated by a white space, followed by a separator line. Each record in the training data set is a line with values separated by white spaces. The data in "golf.dat" file is in the following format.

```
OUTLOOK TEMPERATURE HUMIDITY WINDY PLAY
//****************separator ignored by the algorithm *************
sunny 85 85 false Don'tPlay
sunny 80 90 true Don'tPlay
overcast 83 78 false Play
rain 70 96 false Play
......
```

Fig. 6. VisTree: Initial working window

5.4   GUI for VisTree

The initial working window of VisTree is shown in Fig. 6. The command buttons appear on the left panel and are used to specify meta parameters. The file menu allows the user to convert the attribute file into VisTree ".names" format, load the data file, load and visualize a tree stored as xml file. Command buttons are activated after the file is loaded.

Fig. 7. VisTree: Splitting criteria window

Fig. 8. VisTree: Pruning criteria window

Fig. 9. VisTree: Validation methods window

Fig. 10. VisTree: Tree file

The four command buttons allow the user to specify meta parameters (splitting (Fig. 7) and pruning criterion (Fig. 8)) and desired validation method (Fig. 9). The "Start" button invokes the induction algorithm and the tree appears in a separate Prefuse window (Fig. 11). The performance statistics are shown in the named tab (Fig. 11). Three files are created in the same directory as that of dataset file: one each for the Prefuse view, rule format and performance statistics.

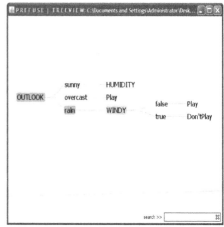

Fig. 11. VisTree: Prefuse window popping up after tree induction

6   Discussion and Conclusion

As mentioned earlier, VisTree is an on-going project and is yet to achieve it's goal-set completely. As a proof-of-concept, the basic functionality for interactive visualization of the induced tree, storage and reloading of the induced tree, variation of meta parameters has been implemented. At the time of writing this paper, VisTree provides three splitting criteria (information gain, gain ratio, gini index), three pruning criteria (cost complexity pruning, reduced error pruning, minimum error pruning) and four validation methods (bootstrap method, k-fold cross validation, leave one out method, hold out method). It also provides functionality to reload multiple trees saved as xml files simultaneously.

It is intended to extend these sets to enrich the quality of exploration that the package can offer. Further, it is planned to supplement the package with more validation methods. Performance statistics based on ROC space [6] are planned to be incorporated to make the package more useful for analytical studies. Addition of capabilities for automatic comparison of the saved trees is also envisaged. The ultimate goal is to make this package available as open source software for data mining community including students, teachers and practitioners.

Acknowledgement

This work was carried out with the support of research grant Dean(R)/R&D/ 2009/465, University of Delhi. We acknowledge with gratitude the review comments provided by the anonymous refree which helped us to improve the presentation of the paper. We also thank the authors of the freewares [15,16,17,18] that have been used for developing VisTree.

References

1. Dawyer, K.: Decision Tree Instability and Active Learning. MSc. thesis, Department of Computing Science, University of Alberta, Edmonton, AB, Canada, Spring (2007)
2. Han, J., Kamber, M.: Data Mining: Concepts and Techniques, 2nd edn. Morgan Kaufman publishers, San Francisco (2008)
3. Shawe-Taylor, J., Christianini, N.: Support Vector Machines and other kernel based learning methods. Cambridge University Press, Cambridge (2000)
4. Haykin, S.: Neural Networks: A Comprehensive Foundation, 2nd edn. Prentice-Hall, Englewood Cliffs (1999)
5. Mitchell, T.: Machine Learning. McGraw Hill, New York (1997)
6. Rokach, L., Maimon, O.: Data Mining with Decision trees: Theory and Applications. Series in Machine Perception and Artifical Intelligence, vol. 69. World Scientific Publishing, Singapore (2008)
7. Cios, K.J., Pedrycz, W., Swiniarski, R.W., Kurgan, L.A.: Data Mining: A Knowledge Discovery Approach. Springer, Heidelberg (2007)
8. Breiman, L., Friedman, J.H., Olshen, R.A., Stone, C.J.: Classification and Regression Trees, 1st edn. Taylor & Francis, Abington (1984)
9. Tan, P.-N., Steinbach, M., Kumar, V.: Introduction to Data Mining. Addison-Wesley, Reading (2005)
10. Duda, R.O., Hart, P.E., Stork, D.G.: Pattern Classification. Wiley, Chichester (1973)
11. Quinlan, J.R.: C4.5: Programs for Machine Learning. Morgan Kaufmann Series in Machine Learning (1992)
12. Quinlan, J.R.: Induction of decision trees. Machine Learning Archive 1(1), 81–106 (1986)
13. Mingers, J.: An Empirical Comparison of Pruning Methods for Decision Tree Induction. Machine Learning archive 4(2), 227–243 (1989)
14. Hall, M., Frank, E., Holmes, G., Pfahringer, B., Reutemann, P., Witten, I.H.: The WEKA Data Mining Software: An Update. SIGKDD Explorations 11(1) (2009)
15. http://java.sun.com/javase/downloads/index.jsp
16. http://netbeans.org/
17. http://prefuse.org/
18. http://imacwww.epfl.ch/Team/Raphael/BookWiley2003/java-illustrations/ID3Standard/ID3.java.html
19. http://www2.cs.uregina.ca/~dbd/cs831/notes/ml/dtrees/c4.5/tutorial.html
20. http://www.cs.waikato.ac.nz/ml/weka/
21. http://weka.wikispaces.com/Explorer+tree+visualization+plugins
22. Available on request. Please email to the corresponding author

Semantics Extraction from Social Computing: A Framework of Reputation Analysis on Buzz Marketing Sites

Takako Hashimoto[1] and Yukari Shirota[2]

[1] Chiba University of Commerce, 1-3-1 Konodai, Ichikawa, Chiba, Japan
[2] Gakushuin University, 1-5-1 Mejiro, Toshima-ku, Tokyo, Japan
takako@cuc.ac.jp,
yukari.shirota@gakushuin.ac.jp

Abstract. Social computing services, which enable people to easily communicate and effectively share the information through the Web, have rapidly spread recently. In the marketing research domain, buzz marketing sites as social computing services have become important in recognizing the reputation of products hold with users. This paper proposes a reputation analysis framework for the buzz marketing sites. Our framework consists of four steps: the first is to extract the topics of the product using natural language processing. The input data comprises consumer messages on buzz marketing sites. Next, important topics on the products are extracted. The third step is to detect emerging consumer needs by identifying new burst topics. Finally, the results are visualized. Based on our framework, product characteristics and emerging consumer needs are extracted and reputations are visualized.

Keywords: Web intelligence, Social Computing, Buzz Marketing, Data Mining.

1 Introduction

Social computing services like blogs, SNSs (social networking services) and buzz marketing sites, which enable people to easily communicate and effectively share the information through the Web, have rapidly spread. We can say that communications in social computing services have generated new consensuses and new intelligence. In buzz marketing sites especially, varied consumers write review messages about a product. They also add their comments on others' messages. These communications affect consumer behavior. Social computing services have become highly-influential in the marketing research domain. In this environment, reputation analysis from messages in social computing services has become significant.

The purpose of this paper is to propose a reputation analysis framework to extract product characteristics and analyze consumer needs from the messages on buzz marketing sites. Our system targets both consumers and marketing planners. It's an application which provides the potential use in marketing and

S. Kikuchi, S. Sachdeva, and S. Bhalla (Eds.): DNIS 2010, LNCS 5999, pp. 244–255, 2010.

an improvement of products by manufactures. Through our framework, they can identify product characteristics and recognize emerging consumer needs related to the products.

Our framework consists of four steps: the first is to extract the topics of the product using natural language processing. The input data comprises consumer messages on buzz marketing sites. Next, important topics on the products are extracted. The third step is to detect emerging consumer needs by identifying new burst topics. Finally, the results are visualized.

The remainder of the paper is structured as follows: Chapter 2 describes the research background and related work. Chapter 3 shows the preliminary survey results on buzz marketing sites. Chapter 4 proposes the reputation analysis framework on buzz marketing sites. Chapter 5 concludes the paper and sets a path for future work.

2   Background and Related Work

A reputation analysis is of special concern in the study of web intelligence. In the advanced networked society which provides different communications through social computing services, a reputation analysis from various information sources like blog and buzz marketing sites, has become a key technology. In general, the main purpose of a reputation analysis is positive/negative comments detectin and their visualization. User utilize the results from a reputation analysis for their decision-making.

There are several reputation analysis systems [1], [2], [3]. Existing services focus on the detection of positive/negative comments, as pointed out above. Our framework also takes a similar approach. In addition, we plan to forecast the emerging needs by finding the trend in the real world. That is to say, our proposed reputation analysis framework are aiming to achieve the more integrative technique.

To estimate the positive/negative degree of a certain message, there is a method to clarify sentences with the sentiment in the document [4]. We plan to use this technique to assess the positive/negative degree. A method to clarify the pros and cons of the sentences is also proposed [5]. This work uses personality characteristics to improve accuracy. Our framework is due to use similar personality characteristics, not only to detect the pros and cons, but also to extract other semantics.

To extract the reputation, several techniques are available to detect positive/negative degree using the dictionary. The key point is to construct an efficient dictionary. Kamps et al. proposed a method to calculate the distance between good and bad for the word [6]. A co-occurrence dictionary is also used to detect positive/negative degree [7], [8]. We also plan to use a co-occurrence dictionary to extract the topics and detect positive/negative degree more precisely. We will evaluate existing techniques to meet our purpose in the next phase of our research.

Regarding document clustering, a method to form the clusters according to the positive/negative degree is proposed [9]. We cluster by clarifying documents based on topics. Reputation analysis is then processed according to the positive/negative degree.

3   Survey of Buzz Marketing Sites

We begin by looking at messages on the buzz marketing sites. From among the buzz marketing sites on the Internet, for this paper, we chose kakaku.com [10] as our example of buzz marketing sites. The kakaku.com site is the most popular 'customer purchasing support site' in Japan. It provides price information on electrical appliances, vehicles, toys, and various other products. Buzz marketing sites and shopping malls are also provided. Around 20 million user accesses per month were recorded as of September 2009.

As a preliminary experiment, we surveyed messages on the buzz marketing sites of kakaku.com. Each site for individual products are opened in the buzz marketing sites on kakaku.com. Consumers communicate with each other by adding their messages on the site. We read these messages and classified them to confirm whether or not product characteristics and emerging consumer needs can be extracted.

3.1   Messages on Front Loading Washing Machines with Automatic Drying System

We first checked the messages on three models of front loading washing machines with automatic drying. Front loading washing machines with automatic drying have become popular with Japanese families in recent years. However, their large size and vibration noise have been problems, and prices are still high. Therefore, many messages from consumers appear on the buzz marketing sites. As a preliminary experiment, we did the following:

1. Read all messages on corresponding machines
2. Extract the main topic of each message
3. Extract the feature of focus in each topic
4. Count the number of messages in each topic

Table 1 lists the topics, the feature terms and the number of messages on the three types of the washing machines we selected. The following eight topics are abstracted according to the messages: *Installation, Noise, Cleaning performance, Dryer performance, Price, Comparison with other models, Troubles* and *Other*. For each topic, we also derived feature terms (Table 1).

Figure 1 shows the number of messages on each topic using a cobweb chart. For the overall products (Model A, Model B and Model C), users seem to talk about *Comparison with other models*. In addition, depending on the machine, the number of messages differs with each topic. For example, for the Model C, there are many messages about *Troubles*. This indicates that users actively talk

Table 1. Topics, feature terms, and the number of messages for front loading washing machines with automatic drying system (Messages were collected on Nov. 12, 2009)

Topic	Feature Terms	Products (appearance time in the market)		
		Model A (Nov. 2008)	Model B (Oct. 2008)	Model C (Jun. 2009)
Installation	Size, Space, Width, Height, Measurement, Self-install	82	50	9
Noise	Noise, Vibration	27	14	20
Cleaning performance	Damage, Rip, Discoloration, Fluff, Rejuvenation, Stain	43	21	56
Dryer performance	Odors, Wrinkles, Shrinking, Fabric Care, Speed	66	35	6
Price	Expenditure, Cheap, Stock, "Shop A", "Shop B", "Shop C"	48	89	44
Comparison with other models	"Company A", gCompany B", "Company C", Predecessor	96	126	37
Troubles	Error Signals, Bugs, No-good	30	45	105
Other		56	37	50
	Total	448	417	327

about Troubles on Model C. On the other hand, with the Model A, messages on *Installation* appear more often. Actually, the size of Model A is smaller than Model B and Model C. Users seems to be interested in Model A's easy installation. Model B has many messeages on *Price*. In fact, the price of Model B is slightly higher than Model A and Model C. Therefore, on Model B, users are concerned with the topic about price. It seems reasonable to suppose that these biases show the features of each machine and become a key for analyzing the product's reputation with consumers.

3.2   Messages on Electronic Air Cleaners

Table 2 lists the topics, feature terms and number of messages for the five types of electronic air cleaners we selected. Two types (Model G and Model H)came to

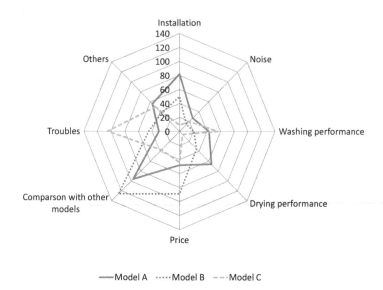

Fig. 1. Number of messages on each topic for corresponding products (Washing machines)

market at around the same time (in 2008) and the other three(Model D, Model E and Model F) entered about a year later (in 2009). Just as with the washing machines, we read all messages, classified the topics, extracted feature terms for each topic and manually counted the number of messages. Figure 2 shows the number of messages on each topic using a cobweb chart.

With the electronic air cleaners, topics such as *Installation, Noise, Cleaning performance, Price, Comparison with other models, Troubles* and *Other* appear just as with the washing machines. For the overall products (Model D, Model E, Model F, Model G and Model H), users seem to talk about *Prices*. In addition, for example, on Model G, there are many messages about *Troubles*. This indicates that users actively talk about troubles on Model G. Model F and Model H especially have many messages on *Price*. On Model F and Model H, users are concerned with the topic about price.

Beyond that, a topic related to *New influenza* appears for models that came to market in 2009. The reason for this burst on the new influenza seems to be that news of the new influenza rapidly grew in the spring of 2009. We believe that the detection of this kind of the bursty topic could be useful in identifying emerging consumer needs.

For burst detection, we propose a framework to check consumer messages periodically in the next chapter.

Table 2. Topics, feature terms, and the number of messages for electronic air cleaners (Messages were collected on Nov. 12, 2009)

Topic	Feature Terms	Products (appearance time in the market)				
		Model D (Sep. 2009)	Model E (Sep. 2009)	Model F (Sep. 2009)	Model G (Nov. 2008)	Model H (Sep. 2008)
Installation	Size, Space	4	0	1	0	0
Noise	Noise	8	4	3	2	0
Air cleaning performance	Humidification, , Setting, Ion	17	21	15	19	19
Price	Expenditure, Cheap, Stock, "Shop A", "Shop B"	25	22	31	8	33
Comparison with other models	"Company A", "Company B"	21	16	12	15	12
Troubles	Feed-water, Tank, Smell	8	0	11	10	26
New influenza	Virus, Prevention	19	23	14	0	0
Other		10	0	6	9	0
	Total	112	86	93	79	74

4 Framework for Reputation Analysis on Buzz Marketing Sites

This chapter discusses a reputation analysis framework to extract product characteristics and analyze emerging consumer needs from messages on buzz marketing sites.

Our framework consists of the following four steps:

1. Topic extraction
2. Important topic detection
3. Emerging needs detection
4. Visualization

The following sections describe each step.

4.1 Topic Extraction

In this step, input data is the set of messages about products, e.g., front loading washing machines on buzz marketing sites. We define one message as one

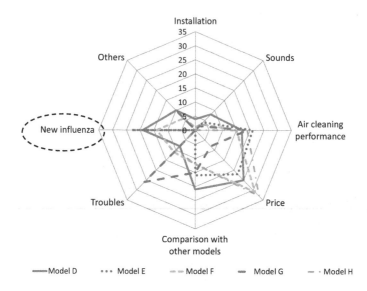

Fig. 2. Number of messages on each topic for corresponding products (Electronic Air Cleaners)

document and extract words using morphological analysis; we then calculate the tf-idf value for each word. We form the document clusters based on vectors of the tf-idf values using the cosine similarity [11]. Each cluster equals the topic of corresponding products. With topic extraction, we plan to use a concept-based co-occurrence dictionary [12] to improve the accuracy of document clustering.

4.2 Important Topic Detection

Important topics fall into two categories. One is important topics for the set of corresponding products, which express characteristics of the overall corresponding products. The other comprises important topics for individual products, which express the characteristics of each product. For example, consider washing machines, the topic related to *Comparison with other machines* is active for overall machines. For example, consider washing machines. The topic related to *Comparison with other machines* is active as a whole. On the other hand, *Troubles* is the bursty topic for Model C.

With reputation analysis, it is important to detect both categories of significant topics. We define the following three parameters to express the important topic for an individual product:

1. Contributing rate of each topic for a set of products P

$$CRPT_j = \frac{\displaystyle\sum_{i=1}^{l} m_{i,j}}{\displaystyle\sum_{j=1}^{n}\left(\sum_{i=1}^{l} m_{i,j}\right)}$$

$CRPT_j$ expresses the important topic on the products as the overall trend. For example, the topic *Comparison with other models* would be the most important topic, because the number of the messages for it was the biggest (Table 1).

Where l is the number of products for a set of products P. P consists of several models (for example, a set of electronic air cleaners). n is the number of topics for P. $m_{i,j}$ is the number of messages about a topic T_j on a product P_i. $CRPT_j$ is the contribution rate of T_j on P. If T_j is important for P, $\sum_{i=1}^{l} m_{i,j}$ tends to be high. That is to say, $CRPT_j$ will be high.

2. Contributing rate of each topic on individual product

$$CRP_iT_j = \frac{m_{i,j}}{\displaystyle\sum_{j=1}^{n} m_{i,j}}$$

CRP_iT_j is the contribution rate of T_j on P_i.
CRP_iT_j expresses the main topic for individual products. We assume that significant topics differ from product to product. For example, for the Model C, there are many messages about *Troubles* (Table 1). The topic related to *Troubles* may express the characteristics of Model C. As a result, CRP_iT_j becomes high. This indicates that *Troubles* is significant on Model C.

Topics with high contribution rates ($CRPT_j$ and CRP_iT_j) are detected as important topics.

3. Positive/negative degree of topic on individual products

To analyze reputation, it is important to identify the positive/negative degree of each topic for the product. For example, the topic related to *Troubles* is significant on Model C. The question is whether the messages about *Troubles* are positive or negative.
We plan to use a dictionary (ontology) consisting of words with positive/negative values and a method to derive the positive/negative degree. Turney et al. proposed the method to derive positive/negative degree according to the semantic orientation of the phrases in the review that contain adjectives

or adverbs[13]. And Takamura et al. provide the dictionary on semantic orientations (positive/negative values) of words through the Web[14]. Based on this dictionary and Turney's method, we will calculate the positive/negative degree as follows :

$$PNP_iT_j = \frac{1}{q}\sum_{k=1}^{q} f(P_i, T_j, w_k)$$
$$q = Q(P_i, T_j)$$
$$w = W(P_i, T_j)$$

Where PNP_iT_j is the positive/negative degree of T_j of P_i. q is the number of words in T_j of P_i. $Q(P_i, T_j)$ is the function to calculate the number of words of ontology in T_j of P_i. The list of ontology words w for T_j of P_i is derived from the function $W(P_i, T_j)$. $f(P_i, T_j, w_k)$ is the function to extract the positive/negative value for each word w_k appearing in T_j of P_i based on the dictionary on semantic orientations (positive/negative values) of words[14]. PNP_iT_j is the mean value of the positive/negative values of words derived from T_j . The positive/negative degree of T_j of P_i depends on the sign of PNP_iT_j. For example, with the Model A, there are actually many negative words about *Dryer performance*(Table 1). Therefore PNP_iT_j of *Dryer performance* for the Model A may be a negative number.

Topics with high positive/negative degrees (PNP_iT_j) are also detected as important topics.

4.3 Emerging Needs Detection

To detect emerging needs, we plan to find changes in topics. In our framework, the messages are acquired periodically (e.g., once a week), and then the document clusters are formed. If a new cluster is generated, we recognize that new changes (or new needs) have been detected. For example, the topic related to *New influenza* can be found for models that came to market in 2009 (Table 2). We can say that the new needs about the new influenza have appeared. We assume that these new needs will be propagated to other products. That is to say, the needs about the new influenza will appear in other product as the new needs in the near future as well.

4.4 Visualization

Based on important topic extraction and emerging needs detection, the results are visualized to users. Figure 1 and Figure 2 are examples of visualization. However, the visualization method has not yet been perfected. We continue to discuss visualization in detail.

4.5 System Structure

Figure 3 shows the system structure of our proposed framework. The system consists of the following four modules and one database: the topic extraction module,

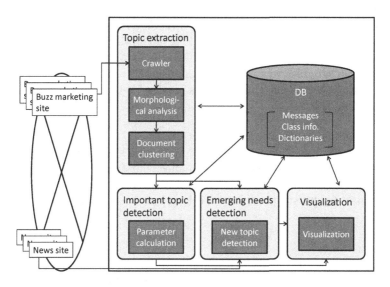

Fig. 3. System structure of our proposed framework

the important topic detection module, the emerging needs detection module, and the visualization module. The database has crawled messages, document class information, and the dictionaries. The topic extraction module acquires messages from the buzz marketing sites. It then extracts words using morphological analysis and forms the document clusters. To improve the accuracy of topic extraction, the module refers to the dictionaries in the database. The main process for the important topic detection module is to calculate the parameters we proposed in Chapter 3. Results of the calculation are stored on the database. The emerging needs detection module detects new burst topics. This module is executed concurrently with the important topic detection module. To precisely detect a new topic, the module refers to the news site and acquires candidates for the new topic. The visualization module visualizes the result of both the important topic detection module and the emerging needs detection module.

4.6 Problems with Reputation Analysis

Based on the above, we have found the following problems with reputation analysis.

1. Ontology provision:
   To improve the accuracy for topic extraction, it is necessary to prepare the high quality ontology. Furthermore, we have to consider the mechanism for ontology updating to detect emerging terms like the new influenza. Ontology generation and update are key issues with reputation analysis.

2. Credibility of the topic:
   With reputation analysis, credibility is the key issue. Even if there is a significant topic for a certain product, it is not always true. For example, the topic on *Troubles* is active in Model C (Table 1), but the main subject of the topic about troubles was misuse by the user, and many emotional messages appeared. In a case like this, it seems reasonable to suppose that credibility of the topic is low.

3. Burst products vs. non-burst products:
   The number of messages on the buzz marketing sites depends on the product. Of course, sales success seems to affect burstiness of the site. However, the degree of sentiment also seems to have an influence. We plan to analyze the reason for bursty products by taking into account emotional characteristics.

4. Roles in discussion:
   It is important to consider what kind of roles should appear in the discussion. Regarding the bursty discussion, both the agitator and the adversary are important. The follower seems to be significant as well. For reputation analysis, we have to look at the roles in the discussion.

5   Conclusion

In this paper, we proposed a framework for reputation analysis on buzz marketing sites. The purpose of our framework is to extract product characteristics and analyze consumer needs. Our framework extracts topics for the products using natural language processing of consumer messages. Important topics for the product are then detected. Emerging consumer needs are also detected. Finally, the results are visualized for the user. Our next step is to experiment with evaluating the effectiveness of our framework. Based on results, we will then improve the framework. At the same time, we will work on solving other issues, such as ontology provision, message credibility, burst discussion detection, and role detection on the buzz marketing site.

References

1. Morinaga, S., Yamanishi, K., Tateishi, K., Fukushima, T.: Mining product reputation on the web. In: Proceedings of the 8th ACM SIGKDD Intenational Conference on Knowledge Discovery and Data Mining (KDD 2002), pp. 341–349. ACM, New York (2002)
2. Yi, J., Niblack, W.: Sentiment mining in webfoutain. In: Proceedings of the 21st International Conference on Data Engineering (ICDE 2005), pp. 1073–1083 (2005)
3. Dave, K., Laurence, S., Pennock, D.M.: Mining the peanut gallery: Opinion extraction and semantic cllasification of product reviews. In: Proceedings of the 12th International World Wide Web Conference (WWW 2003), pp. 519–528. ACM, New York (2003)

4. Wiebe, J., Wilson, T., Bruce, R., Bell, M., Martin, M.: Learning subjective language. Computational Linguistics 30(3) (2004)
5. Gallery, M., Mckeown, K., Hirshberg, J., Shriberg, E.: Identifying agreement and disagreement in conversational speech; Use of bayesian networks to model pragmatic dependencies. In: Proceedings of 42nd Meeting of the Association for Computational Linguistics (ACL 2004), pp. 669–676 (2004)
6. Kamps, J., Marx, M., Mokken, R.J., Rijke de, M.: Using wordnet to measure semantic orientations of adjectives. In: Proceedings of the 4th International Conference on Language Resources and Evaluation, LREC 2004 (2004)
7. Turney, P.D.: Thumbs up? Thumbs down? Semantic orientation applied to unsupervised classification of reviews. In: Proceedings of the 40th Annual Meeting of the Association for Computational Linguistics (ACL 2002), pp. 417–424 (2002)
8. Nasugawa, T., Kanayama, H.: Acquisition of Sentiment Lexicon by Using Context Coherence. In: IPSJ SIG Notes, pp.109–116 (2004) (in Japanese)
9. Kaji, N., Kitsuregawa, M.: Dependency-based Probabilistic Model for Sentiment Classification. In: Proceedings of Data Engineering Workshop 2006 (2006)
10. Kakaku.com, http://corporate.kakaku.com/en/
11. Oguma, J., Utsumi, A.: Document clustering that uses co-occurrence information on word. In: Proceedings of the Annual Conference on JSAI (CD-ROM) (2007)
12. Arita, I., Kikuchi, H., Shirai, K.: Word Clustering Using Concurrent Search Queries. In: IPSJ SIG Notes, pp. 115–120 (2007)
13. Turney, P.D., Littman, M.L.: Unsupervised learning of semantic orientation from a hundred-billion-word corpus: NRC Technical report ERB-1094, Institute for Information Technology,11 pages (2002)
14. Takamura, H., Inui, T., Okumura, M.: Extracting Semantic Orientations of Words using Spin Model. In: Proceedings of the 43rd Annual Meeting of the Association for Computational Linguistics (ACL 2005), pp. 133–140 (2005)

An Approach to Extract Special Skills to Improve the Performance of Resume Selection

Sumit Maheshwari, Abhishek Sainani, and P. Krishna Reddy

Center for Data Engineering,
International Institute of Information
Technology, Hyderabad, (IIIT-H),
Hyderabad, Andhra Pradesh, India - 500032.
pkreddy@iiit.ac.in

Abstract. In the Internet era, the enterprises and companies receive thousands of resumes from the job seekers. Currently available filtering techniques and search services help the recruiters to filter thousands of resumes to few hundred potential ones. Since these filtered resumes are similar to each other, it is difficult to identify the potential resumes by examining each resume. We are investigating the issues related to the development of approaches to improve the performance of resume selection process. We have extended the notion of special features and proposed an approach to identify resumes with special skill information. In the literature, the notion of special features have been applied to improve the process of product selection in E-commerce environment. However, extending the notion of special features for the development of approach to process resumes is a complex task as resumes contain unformatted text or semi-formatted text. In this paper, we have proposed an approach by considering only skills related information of the resumes. The experimental results on the real world data-set of resumes show that the proposed approach has the potential to improve the process of resume selection.

Keywords: Special features, Resume selection.

1 Introduction

In the Internet era, large number of resumes are received on-line, through e-mails or through services provided by companies like Info Edge Limited [3]. For companies, it is a difficult and time consuming process to select the appropriate resume from such a large number of resumes. Research efforts [6][11] are going on to develop the methods for improving the performance of resume selection process.

Normally, resumes share document-level hierarchical contextual structure where the related information units usually occur in the same textual block and text blocks of different information categories usually occur in relatively fixed order [11]. One can observe the hierarchical structure in the resumes. The first layer consists of different sections such as education, experience, skills etc. and the second layer consists of text about corresponding sections.

S. Kikuchi, S. Sachdeva, and S. Bhalla (Eds.): DNIS 2010, LNCS 5999, pp. 256–273, 2010.

Table 1. Sample resume with corresponding sections and their respective features

Education
1. b.tech. (computer science & engineering) iiit, hyderabad (expected may, 2009) 6.66/10 cgpa.
2. senior secondary instrumental school, kota (cbse board 2004) 72%.
3. secondary st. sr. sec. school, ajmer (cbse board 2002) 83%.

Skills
1. programming languages: c, c++
2. operating systems: windows 98/2000/xp, gnu/linux
4. scripting languages: shell, python
5. web technologies: html, cgi, php
6. other tools: microsoft office, latex, gnu/gcc, visual studio 2005/08
7. database technologies: mysql

Experience
1. audio-video conferencing over ip networks:
2. duration: nov. 2007 nov. 2008 team size: 2. technical environment: c++ abstract: the objective of this project was to develop an audio/video conferencing system which enables multiple users to communicate with each other via a global server with improved efficiency in terms of voice clarity and low latency. the system is equipped with resources to facilitate text chat, voice chat and voice/video chat between multiple clients. this client server application was developed using c++ and .net framework in windows environment.
3. windows firewall
4. duration: july-nov 2007 team size: 1 technical environment: c abstract: packets from or to a network are analyzed and according to the users settings actions are taken on how the packets would be handled. various options are provided to the user in accordance to which action is taken ranging from what the packet contains to the source of the packet.
5. document request form automation
6. duration: sep-nov 2006 team size: 2 technical environment: php, mysql abstract: project developed for iiit hyderabad administration. this web-based tool automates the processing of the various documents needed by students.
7. implementation of outer loop join
8. duration: jan-march 2007 team size: 1 abstract: implementation of the above operation as a part of the database management systems course.
9. myshell
10. duration: aug-oct 2006 team size: 1 abstract: developed a program which acted as a shell, starting and running command line arguments as part of our operating systems course.
11. other studies and presentations
12. analysis of animation video viewing
13. using an eye tracker to track which point on the screen were the viewers focusing on, i analyzed various trends in animation video viewing. this study was done as a part of cognitive course.
14. case study in software design
15. i was a part of a six member group involved in the thorough analysis of a software design problem and the task of coming up with a solution pertaining to the problem. this was done as a part of software engineering course.

Achievements
1. secured 1573 air in all india engineering entrance examination, 2005.
2. secured 2216 air in iit-jee screening examination, 2005
3. cleared national talent search examination level 1 in 2002.
4. was among the finalists of the rajasthan state science talent search

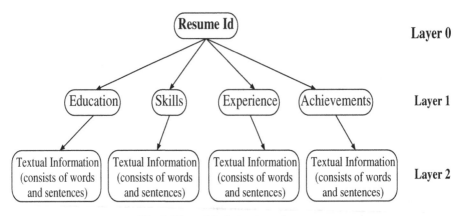

Fig. 1. Hierarchical structure of resume

Table 1 shows the sample resume. Each resume contains different sections and each section contains words and sentences as features. The numbering in each section denotes a feature separated by a delimiter ('newline' in our case). Figure 1 shows the corresponding hierarchical structure for Table 1. The top layer which is termed as "Layer 0" contains resume id. It can be observed that sections like education, experience, skills and achievements form the first layer of the resume. Each section is described by the text containing words and sentences which forms the second layer of the resume. Based on the structure of the content, the text of each section in the second layer can be organized into several layers.

The large number of resumes are reduced to few hundred potential ones based on some filtering techniques or search services [1] [2] [3]. The set of resumes hence obtained are similar to each other as they satisfy the search criteria or requirements for a company. In the current scenario, it is necessary to manually analyze each resume to select appropriate resumes. We define this problem as 'Problem of resume selection from set of similar resumes'. The research issue here is developing an improved approach which could help in selecting appropriate resume by processing similar resumes.

In this paper, we have made an effort to propose an improved approach based on the notion of special information. We consider that there may exist special information in some resumes as compared to others. For example, a resume may contain specialty in education, specialty in experience, specialty in skills or specialty in achievements and so on. Special information may exist in one or more sections of a resume. We assume that identifying such special information and organizing them efficiently enhance the performance of the resume selection process.

In the literature, an approach has been proposed to identify special features to improve the process of product selection in E-commerce environment [9]. They have exploited the fact that every product possesses some specialness, which is exhibited through few special features. Their approach identifies the special features and organizes the features of the product in an effective manner. In this paper we have extended the notion of special features to improve the process of resume selection. However, resume selection process is a more complex problem than product selection process because:

(i) It was observed that the features are standardized in the product selection scenario whereas in case of resumes, features are free-form text which is difficult to process.

(ii) Resume has a hierarchical structure as compared to the product feature descriptions having single layered structure. Each resume contain several sections, and each section contains different types of text. For example, experience section contains long sentences, skills section contains skill type (programming languages) and skill values (c++, java).

So, development of an approach to process the resume dataset is a complex task, as separate approach has to be developed for identifying special information in each type of text and integrating the same appropriately. In this paper, we have proposed an approach to improve the performance of resume selection by considering only the skill related information of the resumes. The development of approach to extract information from other sections of the resume is a part of the future work.

We have conducted experiments on the real world data-set of resumes and the results show that the proposed approach has the potential to improve the process of resume selection.

The rest of paper is organized as follows. In Section 2, we discuss the related work. In Section 3, we explain the notion of "special features" used in the product selection framework [9]. In Section 4, we explain the basic idea of the proposed approach and discuss the algorithm. In Section 5, Experimental results are presented. Discussion is presented in Section 6. Summary and Conclusion are provided in Section 7.

2   Related Work

In [6], a toolkit named as "Learning Pinocchio $(LP)^2$", was applied on resumes to learn Information Extraction rules from resumes written in English. The information identified in their task includes a flat structure of Name, Street, City, Province, Email, Telephone, Fax and Zip code. Learning Pinocchio is an adaptive system for IE, based on a kind of transformation based like rule learning. Rules are learned by generalizing over a set of examples marked via XML tags in a training corpus.

An approach has been proposed in [11] for resume information extraction to support automatic resume management and routing. A cascaded information extraction (IE) framework is designed. In the first pass, a resume is segmented into consecutive blocks attached with labels indicating the information types. Then, in the second pass, the detailed information, such as Name and Address, are identified in certain blocks (e.g. blocks labelled with Personal Information), instead of searching in the entire resume. Based on the requirements of an ongoing recruitment management system which incorporates database construction with IE technologies and resume recommendation (routing), general information fields like Personal Information, Education etc. are defined.

In [9], the problem of 'selecting a product from a group of similar products in E-commerce environment' faced by the customer is investigated. It proposes an improved approach to help a customer select the appropriate product by exploiting the fact that every product possesses some specialness, which is exhibited through few special features. The approach identifies the special features for each product and organizes the features of the products in an effective manner.

In Web mining research area, efforts are being made to identify outliers in a given set of web documents. Web content outliers are documents with varying content compared to similar Web documents taken from the same domain [4]. These approaches concentrate more on classifying objects as outliers rather than mining specialness of each object.

In [8], an effort has been made to discover unexpected information from the competitor's site. It compares the user's Web pages with those of the competitors and finds several kinds of unexpected information.

In data mining research, there have been efforts to mine outliers in numerical dataset. In [5], an approach has been proposed in which each object is assigned a degree of being an outlier, which is called local outlier factor. It is local as the degree depends on how isolated the object is with respect to the surrounding neighborhood. In [7], an approach has been proposed to mine the distance based outliers. The notion of K-nearest neighbor has been used to identify outliers in [10].

In this paper we have made an effort to extend the notion of special features to improve the process of resume selection. It can be observed that the work done on resumes mostly focuses on information extraction of resume or building a classifier to extract the information from resume and storing it in structured manner. Discovering unexpected information approach does not make an effort to exploit the special properties of each object within the whole set of pages including competitors and users set of pages. The outlier algorithms discussed in data mining area deals with numerical data-sets and have not applied in case of Web documents. The outlier algorithms proposed in Web mining area concentrates more on classifying objects as an outlier, rather than mining specialness of each object. None of the above approach tries to extract special information from the given set of resumes.

3   Identification and Organization of Special Feature Knowledge

In this section, we explain the basic framework related to the notion of special features as discussed in [9]. The framework consists of two main approaches: extraction of special features and organization of special features.

3.1   Extraction of Special Features

Before explaining the notion of special features we define the term 'degree of specialness'.

Degree of specialness: Let O be the set of 'n' similar objects, where object 'o_i' $\in$ O and each object 'o_i' possesses set of features '$f(o_i)$'. Let 'F' be set all features such that $F = \cup_{i=1}^{n} f(o_i)$. Each feature in F is denoted by f_j where $0 \leq j \leq |F|$ and $n(f_j)$ denote the number of objects to which feature f_j belongs to. Note that, the set F is a multi-set where each feature is represented as a tuple of $< resumeid, feature >$ and two different tuple may contain the same feature more than once since one feature can belong to multiple objects.

Let f_j be a feature, such that $f_j \in f(o_i)$. The degree of specialness (DS) of a feature f_j is its capability of making the object o_i separate/distinct/unique/special from other objects. The DS value for a feature varies between zero to one (both inclusive). The DS value of the feature f_j is denoted by DS(f_j). Then,

$$DS(f_j) = \begin{cases} 1 & \text{if } n(f_j) = 1 \\ 1 - (n(f_j)/|O|) & \text{otherwise} \end{cases} \tag{1}$$

About special features: Based on the DS values of features, features can be classified as common features, common cluster features and special features. The features that are occurring in all the objects have their DS value as '0' and are called common features. The features that are occurring in very few objects have their DS value close to 1 and are called **special features**. The other features are called common cluster features.

3.2   Organization of Features

After assigning the DS values to all the features in the data-set, the next issue is to organize the features in an effective manner by taking into account the corresponding DS values. On the basis of DS values, features can be categorized into one of the three categories: common features, common cluster features and special features.

Three-Level Feature Organization Approach: The features are distributed into three levels: I-level, II-level and the III-level. The I-level contains the common features, II-level contains common cluster features and III-level contains special features. It can be noted that, for any object o_i, its complete set of features $f(o_i)$ is a combination of (i) the common features at I-level (ii) common cluster features for the cluster in which o_i is a member and (iii) special features of object o_i at III-level.

Figure 2 depicts the organization of four similar objects o_1, o_2, o_3 and o_4, where each object has some set of features. The I-level at the top shows the common features present in all the objects. The II-level shows the common features in each cluster. o_1 and o_2 form one cluster and similarly o_3 and o_4 form another cluster. The III-level shows the special features for each object. The complete set of features for object o_1 is combination of 'common features of all the objects' present in I-level, 'common cluster feature of the cluster containing o_1 and o_2' in II-level and 'special features of o_1' present in III-level.

The procedure to organize the features using three-level approach is given in [9]. Here, we provide a summary for the three-level approach. It is a clustering algorithm that takes the set of objects O, similarity threshold (ST) and feature set F as input and returns common features, common cluster features and special features for each object with formation of clusters as an intermediate step. The similarity measure used in clustering algorithm is described below.

Common Features of all the objects (I-level)		
(II-level) Common Cluster Features	Object Identifier	**(III-level)** Special Features
Common features of O1 and O2	O1	special features of O1
	O2	special features of O2
Common features of O3 and O4	O3	special features of O3
	O4	special features of O4

Fig. 2. Three level representation of resume skill features

The similarity between the objects o_i and CL(i) (where CL(i) denotes the i'th cluster) is denoted by sim(o_i, CL(i)) and is calculated as follows:

$$sim(o_i, CL(i)) = |f(o_i) \bigcap CF(i)|$$

where CF(i) represent the features that are common among all the objects present in the cluster CL(i) and $f(o_i)$ denotes the features present in object o_i.

The approach contains two parts. The summary of the first part is as follows. The first cluster is initialized with some object. Next, the following step is repeated for each object: For each other object o_j, if the similarity of o_j with the existing cluster or clusters is greater than similarity threshold, the object o_j is put into into the cluster with maximum similarity; Otherwise, new cluster is initialized with o_j.

In the second part, the features of each cluster are organized into three-levels. The I-level contains the features of all clusters with DS value as '0'. The II-level contains the common features of each cluster. The III-level contains the remaining special features of each object.

About setting similarity threshold (ST) value: Organizing the features using three-level method is an iterative process. The value of ST should be chosen such that the objects are clustered into a reasonable number of clusters and the number of features shown to user can be reduced. The objective of clustering the objects is to reduce the effort of users by providing them with more convenient view and also less number of features. In case of large number of clusters, it leads to more confusion. Finally, we can set the ST to particular value which gives minimum number of clusters, and minimum number of features to be shown to user. For example, ST could be chosen as fifty percent of the average number of features in an object eliminating the common features. The threshold can be gradually increased, and the number of clusters formed and total number of features shown to user can be observed. If the number of features to be shown decreases significantly, we can increase the threshold and check the same. It can be observed that if the threshold is decreased, the number of common features for each cluster would decrease and consequently number of clusters shown to user would be increased.

4   Proposed Approach

In this section, after explaining the problem definition, we will discuss the basic idea and the proposed approach. We also discuss the overall framework and options for using the proposed framework.

4.1   Problem Definition

It is a difficult and time consuming process to select appropriate resume from a large set of resumes. Currently available filtering techniques or search services [1] [2] [3] filter thousands of resumes to few hundred potential ones. Since these filtered resumes are similar to each other, examining of each resume becomes essential to know about the potential candidates. The problem is defined as follows: Given a set of similar resumes, develop a methodology to help the enterprises/recruiters to improve the performance of resume processing.

We define 'similar resumes' as a set of resumes that are produced as result after filtering through the enterprise's resume management systems. Similar resumes consist of resumes with same experience or applying for the same job. The input to the proposed approach is a group of similar resumes and the output is an organization of resumes based on their specialness.

4.2   Basic Idea

The problem here is to develop an approach to select the appropriate resumes efficiently. It can be observed that there are some common features which are present in all the resumes in the group and also each resume may possess some special features that could differentiate it from rest of the resumes in the group. The intuition here is that if the special features of each resume are identified, the time required to make a decision for selecting an appropriate resume would be reduced in comparison to the time required by considering all the information.

Normally, each resume is described by a text document and the text in the resume is divided into different sections. A special information of a resume implies special information in each section of a resume. For example, there could be specialty in skills, specialty in achievements, specialty in education etc. The problem here is to identify the special information from each resume. Each section contains different types of text. For example experience section contains long sentences, skills section contains skill type (programming languages) and skill values (c++, java), the development of an approach to process the resume dataset is a complex task, as separate approach have to be developed by identifying special information for each type of text and integrate the same appropriately.

In this paper, we explored only the skill section and develop an approach to identify resumes with 'special skills'. The development of approach to identify special information in other sections of the resume is beyond the scope of this paper and would be considered as a part of the future work.

The main issue here is how to measure the specialness of text in skills section for all the resumes. We extend the notion of special features to propose an effective approach for the resume selection problem.

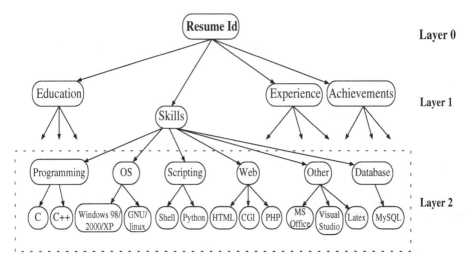

Fig. 3. Hierarchical structure of resume corresponding to Table 1

Table 2. Sample features for skills section

Skill Type features	Skill Values features
1. programming languages	c, c++,java.
2. programming languages	c, c++ java.
3. scripting languages	python, perl, shell
4. scripting languages	python perl shell
5. libraries	opengl, sdl
6. database technologies	mysql mssql
7. operating systems	linux, windows

4.3   Description of Proposed Approach

If we apply the notion of special features on skills section directly the comparison between the features would not be effective. It can be observed that the skill section information (refer Table 1) contains enumerated sequence of text pieces. Each text piece consists of a skill type and its skill values. For example, 'programming languages' is a skill type and 'c, c++, java' are skill values. So, there is a two layer organization in the skill information as shown in Figure 3. The skill information itself forms an hierarchy where skill types form one layer and skill values form another layer. We exploit the inherent organization in the skills information.

We propose an approach by considering that the skill information in the resume is organized into "skill type" and their corresponding "skill values".

Overall the proposed approach consists of a number of steps. First we perform pre-processing on the skill information of the resumes. Then we extract the skill type and skill value features. After extracting the skill type and skill value features, we calculate DS values of the features and organize them.

There are two types of features that could be extracted from skills information. One type is Skill-Type-Feature-Set (STFS) and another is Skill-Value-Feature-Set (SVFSs). Each element in STFS is a two attribute tuple $< ResumeId, Skilltype >$ and each element in SVFS is defined as $< ResumeId, Skilltype, Skillvalue >$. Note that, for a given resume, there exists one STFS and several SVFSs. For the same skill type, same skill values exist, but in different form. For example, in Table 2, the skill values for skill type 'programming languages' are same, except the presence of some special characters (comma in this case). Thus, direct comparison cannot be done. So, both STFS and SVFSs are formed after carrying out the preprocessing steps and then applying the described algorithm (refer Table 3) on the skills information.

Extracting skill types and skill values: The algorithm to extract skill types and skill values is divided into two parts. In the first step, we apply the preprocessing and in the second step we apply an algorithm described in Table 3 to identify the skill type and skill value features. The preprocessing steps are as follows:

   (i) Entire input text is converted to lower case and special characters are removed.
  (ii) Stop words [1] occurring in general purpose stop words list are removed.
 (iii) The skills section in resume is identified with the keyword 'Skills' in the heading irrespective of the position of the Skills section in the resume.
 (iv) The skill type and its skill value(s) are identified and separately stored using a delimiter (: in our case).
  (v) Skill value(s) corresponding to each skill type are sorted lexically and separated by a comma (,). For a skill value having more than one word, the words are concatenated. For example, the skill values for skill feature 'database technologies: ms sql, postgres sql, mysql' would be changed to skill value string: 'mssql, mysql, postgressql'.
 (vi) To resolve human errors like spelling mistakes, typo errors etc., we define a data structure called 'skill values list' with 'skill type' as a hash key and its possible 'skill values' as its values. Each skill value is checked in the skill values list. In case of many partial matches, the skill value is replaced by the skill value from the list with which it has the longest match. In case of no match, the list is manually updated with the skill value after verification. The possible skill values are extracted from the resume dataset.
 (vii) A skill value can have more than one name referring to it. For example mssql and microsoft sql refers to same skill. To resolve such ambiguity we identify the various possible ways of redundant occurrences through data analysis and prepare a hash table with the canonical names as the hash key and various possible names as a list of hash values corresponding to the canonical name. All these different names should be replaced by a common name or canonical name to resolve this issue.

The description of algorithm shown in Table 3 is as follows. The input to the second part of the algorithm is set R consisting of 'n' resumes, dictionary S that contains all

---

[1] Stop words is the name given to words which are filtered out prior to, or after, processing of natural language data (text).

the distinct skill types present in the set R and $|S|$ denotes the cardinality of dictionary S. The output consists of the skill type feature set (STFS) and skill value feature set (SVFS). In STFS each element is a tuple consisting of resume identifier and skill type as its attributes whereas in SVFS each element is a tuple consisting of resume identifier, skill type and skill value. The steps of the algorithm are as follows: We take each resume and repeat the following steps for each resume. (i) Identify the skill section of the resume using the 'Skills' tag. (ii) Process each line of the skill section to identify the skill type and corresponding skill value. (iii) The resume id (r_i) and skill type is stored in $STFS_i$ index of the array of SVFS where as resume id (r_i), skill type (s_j) and skill value is stored are the index $SVFS_{ij}$ of the array SVFS.

Thus after performing the preprocessing steps and applying the above described algorithm we get STFS and SVFSs. The next task is to calculate the specialness values of all the features in STFS and SVFSs and organize the same.

Table 3. Algorithm to calculate STFS and SVFS

Input: R: Set of 'n' resumes; F: set of features for all 'n' resumes; S: dictionary for all the skill types and $|S|$ is number of distinct skill types **Output:** STFS and SVFS

1. Notations used:
    i, j: integers;
    S_{r_i}: skills information for resume r_i
    $STFS_i$: array for skill types for resume r_i,
            where each tuple contain $< r_i, Skilltype >$
    $SVFS_{ij}$: array of skill values for resume r_i and skill type s_j,
            where each tuple contain $< r_i, s_j, Skillvalue >$
2. **for** i=1 to n
3.   Get the skill section features for resume r_i in S_{r_i}
4.   **for** each s_j in S
5.   **if** $s_j \in S_{r_i}$
6.   store the tuple $< r_i, s_j >$ in $STFS_i$
7.   store the tuple $< r_i, s_j, skillvalue >$ in $SVFS_{ij}$
8.   end
9. end

Calculating DS Value and Organizing the Special Skill Types: Given the STFS, the problem is to identify the specialness value and then on the basis of specialness value organize all the features in the set.

Computing Specialness Value for STFS: Let R be a set of 'n' similar resumes, where resume $r_i \in R$. Each resume r_i possess some set of features. Let f(r_i) be set of skill type features for resume r_i. Let F be set of all skill type features for all resumes such that F = $\cup_{i=1}^n f(r_i)$. Each feature F is denoted by f_j where $0 \leq j \leq |F|$ and n(f_j) denote the number of resumes to which feature f_j belongs. The DS value for each feature in STFS is calculated as defined in Equation 1. The input consists of the feature set F and output consists of Feature set F along with the DS values for all the features in the set.

Organization of STFS: We apply the above described three-level organization algorithm on the features in STFS and organize the features as shown in Figure 2. The input to the algorithm consists of feature set F which contains all the features in STFS along with their DS values, threshold ST and set of resumes R and the output of the algorithm consists of three-level organization of STFS features.

Calculating DS Value and Organizing Special Skill Values: Given the skill types and SVFS, the problem is to identify the specialness value and then on the basis of specialness value organize all the features.

Computing Specialness Value for SVFSs: Let S be a set containing distinct skill type features from all the resumes and $s_j \in S$ denotes a particular skill type. Let $f(s_{ij})$ denotes the skill value features for skill type $s_j \in S$ and resume $r_i \in R$. Let $F(s_j)$ be set of all skill value features for skill type s_j for all the resumes such that $F(s_j) = \cup_{i=1}^{n} f(s_{ij})$. The DS value of each feature in SVFSs is calculated using Equation 1. The input consists of the feature set $F(s_j)$ for all $s_j \in S$ and output consists of Feature sets along with the DS values for all the features for each of skill type.

Organization of SVFSs Features: We apply the above described three-level organization algorithm on the features in SVFSs and organize the features as shown in Figure 2. The three-level organization algorithm is run for each distinct skill type $s_j \in S$. The input to the algorithm consists of feature set $F(s_j)$ that consists of features for skill type s_j along with their DS values, threshold ST and set of resumes R and the output of the algorithm consists of three-level organization of skill value features for each skill type s_j.

4.4 Overall Framework

In this section we explain the overall framework. The input to the proposed approach are the resumes stored as text documents where each document contains different sections along with their descriptions (refer Table 1). The steps of proposed framework are discussed below (refer Figure 4).

1. **Identification of features from skills section:** We extract Skill Type Feature Set and Skill value Feature set from skills information for all the resumes.
   - **Identifying Skill Type Feature Set (STFS):** We identify the skill type features for all the resumes and form an STFS.
   - **Identifying Skill Value Feature Set (SVFSs):** We identify the skill value features for each of the skill type and form SVFSs for all the skill types.
2. **Calculating DS Value and Organizing Special Skill Type Features:** We compute the DS value for skill type features on the basis of DS values we organize the skill type features.
   - **Computing DS Value for Skill Type Features:** We compute the DS value for skill type features based on the notion of degree of specialness as defined in Equation 1.
   - **Organization of Skill Type Features:** We organize the skill type features using the three-level organization approach described in Section 3.2.

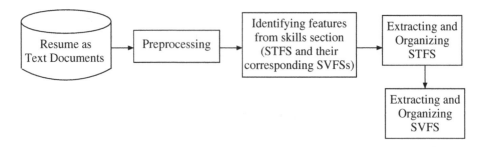

Fig. 4. Flow diagram of the overall framework

3. **Calculating DS Value and Organizing Special Skill Value Features:** We compute the DS value for skill value features for each of skill type and on the basis of DS values we organize the skill value features for each of the skill type.
   - **Computing DS Value for Skill Value Features:** We compute the DS value for skill value features for each of skill type based on the notion of degree of specialness defined in Equation 1.
   - **Organization of Skill Value Features:** We organize the skill value features for each of skill type using the three-level organization approach described in Section 3.2.

4.5   About Using the Proposed Framework

The user can request the system to organize the resumes according to skill type or skill values for the selected skill type. The resumes are organized firstly on the basis of skill type forming the first layer and second layer consists of tables for skill value of each skill type. Now, if the recruiter wants to select a resume only on the basis of skill type, he/she can give only the skill type information as input and browse through the output containing the special skill type information only. And if he/she wants to select a resume based on the skill value for a respective skill type he/she can do so by giving the skill type and skill value information as input and browsing through the special skill types and then special skill values for a respective skill type.

5   Experimental Results

To evaluate the performance, we have applied the proposed framework on real world data-set of resumes. Data-set contains 100 resumes from undergraduate students of computer science department in a University. All the resumes are available in the same format as shown in Table 1. Total number of features in skills section were 643. The skill types present in the data-set are 'programming languages', 'scripting languages', 'operating systems', 'web technologies', 'database technologies', 'libraries', 'other tools', 'compiler tools', 'mobile platforms' and 'middleware technologies'.

We define the performance metric called 'reduction factor' (rf) to measure the performance improvement. The rf denotes the reduction in the number of features that the recruiter needs to browse to select a resume from set of 'n' similar resumes.

Table 4. Organization of skill type features

Common Features (I-level)		
programming languages, operating systems, web technologies, database technologies		
Common Cluster Features (II-level)	**Resume Identifier**	**Special Features (III-level)**
other tools, scripting languages	78	libraries, mobile platforms middleware technologies
	83, 13, 91, 114, 112, 52, 67, 54, 15, 108, 105, 109, 107	compilers, mobile platforms
	25, 5	libraries, mobile platforms
	36, 77, 15	Compiler tools
	43, 44, 70, 71, 76, 69, 40, 39, 50, 57, 62, 59, 101, 95, 65, 68, 84, 90, 110 88, 19, 113, 6, 3, 32, 92, 29, 104, 93	libraries
	9, 98, 86, 73, 79, 8, 66, 28, 115, 103 106, 11, 37, 24, 53, 55, 47, 1, 41	None
libraries, other tools	63, 35	compiler tools
	89, 94, 17, 49	None
scripting languages, libraries	75, 96, 45, 2, 58, 33, 14, 10, 51	None
other tools	82, 27, 99, 38	None
scripting languages	111, 4, 22, 23, 21, 46, 42, 16	None

Table 5. Reduction Factor value for STFS

Feature Type	$|F|$	$\sum_{i=1}^{L} F(i)$	rf
Skill type	643	79	0.88

Let 'F' denote the total number of features for all the resumes, F(i) denote the number of features in 'i'-level and 'L' denotes the number of levels. The 'rf' is defined as,

$$rf = 1 - \frac{\sum_{i=1}^{L} F(i)}{F}$$

Results for Skill Type Features (STFS): The Table 4 shows the reduction factor for skill type features. It can be seen that rf value comes to 88%. The results indicate that 88% reduction in the effort could be achieved in resume selection process. The total numbers of skill types features present were 643 and numbers of features being displayed to user are only 79. The Table 5 shows the organization of corresponding skill type features (set F) using three-level approach. The I-level shows the common skill type, the II-level shows the common skill types for each cluster of resumes and the III-level shows the special skill types for each resume. The resumes can be classified based on their skill type in one click. Since number of resumes share same special feature we have mentioned them in same row separated by delimiter comma (,) for user convenience as well as to reduce space. There is such large reduction in the number of features displayed as large number of features are present as common features so instead of displaying them for each resume, it's been displayed only once. Similarly the cluster features are displayed once for all the resumes present in a cluster instead of separately displaying for each one.

Results for Skill Value Features (SVFSs): The reduction factor in the skill value features for each of the skill type is shown in Table 6. It can be observed that rf values for SVFSs for some skill types is very high, for few skill types low and in some cases medium. The reason for high reduction factor for some skill types is that there are number of resumes that share common skill values for these skill types and thus the clusters formed are uniform. The reason for low reduction factor for skill type such as middleware technologies or mobile platforms is because the number of features in these sets

Table 6. Reduction Factor values for SVFSs

Feature Type	$\lvert F \rvert$	$\sum_{i=1}^{L} F(i)$	rf
database technologies	148	10	0.93
programming languages	283	38	0.86
scripting languages	150	66	0.56
compiler tools	41	7	0.83
mobile platforms	4	3	0.25
middleware technologies	3	3	0
libraries	170	91	0.56
web technologies	354	154	0.34
operating systems	271	16	0.94
other tools	302	208	0.31

Table 7. Organization of skill values for skill type "database technologies"

Common Features (I-level)		
None		
Common Cluster Features (II-level)	**Resume Identifier**	**Special Features (III-level)**
mssql, mysql	10, 101, 105, 107, 114, 115, 14, 16, 19, 24, 25, 29, 3, 32, 33, 35, 36, 39, 4, 40, 41, 42, 46, 47, 5, 50, 53, 54, 55, 57, 58, 59, 6, 62, 63, 64, 65, 66, 67, 68, 69, 70, 71, 73, 75, 76, 77, 78, 84, 86, 88, 90, 92, 93, 98	none
	103, 112	postgresql
	51	oracle
mysql	1, 108, 109, 11, 110, 111, 15, 17, 2, 21, 22, 23, 26, 27, 28, 37, 43, 44, 45, 79, 8, 87, 89, 9, 91, 95, 96	none
	104	msaccess
	113	jdbc, postgresql
mssql	38, 49, 94	none

Table 8. Organization of skill values for skill type "compiler tools"

Common Features (I-level)		
None		
Common Cluster Features (II-level)	**Resume Identifier**	**Special Features (III-level)**
lex, yacc	105, 107, 108, 109, 112, 114, 13, 15, 36, 52, 54, 64, 67, 77, 83	none
	26, 91	phoenix
	35	phoenix, rdk
	63	phoenix

Table 9. Organization of skill values for skill type "programming languages"

Common Features (I-level)		
c, c++		
Common Cluster Features (II-level)	**ResumeId**	**Special Features (III-level)**
javascript, msil	101, 25	none
	112	j2ee
javascript, python	17	latex
	19	.net, vc++
	35	batchscripting, shellscripting
matlab, python	38	perl, php
	62	none
javascript, vb	47	none
	87	symbian
python	1, 27, 68, 82, 89	none
	63	shellscripting
javascript	10, 103, 104, 111, 113, 21, 24, 3, 33, 37, 39, 43, 49, 5, 51, 53, 6, 75, 78, 94, 98	none
	11	perl
	54	j2me, socketprogramming
	99	lisp, vhdl
Nasm	106, 108	none
	109	msil, oz
matlab	14, 4, 90	none
Mips	69, 76, 86, 92	none
	36	socketprogramming
	55	actionscript, mxml
	57	openc++, symbian
	77	prolog

are very less. Thus their is very little scope of clustering the resumes based on common features. Though in most of the cases reduction factor is above 50%. Thus we can say that on average there is 50% reduction in the efforts of HR managers in resume selection process.

For each skill type, its respective skill value features are organized using three-level feature organization. Table 7 shows the organization of skill value for skill type 'database technologies'. The I-level shows the common skill value in 'database technologies', the II-level shows the common skill value in 'database technologies' for each cluster of resumes and the III-level shows special skill value in 'database technologies' for each resume. Table 8 and 9 shows the organization of skill value for skill type 'compiler tools' and 'programming languages' respectively. Similarly skill value features for other skill type like 'scripting languages', 'mobile platform', 'middleware technologies', 'libraries', 'operating systems', 'web technologies' and 'other tools' can be organized in the similar manner.

6   Discussion

There has been little research work related to the issue of resume extraction and selection. Most of the work has been focused on information extraction from resumes. Accordingly, various approaches have been proposed to extract structured information from a given set of resumes. Also most of the companies use a resume management system that helps them to get the selected resumes based on the user query. The resume management systems give hundreds of results for which again user has to manually browse each of the resumes. Secondly the system is dependent on user query.

In this paper we have made an effort to develop an approach to reduce the task of manually browsing each resume by discovering the special features and organizing them in an efficient manner. Also the system is independent of user query and helps the user to discover important information that user might be unaware of.

The evaluation of the proposed approach mentioned in the paper in the form of reduction factor indicates the benefit in the information reduction by comparing the text. But the real evaluation of the proposed approach is yet to be done by observing how it can help the enterprises.

The problem of resume selection and extraction is a problem that has not received much attention so far. The approach proposed in this paper is just the beginning and provides a direction towards the problem of resume selection. More research work is required to address the problem of resume selection and extraction.

7   Summary and Conclusion

Selecting appropriate resume from a group of similar resumes is one of the problems faced by the recruiters in most of the companies. We have extended the notion of special features to extract the special skills from given set of similar resumes. We have proposed an approach to identify resumes by analyzing "skill" related information. The proposed approach has the potential to improve the performance of resume processing by extracting both special skills type and special skill values. With the help of the experimental results we have shown that there is 50-94% reduction in the number of features that the recruiter needs to browse through to select appropriate resumes.

The resume selection process is a more complex problem as resume contains freeform texts which are difficult to compare and also has an hierarchical structure containing different sections. As each resume contains different sections with each section containing different types of text, an integrated approach has to be developed by considering information in each section. In this paper, we have developed an approach by considering only the skills related information of the resume. As a part of future work, we are planning to investigate approaches to extract special information from other sections of the resume and develop an integrated approach for resume processing.

Acknowledgements

This work has been carried out with the support from Nokia Global University Grant.

References

1. Times business solutions limited, July 24 (2000) http://www.timesjobs.com/
2. Flagship brand of monster worldwide, inc., July 27 (2009) http://www.monsterindia.com/
3. Info edge (india) ltd., July 30 (2008) http://www.naukri.com/
4. Agyemang, M., Barker, K., Alhajj, R.S.: Mining web content outliers using structure oriented weighting techniques and n-grams. In: Preneel, B., Tavares, S. (eds.) SAC 2005. LNCS, vol. 3897, pp. 482–487. Springer, Heidelberg (2006)
5. Breunig, M.M., Kriegel, H.-P., Ng, R.T., Sander, J.: Lof: identifying density-based local outliers. SIGMOD Rec. 29(2), 93–104 (2000)
6. Ciravegna, F., Lavelli, A.: Learningpinocchio: adaptive information extraction for real world applications. Nat. Lang. Eng. 10(2), 145–165 (2004)
7. Knorr, E.M., Ng, R.T.: Algorithms for mining distance-based outliers in large datasets. In: VLDB 1998: Proceedings of the 24rd International Conference on Very Large Data Bases, pp. 392–403. Morgan Kaufmann Publishers Inc., San Francisco (1998)
8. Liu, B., Ma, Y., Yu, P.S.: Discovering unexpected information from your competitors' web sites. In: KDD 2001: Proceedings of the seventh ACM SIGKDD international conference on Knowledge discovery and data mining, pp. 144–153. ACM, New York (2001)
9. Maheshwari, S., Reddy, P.: Discovering special product features for improving the process of product selection in e-commerce environment. In: ICEC 2009: Proceedings of the 11th international conference on Electronic commerce, Taipei, Taiwan. ACM, New York (2009)
10. Ramaswamy, S., Rastogi, R., Shim, K.: Efficient algorithms for mining outliers from large data sets. SIGMOD Rec. 29(2), 427–438 (2000)
11. Yu, K., Guan, G., Zhou, M.: Resume information extraction with cascaded hybrid model. In: ACL 2005: Proceedings of the 43rd Annual Meeting on Association for Computational Linguistics, Morristown, NJ, USA, pp. 499–506. Association for Computational Linguistics (2005)

EcoBroker: An Economic Incentive-Based Brokerage Model for Efficiently Handling Multiple-Item Queries to Improve Data Availability via Replication in Mobile-P2P Networks

Anirban Mondal[1], Kuldeep Yadav[1], and Sanjay Kumar Madria[2]

[1] Indraprastha Institute of Information Technology
New Delhi, India
{anirban,kuldeep}@iiitd.ac.in
[2] University of Missouri-Rolla
Rolla, USA
madrias@mst.edu

Abstract. This work proposes EcoBroker, a novel economic incentive-based brokerage model for improving data availability via replication for multiple-item queries in Mobile-P2P networks. In Ecobroker, data requestors need to pay the price (in virtual currency) of their requested data items to data-providers. The main contributions of EcoBroker are two-fold. First, its economic incentive model effectively combats free-riding by incentivizing MPs to become brokers and to host replicated data, thereby improving data availability. Second, its brokerage model facilitates efficient processing of queries involving multiple data items. Our performance evaluation indicates that EcoBroker indeed improves data availability and querying-related communication overhead in Mobile-P2P networks.

1 Introduction

In a Mobile Ad-hoc Peer-to-Peer (M-P2P) network, mobile peers (MPs) interact with each other in a peer-to-peer (P2P) fashion. Proliferation of mobile devices (e.g., laptops, PDAs, mobile phones) coupled with the ever-increasing popularity of the P2P paradigm (e.g., Kazaa [14], Gnutella [7]) strongly motivate M-P2P network applications. Mobile devices with support for wireless device-to-device P2P communication are beginning to be deployed such as Microsoft's Zune [12].

M-P2P applications facilitate mobile users in sharing information with each other *on-the-fly* in a P2P manner. Notably, M-P2P users are often interested in issuing queries involving multiple (possibly related) data items. For example, a music fan Jim, who is moving in a shopping mall, could issue a request for obtaining several songs and video-clips of the Beatles music band. In the same vein, John, who has recently been intrigued by the paintings of Salvadore Dali (at a fine arts exhibition), may want to issue a query for multiple paintings of Dali. Jane, who likes nature photography, may want to request multiple photos of sunsets. Incidentally, the issuance of multiple-item queries is also common in static P2P systems such as Kazaa, where peers often issue queries for downloading entire albums of a specific artist. Notably, P2P interactions among mobile users are generally not freely supported by existing wireless communication infrastructures.

S. Kikuchi, S. Sachdeva, and S. Bhalla (Eds.): DNIS 2010, LNCS 5999, pp. 274–283, 2010.

Our target applications mainly concern slow-moving objects e.g., cars on busy streets, people moving in a market-place or students in a campus. This work does not consider updates to data items. Items shared in our M-P2P application scenarios mostly involve MP3 songs, video-clips and photos, which are typically not updated in practice.

Data availability in M-P2P networks is typically lower than in fixed networks due to frequent network partitioning arising from user movement and users switching 'on'/'off' their mobile devices. The problem of data availability is further exacerbated by free-riding. Free-riding has been known to be rampant in static P2P environments i.e., a large percentage of the peers do not provide any data/services to the network [13]. The adverse effect of free-riding on data availability becomes even more pronounced in M-P2P environments due to the generally limited resources (e.g., energy, memory space, bandwidth) of MPs. Thus, an economic incentive model for enticing MPs to provide data and services becomes a necessity for improving M-P2P data availability.

Recall that our application scenarios call for efficient handling of multiple-item queries. Given a query involving k items, a naive way would be to issue k separate queries, thereby incurring considerable communication overhead as well as significantly taxing the limited resources of the M-P2P network. On the other hand, issuing a single query for k items would provide opportunities for optimizing communication overhead and preserving limited M-P2P resources such as MP energy and bandwidth. Observe that such optimizations would be possible if certain designated MPs acted as **brokers** and hosted replicas of 'hot' items at themselves. This provides a strong motivation for a brokerage model for improving data availability in M-P2P networks, especially in case of multiple-item queries. Hence, we propose **EcoBroker**, a novel economic incentive-based brokerage model for improving data availability via replication for multiple-item queries in M-P2P networks.

The main contributions of EcoBroker are two-fold:

1. Its economic incentive model effectively combats free-riding by incentivizing MPs to become brokers and to host replicated data, thereby improving data availability.
2. Its brokerage model facilitates efficient processing of queries involving multiple data items.

In the incentive-based model of EcoBroker, each data item has a *price* (in *virtual currency*). Data item price depends on access frequency and data quality [16] (e.g., image resolution, audio quality). A query issuing MP pays the *price* of the queried data item to the query-serving MP, a commission to the successful broker and a relay commission to each MP in the successful query path. Thus, EcoBroker provides an incentive for free-riding MPs to provide data as well as brokerage and relay services so that they can earn currency (i.e., broker commissions) for issuing their own requests. **Revenue** of an MP is defined as the difference between the amount of virtual currency that it earns (by providing data, brokerage and relay services) and the amount that it spends (by requesting data). Notably, virtual currency is suitable for M-P2P environments due to high transaction costs of micro-payments in real currency [22]. Secure virtual currency payments have been discussed in [4].

In the brokerage model of EcoBroker, brokers select and obtain the items to replicate at themselves to facilitate them in earning currency. Notably, brokers replicate items at themselves based on the currency-earning potential of the items. Querying in EcoBroker proceeds via broadcast until the queries are intercepted by brokers. When a broker intercepts a multiple-item query, it can quickly serve the query partially from the replicated items existing at itself. For the remaining queried items, whose replicas it does not contain, it uses its index to locate and obtain those items. In this paper, we do not describe the index used by brokers due to space constraints. Instead, we use an existing index for M-P2P environments, namely the CR*-tree index, which we have proposed in our previous work [17]. As such, our focus in this paper is on performing effective replication at brokers for facilitating multiple-item queries.

Notably, each item has only one original owner and brokers are allowed to replicate for brokerage purposes only by obtaining the items from their respective original owners. Hence, brokers require to provide an incentive to the original owners of those items. In the absence of any incentive, the original owners would not want to replicate their items at the broker MPs because they would lose revenues on future accesses to their own items. Thus, broker MPs make a one-time payment to the original owners for the right to replicate their items. Incidentally, replicating their items at the broker MPs does not preclude the original owners from hosting their own items.

In EcoBroker, brokers could be pre-designated in accordance with the application scenario under consideration, and there could be multiple pre-designated brokers. For example, recall our application scenario concerning an M-P2P user (in a shopping mall) searching for multiple songs of the Beatles music band. In this case, the shop-owners or the shopping mall administrators can act as brokers. If songs or movies are shared among M-P2P users in a University campus setting, some of the students (e.g., student organization leaders) can act as brokers.

Our performance evaluation demonstrates that EcoBroker indeed improves data availability and querying-related communication overhead in Mobile-P2P networks. The remainder of this paper is organized as follows. Section 2 provides an overview of relevant existing works, while Section 3 discusses the economic incentive-based model of Eco-Broker. Section 4 presents the brokerage model of EcoBroker, while Section 5 reports our performance evaluation. Finally, we conclude in Section 6.

2   Related Work

Economic models for distributed systems [5,15] primarily focus on resource allocation. These models do not address unique M-P2P issues such as node mobility, mobile resource constraints, frequent network partitioning, free-riding and incentives for peer participation. Economic schemes for resource allocation in wireless ad hoc networks [25] do not consider free-riding. The works in [2,3,21,26] provide incentives to peers for forwarding messages, but they do not incentivize free-riders to host data.

Schemes for combating free-riding in static P2P networks [8,10,13] are too static to be deployed in M-P2P networks as they assume peers' availability and fixed topology. Schemes for improving data availability in mobile ad hoc networks (MANETs) (e.g., the **E-DCG+** approach [11]) primarily focus on replica allocation, but do not consider economic incentive schemes, M-P2P architecture and brokerage model for improving data availability. Interestingly, the proposals in [23,24] consider economic schemes in M-P2P networks. However, they focus on data dissemination with the aim of reaching as many peers as possible, while we consider on-demand services. Furthermore, they do not consider free-riders and M-P2P brokerage models.

P2P replication suitable for mobile environments has been incorporated in systems such as ROAM [19], Clique [20] and Rumor [9]. However, these systems do not incorporate economic schemes and brokerage models.

3   The EcoBroker Economic Incentive-Based Model

The architecture of EcoBroker consists of query-issuing MPs, relay MPs, broker MPs and data-providing MPs. Relay MPs forward messages (e.g., queries, data) in lieu of a *relay commission*. Broker MPs facilitate query-issuing MPs in obtaining their queried data items in lieu of a *broker commission*.

In EcoBroker, the price μ_d of a data item d is computed as follows:

$$\mu_d = \eta_d \times DQ_d \tag{1}$$

where η_d represents the access frequency of d. Notably, data item price increases with increasing access frequency due to frequently accessed items being more important to the network as a whole. DQ_d reflects the quality of d (e.g., image resolution, audio quality). The value of DQ is determined as in our previous work in [16], where we considered three discrete levels of DQ i.e., *high*, *medium* and *low*, their values being 1, 0.5 and 0.25 respectively. Understandably, higher-quality data items command higher prices. As a single instance, an MP requesting for MP3 songs would typically be willing to pay a higher price for obtaining better audio quality. Notably, there can be alternative approaches to defining data item prices, and some of these approaches have been examined in our previous works [17,18].

Broker commission is 5% of the price of each item retrieved by the broker. In Eco-Broker, a broker only earns its commission if it retrieves all the items in a given query, hence partially answered queries do not entail a broker commission. Relay commission is a constant r, which is application-dependent. In particular, the value of r is significantly less than that of the average data item prices, thereby implying that EcoBroker provides more incentives to MPs for hosting data than for relaying messages.

4   Brokerage Model of EcoBroker for Efficiently Handling Multiple-Item Queries

This section presents the brokerage model of EcoBroker. In particular, we discuss how brokers select and obtain the items to replicate at themselves for earning currency.

For selecting the data items to replicate at itself, a given broker MP M keeps track of the queries that pass through itself. This enables M to determine the currency-earning potential for various items in the network. M maintains a list L, each entry of which is of the form $\{d_{id}, \mu_d, size_d, LL\}$, where d_{id} refers to the unique identifier of a given data item d, μ_d is the price of d and $size_d$ is the size of d. LL is a pointer to a list, each entry of which is of the form $\{k, \eta_k\}$. Here, k represents the number of items in each query issued for d and η_k is the access frequency of each k-item query.

Note that the other items in the queries issued for k could differ since η_k only considers the number of k-item queries issued for d as opposed to the actual items in those queries.

Selection of Candidate Data Items for Replication

M uses the information in list L to assign a score λ to each item. As the value of λ for an item increases, its currency-earning potential for M also increases. Hence, M prefers to replicate items with higher values of λ. For each item, M computes λ as follows.

$$\lambda = \sum_{k=1}^{max_k} (\mu \times \eta_k) / (size \times k) \qquad (2)$$

where μ represents the price of the data item d and η_k is the *estimated* access frequency of d for k-item queries. M estimates the value of η_k for d based on the number of queries that had recently passed through it for d. The term $size$ refers to the size of d. k represents the number of queried items in each query (for d) intercepted by M. As a single instance, if a query for 5 items (one of which is d) is intercepted by M, $k = 5$.

In Equation 2, max_k refers to the maximum number of items allowable in a single query. The value of max_k is application-dependent. We found $max_k = 7$ to be reasonable for our application scenarios. Intuitively, if the value of max_k is too high, queries would be likely to fail because it may not be practically feasible to successfully obtain all the queried items.

The value of λ increases with increase in μ and η_k because higher-priced items and items with higher access frequencies imply more currency-earning potential for M. λ decreases with increase in item size due to reduction in M's currency-earning potential per unit of its limited memory space. λ decreases with increase in k because queries with higher values of k entail an increased probability of failure in locating the other items in the query. Recall that partially answered queries entail no earnings for brokers.

M sorts the items in L in descending order of the values of λ and selects those items (from L), whose values of λ exceed λ_{Th}, where λ_{Th} is the average value of λ for all the data items in L. Thus, $\lambda_{Th} = ((1/n_d) \sum_{i=1}^{n_d} \lambda_d)$, where n_d is the number of data items in L and λ_d is the value of λ for a given item d. Items, for which the value of λ falls below λ_{Th}, are then deleted from L. Thus, now list L contains the items, which M wants to replicate at itself. List L is refreshed periodically to ensure that replication is performed based on recent access information. Notably, M may not necessarily be able to obtain and host all the items in L due to M's memory space constraints as well as item price constraints. Now let us see how M determines the maximum price, which it is willing to pay, for obtaining each item in L.

Determination of the Maximum Price for Obtaining a Replica from the Original Data-Provider

Let $cost_{max_d}$ be the maximum price that an MP M is willing to pay for obtaining the replica of a given item d in list L. For each item d, M computes $cost_{max_d}$ as follows:

$$cost_{max_d} = 0.5 \times (\mu \times \eta) / k_{avg} \qquad (3)$$

where μ is the price of a given item d and η is the total estimated access frequency of d. The term ($\mu \times \eta$) is the total estimated (future) currency-earning potential of d. The factor of 0.5 reflects that M is willing to pay 50% of the estimated currency-earning potential of d to d's owner. Thus, the estimated (future) currency from the replicated data item is shared *equally* between M and the owner of d to ensure fairness. This acts as an incentive for d's owner to sell d to M because it would earn currency without having to expend its limited energy and bandwidth for serving queries on d. Additionally, d's owner could use its energy and bandwidth to serve queries on other items that it owns, thereby enabling it to earn more currency.

In Equation 3, k_{avg} is the weighted average value of k for all the queries for d that passed through M. M computes k_{avg} as follows:

$$k_{avg} = (\sum_{k=1}^{max_k} (k \times \eta_k)) / (\sum_{k=1}^{max_k} \eta_k) \qquad (4)$$

For example, if there were 20 three-item queries (in which d was one of the queried items) and 40 such five-item queries involving d, the value of $k = (20 \times 3 + 40 \times 5)/(20 + 40)$ i.e., 4.33. Notably, the value of $cost_{max_d}$ decreases with increasing value of k_{avg} because as the number of *other items* included in queries involving d increases, the probability of M earning currency by hosting d decreases. This is because of increased chances of query failure on any of the other items.

Algorithm for Obtaining the Selected Items for Replication

Now let us examine the algorithm used by M to obtain its desired items. M broadcasts a query request for the items in L. M's broadcast request is a list of the form $\{ d_{id}, cost_{max_d} \}$, where d_{id} is the unique identifier of the item and $cost_{max_d}$ is the maximum price that M is willing to pay for hosting the item.

Recall that each item has only one original owner. Hence, when the owner M_O of an item d intercepts M's broadcast query, it evaluates d's currency-earning potential α_d at itself. If $\alpha_d < cost_{max_d}$, it decides to provide d to M in lieu of a payment of $cost_{max_d}$. M_O computes the value of α_d as follows.

$$\alpha_d = (\mu_d \times \eta_d) / k_{avg} \qquad (5)$$

where μ_d and η_d are d's price and estimated access frequency respectively. The value of k_{avg} is computed by Equation 4. The value of α_d decreases with increase in k_{avg} due to the reasons explained for Equation 3. Notably, the values of η_d and k_{avg} as computed by M_O and M are likely to be different because each MP estimates these values based on the queries that pass through them. Furthermore, M_O might also be willing to provide d to M if its energy is low or if it lacks adequate bandwidth to serve queries on d.

Only the willing owners of the items reply to M. Upon receiving their replies, M sorts the items in descending order of λ. Then, M obtains the replicas from the corresponding owners one-by-one and makes the necessary payments to them. Thus, M keeps filling up its available memory space starting with the item with the highest value of λ, subject to memory space and item size constraints. M terminates the replication procedure when its available memory space for replication is exhausted.

5   Performance Evaluation

This section reports our performance evaluation by means of simulation.

For simulation purposes, we have used OMNeT++ 3.3p1 [6]. OMNeT++ is an object-oriented modular discrete event network simulation framework. It actively supports parallel distributed simulation and mobility, thereby making it a good choice for providing a realistic M-P2P simulation environment.

MPs move according to the *Random Waypoint Model* [1] within a region of area 1000 metres ×1000 metres. The *Random Waypoint Model* is appropriate for our application scenarios, which involve random movement of users. 10% of the peers were data-providers, while the others were free-riders. Thus, a total of 100 MPs comprised 10 data-providers and 90 free-riders (which provide no data). The number of brokers in our experiments was 10 i.e., out of the 90 free-riders, 10 peers decided to act as brokers. Each data-provider owns and hosts 10 items of different sizes. Each query is a request for k data items. 10 such k-item queries/second are issued in the network. We use a highly skewed Zipf distribution with zipf factor of 0.9 to determine the number of such k-item queries to be directed to each MP as well as to decide the frequency with which each individual item should occur in the queries. Communication range of all MPs is a circle of 100 metre radius. Table 1 summarizes our performance study parameters.

Table 1. Performance Study Parameters

Parameter	Default value	Variations
No. of MPs (N_{MP})	100	
No. of items in query (k)	5	2,3,4
Zipf factor (ZF)	0.9	
Queries/second	10	
Bandwidth between MPs	28 Kbps to 100 Kbps	
Probability of MP availability	60% to 80%	
Size of a data item	50 Kb to 350 Kb	
Memory space of each MP	1 MB to 2 MB	
Speed of an MP	1 metre/s to 10 metres/s	
Size of message headers	220 bytes	

Our performance metrics are **data availability (DA)** and **average querying hop-counts (HC)**. DA equals ((N_S/N_Q) × 100), where N_S is the number of successful queries and N_Q is the total number of queries. Thus, DA reflects the percentage of successful queries. Notably, a k-item query is considered to be successful if and only if all the queried items are retrieved. Hence, partially answered queries are considered to be unsuccessful. Queries can fail due to MPs being switched 'off' or due to network partitioning or due to failure in retrieval of at least one of the queried items. HC is the average number of hops per query.

As reference, we adapt a non-economic and non-broker-based model **NB (No_Broker)** since existing M-P2P proposals do not address economic broker-based models. In NB, brokerage is not performed and querying is broadcast-based. As NB does

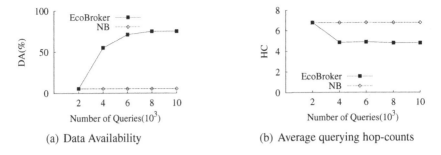

(a) Data Availability   (b) Average querying hop-counts

Fig. 1. Performance of EcoBroker

not provide incentives for free-riders to host data, only a single copy of any given data item d exists at the owner of d.

Performance of EcoBroker

Figure 1 depicts the performance of EcoBroker using default values of the parameters in Table 1. Thus, all the queries in this experiment were 5-item queries. Broker-related replication procedures are initiated only after the first 2000 queries, hence both Eco-Broker and NB initially show comparable performance in terms of both DA and HC. Periodically, every 200 seconds (i.e., for every 2000 queries since there are 10 queries per second), brokers perform replication procedures.

In Figure 1a, DA remains relatively constant over time in case of NB primarily due to the absence of replication in NB, which implies that there is only copy of a given queried item in the network. On the other hand, DA increases significantly in case of EcoBroker due to its effective brokerage-related replication model, which boosts data availability. DA eventually plateaus for EcoBroker due to network partitioning and unavailability of some of the MPs.

As the results in Figure 1b indicate, HC was significantly close to the TTL of 7 hops in case of NB. This is because this experiment involved 5-item queries and in the absence of any replication, the queries needed more hops to retrieve the queried items. In contrast, HC decreased considerably in case of EcoBroker because of its effective economic incentive-based brokerage model, which incentivizes brokers to host replicas, thereby resulting in shorter query paths. Incidentally, after the initial decrease in HC for EcoBroker, HC hits a saturation point because of additional querying hops required for retrieving queried items, whose replicas do not exist at the broker. Since this experiment involved 5-item queries, it can be reasonably expected that brokers were not able to replicate all the queried items at themselves possibly due to memory space constraints as well as due to item price constraints.

Effect of Variations in k

Figure 2 depicts the results of varying k. Recall that k is the number of items per query. In Figure 2a, observe that as k increases, DA decreases for both EcoBroker and NB.

This is because at higher values of k, probability of query failure increases due to failure in the retrieval of at least one of the queried items. Observe that the decrease in DA with increasing k is significantly less pronounced for EcoBroker than for NB. This is because the effect of EcoBroker's economic incentive-based brokerage model becomes more prominent with increasing values of k. In particular, for 4-item queries and for 5-item queries, EcoBroker outperforms NB by upto 70% in terms of DA due to its economic brokerage model.

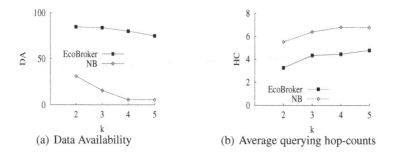

(a) Data Availability   (b) Average querying hop-counts

Fig. 2. Effect of variations in k

As the results in Figure 2b indicate, HC increases with increasing values of k for both the approaches. Understandably, queries involving more items incur more hop-counts. Observe that the increase in HC is most significant as k increases from 2 to 3. This is because 2-items queries have a significantly higher probability of getting answered successfully (possibly at the same MP) as compared to the case for 3-item queries. Eco-Broker outperforms NB in terms of HC due to its brokerage model, which which in-centivizes brokers to host replicas, thereby facilitating queries in being answered within lower number of hops.

6 Conclusion

We have proposed EcoBroker, a novel economic incentive-based brokerage model for improving data availability via replication for multiple-item queries in M-P2P networks. In Ecobroker, data requestors pay the price (in virtual currency) of their requested data items to data-providers. The economic incentive model of EcoBroker effectively combats free-riding by incentivizing MPs to become brokers and to host replicated data, thereby improving data availability. Furthermore, EcoBroker's brokerage model facilitates efficient processing of queries involving multiple data items. Our performance evaluation demonstrates that EcoBroker indeed improves data availability and querying-related communication overhead in M-P2P networks.

References

1. Broch, J., Maltz, D.A., Johnson, D.B., Hu, Y.C., Jetcheva, J.: A performance comparison of multi-hop wireless ad hoc network routing protocol. In: Proc. MOBICOM (1998)

2. Buttyan, L., Hubaux, J.P.: Stimulating cooperation in self-organizing mobile ad hoc networks. ACM/Kluwer Mobile Networks and Applications 8(5) (2003)
3. Crowcroft, J., Gibbens, R., Kelly, F., Ostring, S.: Modelling incentives for collaboration in mobile ad hoc networks. In: Proc. WiOpt (2003)
4. Elrufaie, E., Turner, D.: Bidding in P2P content distribution networks using the lightweight currency paradigm. In: Proc. ITCC (2004)
5. Ferguson, D.F., Yemini, Y., Nikolaou, C.: Microeconomic algorithms for load balancing in distributed computer systems. In: Proc. ICDCS, pp. 491–499 (1988)
6. OMNeT++ for simulation, http://www.omnetpp.org/
7. Gnutella, http://www.gnutella.com/
8. Golle, P., Brown, K.L., Mironov, I.: Incentives for sharing in peer-to-peer networks. In: Proc. Electronic Commerce (2001)
9. Guy, R., Reiher, P., Ratner, D., Gunter, M., Ma, W., Popek, G.: Rumor: Mobile data access through optimistic peer-to-peer replication. In: Proc. ER Workshops (1998)
10. Ham, M., Agha, G.: ARA: A robust audit to prevent free-riding in P2P networks. In: Proc. P2P, pp. 125–132 (2005)
11. Hara, T., Madria, S.K.: Data replication for improving data accessibility in ad hoc networks. IEEE Transactions on Mobile Computing 5(11) (2006)
12. http://www.microsoft.com/presspass/presskits/zune/default.mspx
13. Kamvar, S., Schlosser, M., Garcia-Molina, H.: Incentives for combatting free-riding on P2P networks. In: Kosch, H., Böszörményi, L., Hellwagner, H. (eds.) Euro-Par 2003. LNCS, vol. 2790, pp. 1273–1279. Springer, Heidelberg (2003)
14. Kazaa, http://www.kazaa.com/
15. Kurose, J.F., Simha, R.: A microeconomic approach to optimal resource allocation in distributed computer systems. IEEE Trans. Computers 38(5), 705–717 (1989)
16. Mondal, A., Madria, S.K., Kitsuregawa, M.: CADRE: A collaborative replica allocation and deallocation approach for Mobile-P2P networks. In: Proc. IDEAS (2006)
17. Mondal, A., Madria, S.K., Kitsuregawa, M.: ConQuer: A peer group-based incentive model for constraint querying in mobile-P2P networks. In: Proc. MDM (2006)
18. Mondal, A., Madria, S.K., Kitsuregawa, M.: EcoRep: An economic model for efficient dynamic replication in Mobile-P2P networks. In: Proc. COMAD (2006)
19. Ratner, D., Reiher, P.L., Popek, G.J., Kuenning, G.H.: Replication requirements in mobile environments. Mobile Networks and Applications 6(6) (2001)
20. Richard, B., Nioclais, D., Chalon, D.: Clique: A transparent, peer-to-peer replicated file system. In: Proc. MDM (2003)
21. Srinivasan, V., Nuggehalli, P., Chiasserini, C.F., Rao, R.R.: Cooperation in wireless ad hoc networks. In: Proc. INFOCOM (2003)
22. Turner, D.A., Ross, K.W.: A lightweight currency paradigm for the P2P resource market. In: Proc. Electronic Commerce Research (2004)
23. Wolfson, O., Xu, B., Sistla, A.P.: An economic model for resource exchange in mobile Peer-to-Peer networks. In: Proc. SSDBM (2004)
24. Xu, B., Wolfson, O., Rishe, N.: Benefit and pricing of spatio-temporal information in Mobile Peer-to-Peer networks. In: Proc. HICSS-39 (2006)
25. Xue, Y., Li, B., Nahrstedt, K.: Optimal resource allocation in wireless ad hoc networks: A price-based approach. IEEE Transactions on Mobile Computing (2005)
26. Zhong, S., Chen, J., Yang, Y.R.: Sprite: A simple, cheat-proof, credit-based system for mobile ad-hoc networks. In: Proc. IEEE INFOCOM (2003)

Interface Tailoring by Exploiting Temporality of Attributes for Small Screens

M. Kumara Swamy, P. Krishna Reddy,
R. Uday Kiran, and M. Venugopal Reddy

Center for Data Engineering
International Institute of Information
Technology-Hyderabad (IIIT-H), Hyderabad,
Andhra Pradesh, India - 500 032
pkreddy@iiit.ac.in

Abstract. In the pervasive computing era, mobile phones and personal digital assistants are widely used for data collection. The traditional user interfaces which are employed for data collection in personal computer environments are to be modified appropriately for the mobile environment. Because, it is difficult to display full interface on a single mobile screen due to the limitation of the mobile phone screen size. Interface tailoring methods are investigated in the literature for designing user interface for mobile screens. In the literature, temporality-based approach is proposed for designing efficient user interface for personal computer environment. In this paper, we extend the notion of attribute temporality to interface tailoring methods and propose an improved user interface design approach for small screens. The analysis on the real-world datasets shows that the proposed approach can be used for better user interface design for small screens.

Keywords: user interfaces, context-based approach, interface tailoring, mobile interface forms.

1 Introduction

Mobile phones and personal digital assistants are used for data collection in modern information systems. It is true that mobile phones and personal digital assistants provide significant advantages due to pervasive nature of communication and computing abilities. However, due to small screen size and other factors, the design of UI forms for mobile devices is a challenging task as the approaches used to design UIs for personal computer (PC)-based systems can not be extended for mobile devices in a straightforward manner [5]. So, appropriate methods are to be investigated to device efficient UIs for mobile environment. Normally, the large UI form in PC environment is divided into multiple small UI forms by considering small screen size in the mobile environment. As a result, user has to go through multiple screens which increases navigational burden. In addition, it also increases both cost of data-entry and maintenance. So, developing efficient UI methods for mobile environment is an ongoing research problem.

S. Kikuchi, S. Sachdeva, and S. Bhalla (Eds.): DNIS 2010, LNCS 5999, pp. 284–295, 2010.

In the literature, interface tailoring approaches are investigated to develop UI forms for mobile devices [1,4]. In these approaches, the PC-based UI is divided into multiple screens (or high-level components) in such a way that the UI can accommodate in a mobile screen. The research issue is how to divide a large UI form into small UI forms (or screens) to reduce the navigational burden and improve the efficiency to the extent possible.

For PC-based systems, context-based approach (CA) [11] is followed for designing the UI forms. Recently, context-based approach with attribute temporality (CAAT) [7] has been proposed to reduce the navigational burden in UI forms for PC-based systems by exploiting the notion of "attribute temporality". The attribute temporality value indicates the active period of the attribute during which the attribute receives values. The attributes of a given context are clustered into several active-contexts based on the temporality of attribute and UIs are designed for each active-context. It was found out that the CAAT approach improves the UI performance by reducing the navigational burden significantly.

It was observed that there is a scope to design efficient UI by reducing the size of UI for small screens by extending the notion of attribute temporality to mobile environments. In this paper, we made an effort to propose an improved UI approach for small screens by extending the notion of attribute temporality to interface tailoring approaches. We have carried out analysis by considering real world dataset. The analysis results show that the proposed approach has a potential to improve the performance for mobile environments.

The rest of the paper is organized as follows. In the next section, we discuss the related work. In section 3, we present the overview of the existing approaches. In section 4, we explain the proposed approach. The analysis results and discussion are provided in Section 5. The last section consists of summary and conclusions.

2   Related Work

In this section, we discuss the related work in the area of UI design tools and approaches in both mobile and PC-based environments.

2.1   UI Design Tools for Mobile-Based Systems

Multi-User Publishing Environment (MUPE) [2] is an open source platform developed to enable multi-users to publish and communicate with each other. MUPE allows a user to generate a mobile UI form interactively and send this form by expecting the data from his/her contact list. This application will also generate an integrated view of the data received by the mobile UI form.

A template-based approach to generate UI for mobile phone is presented in [3]. In this work, a UI design templates generator is proposed for UI designers to easily and quickly create the UI templates for mobile phone. The developed UI templates can be fine tuned with a visual UI authoring tool to generate the UI prototype for the target mobile phone system.

2.2 UI Design Approaches for Mobile-Based Systems

The problem of division of large interface into numerous small screens is identified in [1,4]. User interface tailoring [4] is a solution to such problem, which tailors information based on the mobile devices screen. In this approach, the information is tailored to display only the important information on the interface. This approach does not guarantee the integrity of information [1]. Another approach is dividing large PC interface into smaller ones such that it can match the size of the mobile phone screen [1]. In this method, the large interface is divided into small screens. The issue is to investigate improved approaches to reduce the navigational burden as far as possible.

2.3 UI Design Tools in PC-Based Systems

Brad Myers [15] surveyed the state of the art of UI software tools. The existing UI design tools are classified into categories such as language-based tools, interactive graphical specification tools and model-based generation tools. All these models are based on the dynamic behavior of a UI as well as the way as to how the designer can specify the layout. There exist tools which allow to design UI interactively, or which automatically generate the UI from a high-level model or specification. These tools also enable the UI developer to customize their design-output.

User Interface Management Systems (UIMSs) consist of UI construction tools which are used mainly to construct the dialogue component of the interface. The UIMS significantly reduces the time required to develop an interface and also the chances of errors to a minimum because the designer is dealing with a compact and higher level of specification. A high-level UIMS which automatically generates the lexical and syntactic design of GUIs is presented in [19] based on the given description and user preferences. Similar approaches have been used in the Menu Interaction Kontrol Environment (MIKE) [16] and User Interface Design Environment (UIDE) [10]. MIKE automatically generates textual interfaces from a description of the semantic commands supported by the application. The designer then adds the graphical interaction facilities to the interfaces generated by MIKE. UIDE provides a high-level conceptual design tool in which the designer describes the UI as a knowledge base. UIDE can algorithmically transform the knowledge base into a number of functionally equivalent interfaces, each of which is slightly different from the original interface. The transformed interface definition can then be input to a UIMS which implements the UI.

Egbert Schlungaum [17,18] summarizes the area of model-based UI software tools in order to prepare a basis for automatic UI generation. Efforts are made to create a common declarative model to specify the necessary information for automatic UI generation by combining various model-based UI approaches. A system called TADEUS was developed that takes requirement analysis as the input and generates UIs.

2.4 UI Design Approaches in PC-Based Systems

A survey was conducted on context-aware system in [12]. This survey presented common architecture principles of context-aware systems and derived a layered conceptual design framework to explain the different elements common to most context-aware architectures. These design principles are used to introduce various existing context-aware systems focusing on context-aware middleware and frameworks, which ease the development of context-aware applications.

A survey was conducted in [11] to understand context and context-aware applications. The concepts such as context and context-aware computing are defined. These definitions assist to understand the boundaries of context-aware computing, and facilitate application designers for selecting the context to use, structuring the context in applications, and in deciding what context-aware features to implement.

For improved UI design, the notion of attribute temporality was exploited in [7]. Given a context and temporal values, this approach divides context into several active-contexts. UIs are designed for all active-contexts.

About the proposed approach: In the mobile environment, interface tailoring approaches are investigated to tailor the PC interface to fit in small mobile screens. However, it was observed that the notion of attribute temporality can improve the performance by further reducing the navigational burden. In this paper, we have made an effort to improve the performance of interface tailoring methods to design UI forms for small screens by extending the notion of attribute temporality.

3 Overview of Context-Based and Interface Tailoring Approaches

In information systems, UI is a place where the user interacts with the database systems to populate the data. In this section, we briefly discuss the context-based approach with attribute temporality and interface tailoring approaches.

3.1 Context-Based Approach with Attribute Temporality (CAAT)

Normally, context-based approach (CA) [11] is followed for designing UIs in PC-based systems. In context-based approach, the UI forms are designed for each context. The context refers to a particular task or activity of an information system.

In information systems, it can be observed that some attributes receive data for certain time period and they do not receive values during the remaining time period. This notion is called attribute temporality. By extending the notion of attribute temporality on CA, it is possible to design better UIs. In this approach, the notion of active-context is defined which is set of attributes that receives values for a fixed time period. In the context-based approach with attribute temporality (CAAT) [7], the attributes of a given context are divided into several small active-contexts and UIs are designed for each active-context.

The CAAT approach divides the context into several active-contexts. Based on the timing of the data-entry, appropriate active-context is displayed to the user. In CA, all the attributes are shown to the user for the data-entry. Whereas, in CAAT, only the attributes of corresponding active-contexts are shown. Since the number of attributes in active-context is small as compared to the number of attributes in entire context, the performance is improved.

Table 1. Sample data consisting of attribute names and the corresponding durations

Name of the attribute	Duration of the attribute (days)
a	$1 - 5$
b	$2 - 14$
c	$5 - 12$
d	$12 - 15$
e	$1 - 15$
f	$9 - 14$
g	$1 - 15$
h	$10 - 15$
i	$2 - 8$
j	$8 - 10$

The method to divide the given context into active-contexts is as follows. The algorithm is given in [7]. The Jaccard coefficient [20] similarity criteria is employed to find the similarity between the two active-contexts. Initially, the attributes of a given context are clustered into day-wise active-contexts (the attributes valid on a day) at 'Level-0'. The merging process starts at 'Level-0' with the first two contiguous active-contexts. When the similarity value of these two contiguous active-contexts is greater than or equals to the given similarity threshold, they are merged into a single active-context. The newly generated active-context is again merged with the next contiguous active-context if the similarity value exceeds the threshold value. Otherwise, the same process is repeated from the next consecutive active-context. All these newly generated active-contexts form the active-contexts at 'Level-1'. The process is repeated for the active-context in 'Level-1' to form higher level active-context. In this way, finally, all the active-contexts are merged into a single largest active-context at 'Level-n'. The appropriate active-contexts are selected for designing UI.

We explain the process of finding the active-contexts through an example.

Example 1: Consider the sample data in Table 1. In this table, column one contains the name of the attributes and the other column contains active duration of the corresponding attributes in days. In this column, the starting day and ending day, during which the attribute receives the values, are shown. The

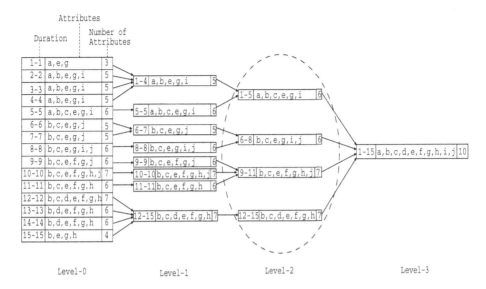

Fig. 1. Clustering processes to find active-contexts in CAAT approach

period which the attribute receives the values is called duration. The maximum duration for this sample data is 15 days. The attributes are clustered into different active-contexts. The hierarchy of derived active-contexts is shown in Figure 1. In this figure, each node (active-context) is consisting of three values. The first value shows the duration, the second value shows the attributes and the final value indicates the number of attributes for the corresponding duration. The 'Level-0' shows the initial active-contexts, 'Level-1', 'Level-2' and 'Level-3' are the active-contexts at different levels. The final hierarchy is generated based on the above procedure. Initially, day-wise active-contexts are generated, the attributes active in a day are grouped together based on the durations. As a result, 15 active-contexts are generated. These day-wise active-contexts are merged to higher level active-contexts based on the similarity between the active-contexts. Let us use Jaccard coefficient as the similarity metric to merge the two active-contexts. Let the similarity threshold value be 0.9. The merging process is repeated to merge all the active-contexts into a single large active-context. In Figure 1, 'Level-0' contains day-wise active-contexts and 'Level-1' to 'Level-3' contain the corresponding active-contexts. At 'Level-3', all the attributes are merged into a single large active-context. The final active-contexts are extracted from 'Level-2', where the temporality of each active-context is distinct and average number of attributes in each active-context is minimum. The dotted circle in the figure shows the layer of active-context. As a result, four active-contexts are extracted for the durations 1-5, 6-8, 9-11 and 12-15. For these active-contexts the UI forms are designed.

3.2   Interface Tailoring Approach

Interface tailoring refers to the ability to customize and optimize an interface according to the context in which it is used [6]. Tailoring of the interface design includes the capability of adaptation of content delivery to various devices, which preserve consistency and usability of the service [4]. One of the interface tailoring approaches namely, a constrained-based UI method [1] is proposed to tailor the PC-based UI such that it matches the mobile screen using component-tree.

In this approach, the attributes are divided into several components by considering the constraints of mobile screen. Suppose, we take number of attributes to be displayed in the screen as a constraint. Based on this constraint, a component tree is built starting from high-level components to attribute levels. The attributes in the component-tree can be displayed in mobile UI using either depth-first principle or breadth-first principle.

In summary, using interface tailoring approaches, the attributes of given UI form for PC-based systems are divided into multiple high-level components such that each component or sub-component can fit into the mobile screen.

4   Proposed Interface Tailoring with Attribute Temporality (ITAT) Approach

The notion of "attribute temporality" can be exploited to interface tailoring approach to improve the performance. So, given the context, we apply attribute temporality and select appropriate active-contexts which match the screen size of mobile phone. If number of attributes are more in an active-context such that it can not fit into a mobile phone, interface tailoring approach is followed to divide the active-context into small screens.

We call the proposed approach as "Interface Tailoring with Attribute Temporality (ITAT)". It contains two phases. In the first phase, we identify the active-contexts from the given context and in the second phase, we apply interface tailoring approach for such active-contexts whose size is more than the user-specified mobile screen size.

- **First phase of ITAT:** To find the active-contexts, we modify the CAAT algorithm [7] by including the notion of active-context tree (AC_tree) which contains all the extracted active-contexts. The algorithm takes the context 'C' as the input and generates AC_Tree as the output. Initially, smaller day-wise ACs are generated. These day-wise ACs are merged into larger ACs at various levels and finally they form a single largest AC. All the extracted ACs, starting from day-wise to the single large AC, are inserted into AC_tree. Next, we find set of ACs which fit in a mobile screen by traversing the AC_tree. The tree traversal starts from the large AC. If the number of attributes in a node either equals or less than the number of attributes that can fit in the mobile screen, the node is selected as an AC for mobile screen and all the other child nodes are not processed further. If the number of attributes in a node exceeds mobile screen size, the lower level nodes are further traversed. This process continues until the entire *AC_tree* is covered.

- **Second phase of ITAT:** In the phase, we find the large *ACs* from a set of the selected ACs for applying the interface tailoring approach. The algorithm reads ACs from the selected ACs sequentially starting from the beginning. If the number of attributes are less than or equal to the mobile screen size then normal UI is generated. Otherwise interface tailoring approach is followed to generate UI. We take the mobile screen size as a constraint, that is number of attributes to be displayed on the mobile screen as a constraint. Based on this constraint, an attribute-tree is built starting from high-level components to attribute levels. The attributes in the attribute-tree can be displayed in mobile UI using breadth-first principle.

The proposed approach is explained with the following example.

Example 2: Continuing with the Example 1, let the mobile screen size, (MSS) be 6. It means a mobile screen can accommodate 6 attributes for taking input. We explain the method to extract active-contexts, later we explain the interface tailoring approach.

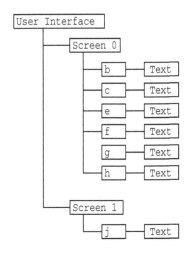

Fig. 2. A sample attribute-tree for active-context at duration "10-10"

An *AC_Tree* is built as shown in Figure 1. However, the active-context lies between 'Level-0' to 'Level-3' that is from day-wise active-context to final single active-context. The tree traversal starts from 'Level-3' to 'Level-0'. The node at 'Level-3' consists of 10 attributes. This node cannot fit in the mobile screen. The traversal goes to the first node in 'Level-2', start reading node by node at this level. The nodes having the durations 1-5 and 6-8 are selected as the *AC* by discarding the child nodes of the selected nodes, that is, child nodes at 'Level-0' and 'Level-1'. The next two nodes are having 7 attributes which cannot fit in the mobile size. Now, the control goes to the corresponding child nodes at the lower level that is at 'Level-1'. Here nodes having durations 9-9 and 11-11 are selected as ACs. At the end, the control goes to 'Level-1' to the final leaf nodes, find the rest of the nodes having durations 10-10, 12-12, 13-13, 14-14 and 15-15, and are selected as ACs as these are the final leaf nodes. After completion of this phase, we have the durations 1-5, 6-8, 9-9, 10-10, 11-11, 12-12, 13-13, 14-14 and 15-15 as the active-contexts.

We apply the second phase on the selected active-contexts. Among the selected active-contexts, the active-contexts at the durations 10-10 and 12-12 cannot fit in the mobile screen as they have 7 attributes in each of the ACs. For these two ACs interface tailoring approach is followed and for the other ACs normal approach is followed to generate UI. The attribute-tree is built with the attributes in active-contexts for duration 10-10 and 12-12 nodes. The sample attribute is shown in the Figure 2 for active-context at duration 10-10. The attributes in this figure are divided into two 'Screens', 'Screen 0' consisting of 6 attributes and the 'Screen 1' has one attribute.

5   Performance Analysis and Discussion

5.1   Performance Analysis

We have conducted performance analysis by considering the UI form data for eSagu [9,13] system. In eSagu system, the farmer communicates about the crop problem by capturing photographs and filling in the corresponding crop observation (UI) form and sends it to agricultural experts through eSagu portal. Agricultural experts deliver the expert advice based on the crop photographs and data received through the crop observation form. Each crop observation form consists of about 120 attributes on an average. All these attributes are spread over to the entire crop season that starts from sowing stage to harvesting stage. Individual UIs are created by considering each crop as a context.

We have taken contexts of five crops and applied IT approach and calculated the number of mobile screens, by considering mobile screen size as 5 attributes. We applied the proposed approach and calculated the number of mobile screens required. It was found out that the proposed approach reduced the number of mobile screens required as compared to IT approach. The results of the analysis for the five crops are shown in Tables 2. In this table, the first column indicates serial number, the second column shows the crop name (context), and the third column shows the number of mobile screens in IT and the final column shows the number of mobile screens in ITAT approach. It can be observed that, there is a reduction of 7, 6, 7, 13, and 12 screens are reduced for cotton, chilli, rice, maize and sunflower data-entry forms respectively.

We now discuss how reduction in number of mobile screens are obtained in case of UI form for cotton crop. The cotton crop has 106 attributes with 180 days cycle. The temporality values of the attributes are collected with the help

Table 2. Performance of CA, IT and ITAT CA approaches in eSagu^{TM} system

S.No.	Crop Name	Total Attributes	No.of Screens in IT	No.of Screens in ITAT
1	Cotton	106	22	15
2	Chilli	108	22	16
3	Rice	111	23	17
4	Maize	105	21	14
5	Sun Flower	106	22	14

of domain experts. In IT approach, all 106 attributes are used to generate mobile UI. We assumed the screen size as 5, that means the mobile screen can accommodate 5 attributes on one screen. As a result, these attributes are divided into 22 screens. In ITAT approach, 106 attributes are divided into 70 active-contexts where each active-context contains about 75 attributes. So, for 75 attributes, 15 mobile screens are required. Based on the time of the data-entry, appropriate active-context is displayed to the user. As a result, at a given point of time only 75 attributes are required with 15 screens in ITAT. Whereas, in IT approach, 106 attributes are shown all the days with 22 screens. With ITAT approach, there is a saving of 7 screens for any given time period, that means there is a saving of 35 attributes for each active-context.

Similar kind of analysis has been carried out for UI forms of other four crops. The results are shown in the Table 2.

5.2   Discussion

The proposed approach provides opportunity to improve the performance of UI for mobile screens. However, the performance improvement depends upon several factors which are discussed as follows.

1. **Performance improvement is application dependent:** The proposed approach depends on the notion of temporality. But UI attributes of only certain class of applications exhibit attribute temporality. Also, the proposed approach performs better if more number of attributes in UI form exhibit temporality property.

   In our analysis on real world dataset, it was observed that there are about 30 attributes which could not exhibit temporality property and appeared in all active-contexts. So, the performance could be higher, if more number of attributes exhibit temporality property. Still, the results with the proposed approach are encouraging.

2. **Requires extra effort to extract temporality values:** The proposed approach requires extra effort in identification of attribute temporality in a proper manner. Temporality values can be extracted in two ways: one is during requirement analysis phase and the other is by analyzing the past data-entry patterns.

3. **Performance depends on mobile screen size:** Currently, mobile devices are available with varying screen sizes. If the screen size increases, more active-contexts can be fit into one mobile screen. As a result, performance could be improved.

6 Summary and Conclusions

For mobile devices, UIs design is a challenging task due to small screen size and other constraints. Interface tailoring methods are employed to design UIs for small screens. In this paper, we have proposed an improved approach to design UI forms for small screens by extending the notion of attribute temporality to interface tailoring approaches. The analysis results on the real dataset show that the proposed approach has the potential for improving the UI performance in mobil environment.

As a part of future work, we are planning to investigate the index-based approaches using the notion of attribute temporality for designing improved UI for mobile environments.

Acknowledgments

This work has been carried out with the support from Nokia Global University Grant.

References

1. Niu, Y., Li, X., Meng, X., Sun, J., Dong, H.: A Constraint-based User Interface Design Method for Mobile Computing Devices. In: 1st International Symposium on Pervasive Computing and Applications, August 3-5 (2006)
2. Chade, S., Koivisto, A.: Mobile Form-Editor: A push based group communication tool. In: Mobility 2006, Bangkok, Thailand, October 25-27 (2006)
3. Tsai, M.-J., Chen, D.-J.: Generating User Interface for Mobile Phone Devices Using Template-Based Approach and Generic Software Framework. Journal of Information Science and Engineering 23, 1189–1211 (2007)
4. Menkhaus, G., Pree, W.: User Interface Tailoring for Multi-Platform Service Access. In: International Conference on Intelligent User Interfaces: IUI 2002, San Francisco, CA, January 13-16, pp. 208–209 (2002)
5. Luyten, K., Coninx, K.: An XML-based runtime user interface description language for mobile computing devices. Springer, Heidelberg (2001)
6. Szekely, P.: Retrospective and Challenges for Model-Based Interface Development. In: 3rd Int. Workshop on Computer-Aided Design of User Interfaces CADUI 1996, June 5-7. Presses Universitaires de Namur, Namur (1996)
7. Kumara Swamy, M., Krishna Reddy, P.: An Efficient Context-based User Interface by Exploiting Temporality of Attributes. In: 16th Asia-Pacific Software Engineering Conference (APSEC 2009), Penang, Malaysia, December 1-3, pp. 205–212 (2009)
8. Fletcher, L.A., Erickson, D.J., Toomey, T.L., Wagenaar, A.c.: Handheld Computers a Feasible Altgernative to Paper Forms for Field Data Collection. Evaluation Review 27(2) (April 2003)

9. eSagu: IT-based Personalized Agro-Advisory system (January 2010),
   http://www.esagu.in
10. James, F.D., Christina, G., Won, C.K., Srdjan, K.: A Knowledge-Based User Interface Management System. In: CHI 1988 Human Factors in Computer Systems, Washington, D.C, May 15-19, pp. 67–72 (1988)
11. Dey, A.K., Abowd, G.D.: Towards a better understanding of context and context-awareness. In: Conference on Human Factors in Computing Systems (CHI 2000), The Hague, Netherlands (April 2000)
12. Baldauf, M., Dustdar, S., Rosenberg, F.: A Survey on Context-aware systems. International Journal of Ad Hoc and Ubiquitous Computing 2(4) (2007)
13. Krishna Reddy, P., Ramaraju, G.V., Reddy, G.S.: eSagu^{TM}: A Data Warehouse Enabled Personalized Agricultural Advisory System. In: Proceedings SIGMOD 2007, Beijing, China (2007)
14. Myers, B.A.: Challenges of HCI Design and Implementation. Interactions 1(1), 73–83 (1994)
15. Myers, B.A.: User Interface Software Tools. ACM Transactions on Computer-Human Interaction 2(1), 64–103 (1995)
16. Dan, O.R.: MIKE: The Menu Interaction Kontrol Environment. ACM Transactions on Graphics 4(12), 33–42 (1984)
17. Schlungbaum, E., Elwert: Automatic user interface generation from declarative models. In: CAD UZ 1996, pp. 3–18. Namur University Press, Namur (1996)
18. Schlungbaum, E.: Individual User Interfaces and Model-based User Interface Software Tools. In: IUI 1997, Orlando Florida, USA (1997)
19. Singh, G., Green, M.: A High-level User Interface Management System. In: Proceedings SIGCHI 1989, April 1989, pp. 133–138 (1989)
20. Guha, S., Rastogi, R., Shim, K.: ROCK: ARobust Clustering Algorithm for Categorical Attributes. In: 15th Int. Conf. on Data Engineering (1999)

Towards K-Nearest Neighbor Search in Time-Dependent Spatial Network Databases*

Ugur Demiryurek, Farnoush Banaei-Kashani, and Cyrus Shahabi

University of Southern California
Department of Computer Science
Los Angeles, CA 90089-0781
{demiryur,banaeika,shahabi}@usc.edu

Abstract. The class of k Nearest Neighbor (kNN) queries in spatial networks is extensively studied in the context of numerous applications. In this paper, for the first time we study a generalized form of this problem, called the Time-Dependent k Nearest Neighbor problem (TD-kNN) with which edge-weights are time variable. All existing approaches for kNN search assume that the weight (e.g., travel-time) of each edge of the spatial network is constant. However, in real-world edge-weights are time-dependent (i.e., the arrival-time to an edge determines the actual travel-time on that edge) and vary significantly in short durations. We study the applicability of two baseline solutions for TD-kNN and compare their efficiency via extensive experimental evaluations with real-world data-sets, including a variety of large spatial networks with real traffic-data recordings.

1 Introduction

With the ever growing popularity of online map services (e.g., Google Maps) and their wide deployment in hand-held devices (e.g.,iPhone) and car-navigation systems, more and more users search for geographical points of interests (e.g., restaurants, hospitals) and the corresponding directions and travel-times to these locations. Consequently, many recent research studies (e.g., [23,5,18,16,22,2,12,13,17,24]) focus on developing techniques to accurately and efficiently compute the distance and route between objects in large road-networks. However, a majority of these studies rely on pre-computation of distances in the network and assume that the cost of traveling each edge of the road-network is constant (e.g., corresponding to the length of the edge).

On the other hand, we are witnessing an increase in the instrumentation of roads in major cities for collecting real-time traffic data. For example, we are working with LA METRO [1] from whom we are receiving real-time traffic data from more than 6500 sensors on various freeways and artillery roads in Los Angeles (LA) county. Studying this

* This research has been funded in part by NSF grants IIS-0534761 and CNS-0831505 (CyberTrust), the NSF Center for Embedded Networked Sensing (CCR-0120778) and in part from the METRANS Transportation Center, under grants from USDOT and Caltrans. Any opinions, findings, and conclusions or recommendations expressed in this material are those of the author(s) and do not necessarily reflect the views of the National Science Foundation.
[1] http://www.metro.net/

S. Kikuchi, S. Sachdeva, and S. Bhalla (Eds.): DNIS 2010, LNCS 5999, pp. 296–310, 2010.

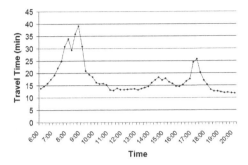

Fig. 1. Real-world travel-time for a weekday on a segment of I-405 in Los Angeles County

real-world traffic data, we observe that the actual travel-time on a road heavily depends on the traffic congestion on the edges and is a function of the time of the day. To illustrate, consider Figure 1 that shows the graph of real-world travel-time on a segment of I-405 freeway in LA between 6am and 8pm on a weekday. Some interesting observations can be made from this figure. First, the travel-time of a segment is a function of the time of the day, hence, the term *time-dependent* travel-time. Second, the change in travel-time is significant, for example from 9:00am to 9:30am, the travel-time of this segment changes from 40 minutes to 20 minutes (100% change). Note that certain network segments (e.g., bridges) can be unavailable during during certain instants of time. Hence, the fastest path from a source to a destination may vary significantly depending on the time of the day. Third, the duration of the change in travel-time is rather short, e.g., within 30 minutes, and the change is continuous and not abrupt. Therefore, the travel-time of an (future) edge may change during a trip. These simple observations have a major computation implication: the time that one arrives at the segment entry determines the travel-time on that segment. Hence, to compute the fastest path from a source to a destination, all combinations of arrival-times at all possible segment entrances must be considered. We call this phenomenon *"arrival-dependency"* and we observe that because of this fact, naive approaches may find incorrect shortest paths (thus incorrect nearest neighbors), especially at the boundaries of traffic rush-hours. Fourth, we remark that there are only a handful of unique travel-time graphs for a given segment (e.g., weekday-graph, weekend-graph, holiday-graph) and hence we can assume that at any given time for any given segment we know the travel-time a priori[2].

Moreover, in addition to the time-dependent traffic data collected by governmental agencies (such as the data we receive from LA Metro), recently an increasing number of navigation companies are also releasing time-dependent travel-time information for road networks. For example, NAVTEQ [19], a leading provider of navigation services, offers a Predictive Flow Service that provides time-dependent travel-times up to one year. Also, INRIX [14] recently announced its new service that provides future traffic (at the temporal granularity of five minutes) computed based on the historical traffic data and local information like weather, school schedules, and sporting events.

---

[2] Traffic prediction is an active area of research and beyond the scope of this paper.

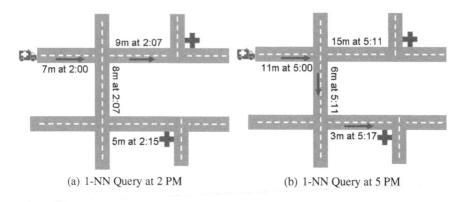

(a) 1-NN Query at 2 PM   (b) 1-NN Query at 5 PM

Fig. 2. Time-dependent 1-NN search

Figure 2 illustrates an example of time-dependent k nearest neighbor search. With this example, an ambulance is looking for the nearest hospital at 2 PM and 5 PM on the same day on a particular road network. The time-dependent travel-time (in minutes) and the arrival time for each edge are shown on the edges. Note that the travel-times on the edges change with arrival time to the edges in Figures 2(a) and 2(b). Therefore, the query launched by the ambulance at 2pm and 5pm would return different results.

From Figure 2, one can come up with a naive approach to answer time-dependent kNN problem by applying an existing kNN search algorithm (e.g., [22,16,2,12]) on different snapshots of the graph generated at discrete times. However, there are fundamental shortcomings with such a naive approach. First, the naive approach needs to update the edge weights and hence, the pre-computation (if any) for every snapshot. This is not realistic for real-world scenarios where the spatial network is large. Second, the naive approach can provide inaccurate results since the computations are done on discrete times rather than in continuous time. Specifically, the shortest path between the objects is derived based on the weights known at the query time for all network edges, disregarding the probably changing weight information during the trip (recall that the network edge weights change over the time). For example, consider the travel-time of all the edges are equal to five minutes at 5 PM in Figure 2(b). Finally, it is very hard to decide on the effective choice of snapshot intervals for real-world applications.

Considering afore-mentioned observations on the impact of time-dependency on fastest path computation and the availability of time-dependent travel-time information for road networks, the need for computational techniques that can answer time-dependent spatial network queries (e.g., time-dependent kNN query) is apparent and immediate. Unfortunately, once we consider the road networks with time-varying edge weights, all the techniques assuming constant edge-weights and/or relying on distance pre-computation would fail to answer k nearest neighbor queries in time-dependent road networks.

In this paper, for the first time, we study the problem of *Time-dependent k Nearest Neighbor* (TD-kNN) search. TD-kNN finds the k static nearest neighbors of a query object which is moving on a *time-dependent network* (i.e., a network where edge weights are variable functions of time). We discuss and compare two different baseline methods to answer time-dependent k nearest neighbor queries in both discrete and continuous time. With the first approach, we use time-expanded graphs to model the time-dependent network. This allows us to exploit previously developed techniques for static networks to solve TD-kNN problem with approximate results. With the second approach, we adopt incremental network expansion by generalizing it to time-dependent networks.

The remainder of this paper is organized as follows. In Section 2, we review the related work on both kNN queries and time-dependent shortest path algorithms. In Section 3, we formally define the time-dependent k nearest neighbor query in spatial networks. In Section 4, we introduce two different baseline approaches for time-dependent kNN queries. In Section 5, we present the results of our experimental evaluation of our proposed approaches with a variety of spatial networks with large number of data and query objects. Finally, in Section 6 we conclude and discuss our future work.

2   Related Work

In this section we review previous studies on kNN query processing in road networks as well as time-dependent shortest path computation.

2.1 kNN Queries in Spatial Networks

In [22], Papadias et al. introduced Incremental Network Expansion (INE) and Incremental Euclidean Restriction (IER) methods to support kNN queries in spatial networks. While INE is adaption of Dijkstra's algorithm, IER exploits the Euclidean restriction principle in which the results are first computed in Euclidean space and then refined by using the network distance. In [16], Kolahdouzan and Shahabi proposed first degree *network Voronoi diagrams* to partition the spatial network to network Voronoi polygons (NVP), one for each data object. They indexed the NVPs with a spatial access method to reduce the problem to a point location problem in Euclidean space and minimize the on-line network distance computation by precomputing the NVPs. Cho et al. [2] presented a system UNICONS where the main idea is to integrate the precomputed k nearest neighbors into the Dijkstra algorithm. Hu et al. [4] proposed a distance signature approach that precomputes the network distance between each data object and network vertex. The distance signatures are used to find a set of candidate results and Dijkstra algorithm is employed to compute their exact network distance. Huang et al. addressed the kNN problem using *Island* approach [12] where each vertex is associated to all the data points that are centers of given radius r (so called islands) covering the vertex. With their approach, they utilized a restricted network expansion from the query point while using the precomputed islands. In [13], Huang et al. introduced *S-GRID* where they partition the spatial network to disjoint sub-networks and precompute the shortest path for each pair of border points. To find the k nearest neighbors, they first perform a network expansion within the sub-networks and then proceed to outer expansion between the border points by utilizing the precomputed information. Recently Samet et al. [23] proposed a method where they associate a label to each edge that represents all nodes to which a shortest path starts with this particular edge. They use these labels to traverse *shortest path quadtrees* that enables geometric pruning to find the network distance. With all these studies, the network edge weights are assumed to be static (i.e., travel-time functions of all edges are constants) and hence the shortest path computations and precomputations are invalidated with time-varying edge weights. Unlike the previous approaches, we make a fundamentally different assumption that the weight of the network edges are time-dependent rather than fixed. Our assumption yields a much more realistic scenario and versatile approach.

2.2 Time-Dependent Shortest Path Studies

Cooke and Halsey [3] introduced the first time-dependent shortest path (TDSP) solution where they formulated the problem in discrete time and use dynamic programming. Dreyfus [8] proposed a generalization of Dijkstra algorithm, but his algorithm is showed (by Halpren [11]) to be true in only FIFO networks. If the FIFO property does not hold in a time-dependent network, then the problem is NP-Hard as shown in [20]. In [1], Chabini proposed a discrete time one-to-all and all-to-one TDSP algorithm that allows waiting at network nodes. In [9], George and Shekhar proposed a time-aggregated graph where they aggregate the travel-times of each edge over the time instants into a time series. Their model has less storage requirements than the time-expanded networks. All these studies assume that the edge weight functions are defined over a finite discrete window of time $t \in t_0, t_1, .., t_n$, where t_n is determined by the total duration of time interval under consideration. Therefore, the problem is reduced the problem of computing, for each time window, minimum-weight paths through the static network where one can apply any of the well-known shortest path algorithms. Although discrete-time algorithms are easy to implement, they have numerous shortcomings mainly on storage (discussed in Section 1). Orda and Rom [20] proposed a Bellman-Ford based solution where edge weights are piece-wise linear functions. In [4], Dean proposed a label-setting algorithm where arrival times are considered as the labels of the nodes. In [7], Ding et al. also used a variation of label-setting algorithm to solve the time-dependent shortest path problem. Their algorithm decouples the path-selection and time-refinement by scanning a sequence of time steps of which the size depends on the values of the arrival time functions. The focus of this algorithm is to find a fastest path in a time-dependent graph for a given start time interval (e.g., between 7:30 AM and 8:30 AM). In [15], Kanoulas et al. introduced a Time-Interval All Fastest Path (allFP) algorithm based on A* algorithm in time-dependent networks. Instead of sorting the priority queue by scalar values, they maintain a priority queue of all paths to be expanded. Therefore, they enumerate all the paths from the source to a destination node which incurs exponential running time in the worst case. In addition, their algorithm is efficient when estimation (heuristic function in A*) can enable effective search space pruning. It is difficult to find such estimation in time-dependent graphs.

3   Problem Definition

In this section, we will formally define the problem of time-dependent kNN search in spatial networks. We assume a spatial network (e.g. the Los Angeles road network) containing a set of static data objects (i.e., points of interest such as restaurants, hospitals) as well as moving query objects searching for their kNN. We model the spatial network as a time-dependent weighted graph whose edge and node properties as well as topological structure are time varying. We assume that the non-negative edge weights are time-dependent travel-times between the nodes and the time-dependent edge weights are given priori. We assume both data and query objects lie on the network edges and all relevant information about the objects is maintained by a central server. We model the data objects as the network nodes. As a query object moves, the central server is updated with the new location of the object. Below, we formally define our terminology.

Definition 1. *Time-dependent Graph*
A Time-dependent Graph (G_T) is defined as $G_T(V, E)$ where $V = \{v_i\}$ is a set of nodes representing the intersections and terminal points, and E ($E \subseteq V \times V$) is a set of edges representing the network segments each connecting two nodes. Each edge e is represented by $e(v_i, v_j)$ where v_i and v_j are starting and ending nodes, respectively, and $v_i \neq v_j$. For every edge $e(v_i, v_j) \in E$, there is an edge travel-time function $c_{i,j}(t)$, where t is the time variable in time domain T. An edge travel-time function $c_{i,j}(t)$ specifies how much time it takes to travel from v_i to v_j starting at time t. For example, Figure 3 depicts a road network modeled as a time-dependent graph $G_T(V, E)$. Figure 3(a) shows the graph structure with five edges and corresponding time-dependent travel-times (as piece-wise linear functions) for each edge.

Definition 2. *Travel-Time*
Let $\{s = v_1, v_2, ..., v_k = d\}$ represent a path which contains a sequence of nodes where $e(v_i, v_{i+1}) \in E$, $i = 1, ..., k - 1$. Given a time-dependent graph G_T, a path $(s \rightsquigarrow d)$, and a departure-time from the source t_s, the time-dependent travel time $TT(s \rightsquigarrow d, t_s)$ is the time it takes to travel along the path. Since the travel-time of an edge varies depending on the arrival-time to that edge (i.e., arrival-dependency), the travel time is computed as follows:

$$TT(s \rightsquigarrow d, t_s) = \sum_{i=1}^{k-1} c_{(v_i, v_{i+1})}(t_i) \text{ where } t_1 = t_s, \ t_{i+1} = t_i + c_{(v_i, v_{i+1})}(t_i), \ i =$$

$1, .., k$.
   For example, in Figure 3 the travel-time of path $\{(v_1, v_2, v_3, v_5)\}$ with departure-time $t = 5$ is $TT(v_1 \rightsquigarrow v_5, 5) = 45$.

Definition 3. *Time-dependent Shortest (Fastest) Path*
Given a G_T, a source $s \in V$, a destination $d \in V$, and a departure-time t_s from the source, the time-dependent shortest path $TDSP(s, d, t_s)$ is a path with the minimum travel-time among all paths from s to d. Since we consider the travel-time between nodes as the distance measure to find the shortest path, we refer to $TDSP(s, d, t_s)$ as time-dependent fastest path $TDFP(s, d, t_s)$ and use them interchangeably in the rest of the paper. In a time-dependent graph, the fastest path from s to d changes based on

the departure-time from the source. For instance, in Figure 3, suppose a query looking for the fastest path from node v_1 to v_5 for $t_s = 5$. In this case, $TDFP(v_1, v_5, 5) = \{v_1, v_2, v_3, v_5\}$. However, the same query will return $TDFP(v_1, v_5, 10) = \{v_1, v_2, v_4, v_5\}$ for $t_s = 10$. This is why all the methods assuming constant edge-weights and/or relying on distance pre-computation would fail to answer k nearest neighbors in time-dependent road networks. Obviously, with constant edge weights (i.e., time-independent), regardless of the query time the query would always return the same path (and path travel-time) as the result.

Definition 4. *Time-dependent k Nearest Neighbor Query (TD-kNN)*
A time-dependent k nearest neighbor query on spatial networks is defined as a query that finds the k nearest neighbors of a query object moving on a time-dependent network G_T. Considering a set of n data objects $P = \{p_1, p_2, ..., p_n\}$, the TD-kNN query with respect to a query point q finds a subset $P' \subseteq P$ of k objects with minimum time-dependent travel-time to q, i.e., for any object $p' \in P'$ and $p \in P - P'$, $TDFP(q, p', t) \leq TDFP(q, p, t)$.

In the rest of this paper, we assume that the edge travel-time functions are given as positive piece-wise linear functions of time and all piece-wise functions have a finite number of pieces. This is consistent with how traffic trends are reported for a given edge in real-world road networks. We also assume that the spatial network G_T satisfies the First-In-First-Out(FIFO) property [4]. This property suggests that moving objects exit from an edge in the same order they enter the edge. Finally, with our algorithm, we do not allow objects to wait at a node, because, in most real-world applications, waiting at a node is not realistic as it requires the moving object to get out of the current road (e.g., the exit freeway) and find a place to park and wait.

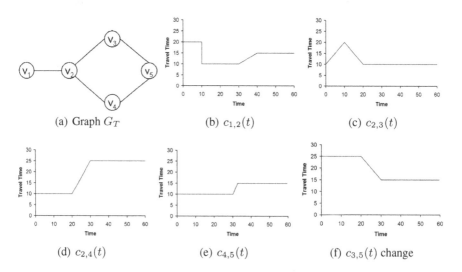

(a) Graph G_T (b) $c_{1,2}(t)$ (c) $c_{2,3}(t)$

(d) $c_{2,4}(t)$ (e) $c_{4,5}(t)$ (f) $c_{3,5}(t)$ change

Fig. 3. A Time-dependent Graph $G_T(V, E)$

4   Baseline TD-KNN Algorithms

In this section, we explain two different algorithms to evaluate time-dependent k nearest neighbor queries in spatial networks. With the first approach, we model the time-dependent road network as a time-expanded graph [21] that approximates the time-dependent network with a snapshot of the network in each time interval. With the second approach, we exploit a generalization of incremental network expansion [22] method where we use time-dependent arrival times as the labels of the nodes to form the greedy search. While the former enables us to use existing nearest neighbor algorithms but with approximate results, the latter allows for obtaining exact results in FIFO time-dependent networks.

4.1   TD-kNN with Time-Expanded Networks

Given a time-dependent graph $G_T(V, E)$, a time-expanded model discretizes the time domain $T = [t_0; t_n]$ into n points of time, and constructs a static graph $G(V, E)$ by making n copies of each node and each edge, respectively. Specifically, time-expanded network replicates the original network for each discrete time unit $t = 0, 1, ..., t_n$, where t_n is determined by the total duration of the time interval under consideration. This model connects a node and its copy at the next instant in addition to the edges in the original network, replicated for every time instant. The weight of an edge in time-expanded network is the time difference between the time events associated with its endpoints. Therefore, a time-varying edge cost can be interpreted as a static flow in the corresponding time-expanded network. Figure 4 shows four consecutive snapshots and the corresponding time-expanded model of the time-dependent network in Figure 3. In this figure, for example, the weight (i.e., travel-time) of edge (v_1, v_2) at $t = 0$ is represented by connecting the copy of node v_1 at $t = 0$ to the copy of node v_2 at $t = 20$.

The time-expanded network approach enables time-dependent k nearest neighbor problem to be solved by applying techniques developed for static networks (e.g., INE). However, there are two drawbacks with any solution based on time-expanded networks. First, since the original network is replicated across time instants, the size of the network increases hence, resulting in high storage overhead and slower response time. The storage requirement for a time expanded-network is $O(|V|T) + O(|V| + |E|T)$, where T is the total number of snapshots. Second, the difference between the shortest path obtained using time-expanded model and the optimal shortest path is very sensitive to the parameter n, and is unbounded. Because, the query time (or the arrival-time to an edge) can always be between any two time points (e.g., between t_0 and t_1), but the edge weights are only captured in either of the time points. For example, consider a shortest path query executed at $t = 12$ in Figure 4, and an error ϵ between the optimal path and the path found using time-expanded network model. In this case, the network snapshot at $t = 10$ is used to compute the shortest path for $t = 12$ and ϵ is accumulated on each edge along the path. Our experiments show that the error rate is especially high during rush hours (see 5.2), hence causing time-expanded models to generate inaccurate results.

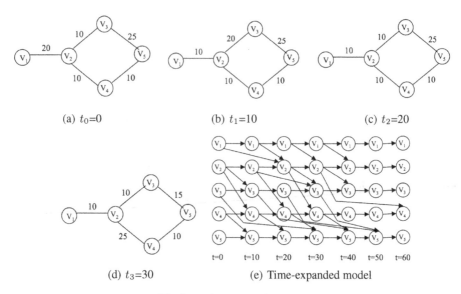

(a) $t_0=0$ (b) $t_1=10$ (c) $t_2=20$

(d) $t_3=30$ (e) Time-expanded model

Fig. 4. A Time-expanded graph

4.2   TD-kNN with Network Expansion

In this section, we propose an algorithm that generalizes the incremental network expansion method [22] (originally proposed for static road networks) to answer k nearest neighbor queries in time-dependent road networks. With this algorithm, starting from the query object q all network nodes reachable from q in every direction are visited in order of their proximity (i.e., time-dependent travel-time) to q until all k nearest data objects are located. We use four main data structures to enable the network expansion solution. The *adjacency component* captures the network connectivity. The *edge component* includes the poly-line representation of each network edge (u, v), length of the edge, and a pair of pointers to the disk pages containing the adjacency lists of its endpoints u, v. The *travel-time component* includes the travel-time functions of each network edge. We use hash table to associate travel-time functions to network edges. The last component is R-tree [10] that indexes the edges' MBRs. Each leaf entry of R-tree contains a pointer to the disk page storing the corresponding edge.

We outline our proposed approach in Algorithm 1. The algorithm takes three parameters as the input, i.e., query location q, number of desired nearest neighbors k, and query time t_q. We first execute a $findEdge(q)$ operation to find the edge that contains q by performing a point location query on R-tree index. Then, we expand the network based on the time-dependent travel-time to each node around q until k objects are found. Specifically, we keep track of the arrival time at the end node of an edge e and use this arrival time to determine the (time-dependent) travel cost of the adjacent edges of edge e. In Algorithm 1, let $t(v)$ denote the time taken to travel from q to node v along the fastest path in a time-dependent network, i.e., time to travel for $TDFP(q, v, t_q)$. Analogous to shortest path distances, for every edge $e(u, v)$, we have $t(v) \leq f_e(t(u))$ where $f_e(t(u))$ is the time taken to travel from q to u plus the time travel from u to v (see Section 3 for time-dependent travel-time computation). The core idea (as initially shown in [8,4]) with this algorithm is that, analogous to shortest path distances, we now have $t(v) \leq f_e(t(u))$ and this will form the basis of our greedy algorithm. We use a priority queue S to keep track of the nodes to be examined. With S, we maintain the set *explored* nodes (which includes the nodes for which we have calculated the actual $t(v)$) as well as a label $l(v)$ for each node not in S, where the label is our current estimate for the least time it takes to reach v from s. We update $l(v)$ by $min(l(v), f_e(t(u)))$ (Line 8) only considering the edges (u, v) where $u \in S$. Finally, we pick the node $w \notin S$ with smallest $l(.)$ value and add it to the set S. If the recently added node is a data object (recall that data objects are modeled as network nodes), we add that data object to nearest neighbor array NN and accordingly compute its travel time (Line 11-13). This process is repeated until the algorithm finds k data objects. It is important to note that Algorithm 1 holds for FIFO networks in which greedy property is maintained.

Algorithm 1. TDFP(q,k,t_q)

1: // S: set of nodes, q: query location, dt: departure-time from q
2: // tt: travel-time of the fastest path, v_i: last node added to S
3: // NN: array of current NNs
4: Initialize $S = \{q\}, t(q) = 0, l(v) = \infty$ for all $v \notin S$
5: $v_i = q$
6: **while** $S \neq V$ **do**
7:     $for\ each\ e(v_i, v_j) \in E\ where\ v_j \notin S$
8:     $l(v_j) = min(l(v_j), f_e(t(v_i)))$
9:     $Let\ w \notin S\ such\ that\ l(w) = min_{v_j \notin S} l(v_j)$
10:     $S = S \bigcup \{w\}, t(w) = l(w), v_i = w$
11:     If v_i *is a dataObject* Then
12:     add v_i to NN;
13:     $tt = t(v_i) - t_q$; //compute travel-time to v_i
14:     End If;
15:     If $NN.size() = k$ Then break;
16: **end while**
17: $return\ NN$ and travel times

---

5 Experimental Evaluation

5.1 Experimental Setup

We conducted several experiments with different spatial networks and various parameters (see Table 1) to evaluate the performance of both TD-kNN algorithms. As our dataset, we used Los Angeles (LA) and San Joaquin County (SJ) road network data with 304,162 and 24,123 road segments, respectively. We obtained these datasets from TIGER/Line [1]. Both of these datasets fit in the memory of a typical machine with 4GB of memory space.

To create realistic time-dependent edge weights, it is necessary to collect huge amounts of data about the network edge behaviors. Towards that end, for the past 1 year, we have

---
[1] http://www.census.gov/geo/www/

been continuously collecting speed, occupancy, volume sensor data from a collection of approximately 6000 sensors located on the road network of Los Angeles County. The sampling rate of the data is 1 reading/sensor/min. Currently, our database consists of about 750 million sensor reading representing speed profiles on the road network segments of LA. We used the historical sensor data to create travel-time functions for LA network. In order to create the time-dependent edge weights of SJ network, we developed a traffic model [6] that synthetically generates time-dependent edge weights for SJ.

Table 1. Experimental parameters

Parameters	Default	Range
Number of objects	10 (K)	1,5,10,15,20(K)
Number of queries	3 (K)	1,2,3,4,5 (K)
Number of k	20	1,10,20,30,40,50
Object Distribution	Uniform	Uniform, Gaussian
Query Distribution	Uniform	Uniform, Gaussian

We generated the parameters represented in Table 1 using a simulator prototype developed in Java. We conducted our experiments on a workstation with 2.7 GHz Pentium Core Duo processor and 12GB RAM memory.

We computed the time-expanded network model of both LA and SJ networks by discritizing the networks for each 15 minutes. Similar to Algorithm 1 (denoted by TD-NE), we implemented a network expansion method to find k nearest neighbors in time-expanded networks (denoted by TE). We continuously monitored each query for 50 timestamps in both of the implementations. For each set of experiments, we only vary one parameter and fix the remaining to the default values in Table 1.

Since the experimental results with both LA and SJ networks differ insignificantly and due to space limitations, we only present the results from LA dataset.

5.2 Results

Correctness and Impact of k. With this experiment, we compare the correctness of the two algorithms (i.e., percentage of correctly identified nearest neighbors). Figure 5(a) plots the correctness versus time ranging from 6 am to 6 pm, while using default settings in Table 1 for all other parameters. As shown, while TD-NE returns correct results all the time, TE's correctness is substantially low around rush hours (i.e., 7-9 am, 4-6 pm). This is because time-dependent weights of each network segment change rapidly especially at the boundaries of the traffic peak periods, resulting the error accumulating along the path.

Next, we compare the performance of the two algorithms with regard to k. Figure 5(b) shows the average query efficiency versus k ranging from 1 to 50. The results indicate that TD-NE outperforms TE with all values of k and the response time of both algorithms increases with the large values of k. Note that the slower response time of TE in this and the following experiments is due to increased size of the network because of replication. One can implement pre-computation techniques to accelerate the response time of TE.

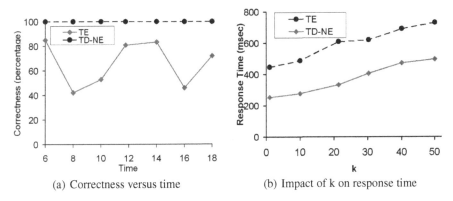

(a) Correctness versus time   (b) Impact of k on response time

Fig. 5. Correctness and impact of k

Impact of Object/Query Distribution and Network Size. With this experiment, we study the impact of object and query distribution as well as network size. Figure 6 shows the response time of both algorithms where the objects and queries follow either uniform or Gaussian distributions. As illustrated, TD-NE yields better performance for queries with Gaussian distribution. This is because as queries are clustered in the spatial network with Gaussian distribution, their nearest neighbor would overlap; hence, allowing TD-NE to save computation.

In addition, we measured the performance of both algorithms with respect to the network size. In order to evaluate the impact of network size, we conducted experiments with the sub-networks of LA dataset ranging from 50K to 250K segments. Figure 6(b) illustrates the response time of both algorithms with different network sizes. In general, with the default parameters in the Table 1, the response time increases for both algorithms as the network size increases.

Impact of Object and Query Cardinality. With this set of experiments, we compare the performance of the two algorithms by varying the cardinality of the data objects

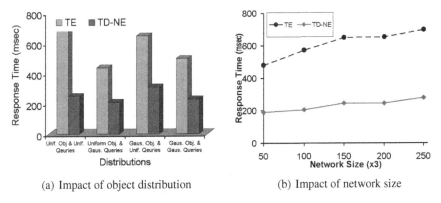

(a) Impact of object distribution   (b) Impact of network size

Fig. 6. Response time versus distribution and network size

(P) from 1K to 20K while using default settings in Table 1 for all other parameters. Figure 7(a) illustrates the impact of the growing object cardinality on response time. The results indicate that the response time linearly increases with the number of data objects in both methods, where TD-NE outperforms TE for all cases. From P=1K to 5K, the response time is slower. Because, since the objects are sparsely distributed when P is small, network expansion visits more redundant network nodes causing extra processing time. Figure 7(b) shows the impact of the query cardinality (Q), ranging from 1K to 5K, on response time. As shown, TD-NE scales better with large number of Q and the performance gap between the approaches increases as Q grows.

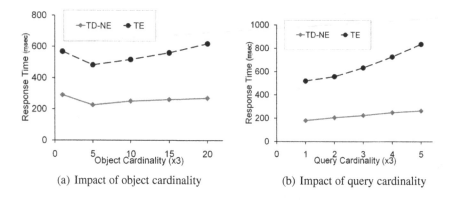

(a) Impact of object cardinality   (b) Impact of query cardinality

Fig. 7. Response time versus query/object cardinality and agility

6   Conclusion and Future Work

In this paper, for the first time we studied the problem of time-dependent k nearest neighbor search (TD-kNN) in spatial networks. We formulated a generalized type of k nearest neighbor query where we, unlike the existing studies, assume the edge weights of the network are time varying rather than fixed. We studied two baseline solutions exploiting time-expanded network and network expansion frameworks and compared their efficiency with real-world data-sets, including a variety of large spatial networks with real traffic-data. Although time-expanded network framework provides a mechanism to use existing kNN algorithms for static networks, the experimental results suggest that the error rate (incorrectly identified nearest neighbors) of this approach is very high especially during traffic peak hours. On the other hand, while network expansion yields correct results at all times, the overhead of executing network expansion is very high particularly in large networks with a sparse set of data objects, hence the need for efficient time-dependent search algorithms.

We intend to pursue this study in three different directions. First, we intend to investigate new data models for effective representation of spatiotemporal road networks. This is critical in supporting development of efficient and accurate time-dependent algorithms (e.g., shortest path), while minimizing the storage and cost of the computation. Second, given that online nearest neighbor queries require near real-time response time, we plan to develop novel preprocessing and indexing techniques that can be used to accelerate kNN computation in time-dependent networks. Third, we plan to study a variety of other spatial queries (including range queries and skyline queries) in time-dependent road networks.

Given the importance of time-dependency for accurate and realistic spatial query processing in road networks, as well as increasing use of traffic sensors and, hence; availability of time-varying traffic data, we predict rapid growth of interest (at academia and industry) in developing various query processing solutions for spatial queries in time-dependent road networks.

References

1. Chabini, I.: The discrete-time dynamic shortest path problem: Complexity, algorithms, and implementations. In: Transportation Research Record (1999)
2. Cho, H.-J., Chung, C.-W.: An efficient and scalable approach to cnn queries in a road network. In: VLDB (2005)
3. Cooke, L., Halsey, E.: The shortest route through a network with timedependent internodal transit times. Journal of Mathematical Analysis and Applications (1966)
4. Dean, B.C.: Algorithms for minimum cost paths in time-dependent networks. Networks (1999)
5. Demiryurek, U., Banaei-Kashani, F., Shahabi, C.: Efficient continuous nearest neighbor query in spatial networks using euclidean restriction. In: Mamoulis, N., Seidl, T., Pedersen, T.B., Torp, K., Assent, I. (eds.) Advances in Spatial and Temporal Databases. LNCS, vol. 5644, pp. 25–43. Springer, Heidelberg (2009)
6. Demiryurek, U., Pan, B., Kashani, F.B., Shahabi, C.: Towards modeling the traffic data on road networks. In: GIS-IWCTS (2009)
7. Ding, B., Yu, J.X., Qin, L.: Finding time-dependent shortest paths over large graphs. In: EDBT (2008)
8. Dreyfus, S.E.: An appraisal of some shortest-path algorithms. Operations Research 17(3) (1969)
9. George, B., Kim, S., Shekhar, S.: Spatio-temporal network databases and routing algorithms: A summary of results. In: Papadias, D., Zhang, D., Kollios, G. (eds.) SSTD 2007. LNCS, vol. 4605, pp. 460–477. Springer, Heidelberg (2007)
10. Guttman, A.: R-trees: A dynamic index structure for spatial searching. In: SIGMOD (1984)
11. Halpern, J.: Shortest route with time dependent length of edges and limited delay possibilities in nodes. In: Mathematical Methods of Operations Research (1969)
12. Huang, X., Jensen, C.S., Saltenis, S.: The island approach to nearest neighbor querying in spatial networks. In: Bauzer Medeiros, C., Egenhofer, M.J., Bertino, E. (eds.) SSTD 2005. LNCS, vol. 3633, pp. 73–90. Springer, Heidelberg (2005)
13. Huang, X., Jensen, C.S., Saltenis, S.: S-grid: A versatile approach to efficient query processing in spatial networks. In: Papadias, D., Zhang, D., Kollios, G. (eds.) SSTD 2007. LNCS, vol. 4605, pp. 93–111. Springer, Heidelberg (2007)
14. Inrix, http://www.inrix.com (last visited January 2, 2010)
15. Kanoulas, E., Du, Y., Xia, T., Zhang, D.: Finding fastest paths on a road network with speed patterns. In: ICDE (2006)
16. Kolahdouzan, M., Shahabi, C.: Voronoi-based k nearest neighbor search for spatial network databases. In: VLDB (2004)
17. Lauther, U.: An extremely fast, exact algorithm for finding shortest paths in static networks with geographical background. In: Geoinformation and Mobilitat (2004)

18. Mouratidis, K., Yiu, M.L., Papadias, D., Mamoulis, N.: Continuous nearest neighbor monitoring in road networks. In: VLDB (2006)
19. Navteq, http://www.navteq.com (last visited January 2, 2010)
20. Orda, A., Rom, R.: Shortest-path and minimum-delay algorithms in networks with time-dependent edge-length. J. ACM (1990)
21. Pallottino, S., Scutellà, M.G.: Shortest path algorithms in transportation models: Classical and innovative aspects. In: Equilibrium and Advanced Transportation Modelling (1998)
22. Papadias, D., Zhang, J., Mamoulis, N., Tao, Y.: Query processing in spatial network databases. In: VLDB (2003)
23. Samet, H., Sankaranarayanan, J., Alborzi, H.: Scalable network distance browsing in spatial databases. In: SIGMOD (2008)
24. Wagner, D., Willhalm, T.: Geometric speed-up techniques for finding shortest paths in large sparse graphs. In: Di Battista, G., Zwick, U. (eds.) ESA 2003. LNCS, vol. 2832, pp. 776–787. Springer, Heidelberg (2003)

Spatial Query Processing Based on Uncertain Location Information

Yoshiharu Ishikawa

Information Technology Center, Nagoya University
ishikawa@itc.nagoya-u.ac.jp
http://www.itc.nagoya-u.ac.jp/~ishikawa

Abstract. Location information acquired by sensors and other devices is not necessarily accurate and has vagueness. In the research field of spatial databases, query processing techniques based on uncertain location information are highly interested in recent years. In this paper, we overview the trend of this field and describe our related projects and future prospects.

1 Introduction

In recent years, due to the development of mobile network technology and GPS devices, information services and applications based on location information are repidly increasing. Location information acquired by GPS usually contains noise and we may not be able to obtain accurate location information. Uncertainty of location occurs not only in mobile computing but also in a location estimation of a mobile robot. In robotics, the location of a mobile robot is often estimated by sensors and movement histories [1]. Based on such a background, research on spatial database queries based on uncertain location information has become an active research field.

Our research group has been investigating query processing techniques for spatial databases based on uncertain information [2,3]. The feature is to use the *Gaussian distribution* to express location vagueness. Gaussian distribution is one of the basic probability distributions and widely used in statistics and pattern recognition [4]. Our research mainly focuses on the situation when the user's position obeys a Gaussian distribution and we have proposed algorithms for range queries [3] and nearest neighbor queries [2]. In this paper, we overview related studies in spatial query processing for uncertain location information then briefly introduce the research activities of our group.

2 Overview of Related Work

There are various ways for representing uncertainty of locations. The simplest approach is to assume the location of an object obeys a *uniform distribution* (e.g., [5]). Another approach is to assume a location is generally described by some *probabilistic density function* (*PDF*). The assumed types of PDFs are different

S. Kikuchi, S. Sachdeva, and S. Bhalla (Eds.): DNIS 2010, LNCS 5999, pp. 311–316, 2010.
© Springer-Verlag Berlin Heidelberg 2010

depending on the studies. Most of the approaches [6,7,8] treat a PDF as a black box and do not consider its details. Their approaches aim at generic algorithms which work well for arbitrary PDFs. On the other hand, some other approaches consider special types of PDFs [2,3]. Using specific properties of a PDF, we would be able to develop efficient and effective algorithms.

We can classify types of spatial queries. There are various types of queries in spatial databases [9]. We can also consider several types of queries for uncertain location information. Most of the proposals focus on the algorithms for range queries [7,10,3,5,8] and nearest neighbor queries [6,11,10,12]. Some studies focus on clustering methods based on uncertain location information [13].

We can consider other criteria by considering which objects are uncertain. Roughly speaking, there are three cases: data objects are uncertain [11,10,5,8], query objects are uncertain [3], and both of the objects are uncertain [6,7,12]. In addition, the proposed algorithms are different in several factors such as the types of object shapes (e.g., points, rectangles) and the number of dimensions (e.g., 1-D, 2-D, arbitrary dimensions).

The concept of an *uncertainty region* is often used in the algorithms for spatial querying based on uncertain location information. An uncertainty region for an object is a region such that the region in which the object is located with the specified probability. It roughly represents the area in which the object locates and used for fast processing when a spatial query is given. An uncertainty region may be contained in the description of the object itself, or it may be derived from the properties of the object. The latter approach is often taken when it is costly to treat the vague representation of an object directly.

When the location of an object is given by a PDF, integration of the PDF is often required to process queries. If the PDF cannot be integrated analytically, it is necessary to use numerical integration such as the Monte Carlo method, but the processing cost is quite huge. In this case, a filtering process to decrease the number of target objects for integration becomes extremely important, and the concept of an uncertainty region is especially effective.

In addition to these approaches, some studies incorporated sampling methods into query processing algorithms [12]. Moreover, [14] extended the notion of the *Voronoi diagram* [15] for uncertain locations to process nearest neighbor queries.

3   Summary of Our Research

3.1   Uncertain Query Locations

In our research, we assume that the Gaussian distribution is used for representing uncertainty of locations. In the previous studies [2,3], we assume that the location of a query object is imprecisely specified by a Gaussian distribution. The location of a query object is formally defined as follows.

Definition 1. *Assume that $\boldsymbol{x}$, the location of a query object q, is represented by a d-dimensional Gaussian distribution [4]*

$$p_q(\boldsymbol{x}) = \frac{1}{(2\pi)^{d/2}|\boldsymbol{\Sigma}|^{1/2}} \exp\left[-\frac{1}{2}(\boldsymbol{x} - \boldsymbol{q})^t \boldsymbol{\Sigma}^{-1}(\boldsymbol{x} - \boldsymbol{q})\right], \tag{1}$$

where q is the average of the distribution, Σ is a $d \times d$ covariance matrix, and $|\Sigma|$ represents its determinant.

As described in the introduction part, the Gaussian distribution is quite popular in statistics and pattern recognition [4] and is also used in estimating the location of moving robots [1]. We have developed efficient spatial query processing algorithms considering the specific properties of a Gaussian distribution.

3.2 Probabilistic Range Queries

As an example, let us consider a neighborhood information retrieval for a moving robot. We assume that a moving robot sequentially estimates its location based on sensor information and own movement histories (Fig. 1). If we represent and track the movement by using a Kalman filter [1] assuming a Gaussian probabilistic process, the location of the robot in every moment is be vaguely represented by a Gaussian distribution.

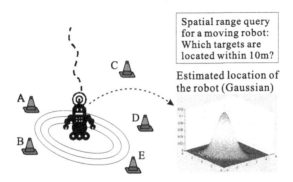

Fig. 1. Spatial range query issued by a moving robot

Consider that we want to retrieve the information of the target objects (e.g., obstacles or items to be obtained) within 10m from the robot. We assume that the data for the target objects (their locations are represented by points) are maintained in the database embedded within the robot. This query cannot be processed as a normal spatial range query—since the location of the robot is uncertain, the distance between the robot and each target is also uncertain and is given probabilistically.

To solve this problem, we have extended the concept of a spatial range query [3]. We defined a *probabilistic range query (PRQ)* such as "retrieve the target objects such that the probabilities such that the locations of the objects are within 10m from the moving robot are less than the threshold". It is formally given as follows.

Definition 2. *Given the probability density function $p_q(x)$, the distance threshold δ $(\delta > 0)$, and the probability threshold θ $(0 < \theta < 1)$, a probabilistic range* query $PRQ(q, \delta, \theta)$ *returns all the objects such that the probabilities that their distances from the query object q are less than or equal to δ are greater than or equal to θ. It is formally defined as follows:*

$$PRQ(q, \delta, \theta) = \{o \mid o \in \mathcal{O}, \Pr(\|\boldsymbol{x} - \boldsymbol{o}\|^2 \leq \delta^2) \geq \theta\}, \tag{2}$$

where $\mathcal{O}$ is the set of the target objects, $\|\cdot\|$ is the length of a vector, and $\|\boldsymbol{x} - \boldsymbol{o}\|^2$ represents the squared Euclidean distance between $\boldsymbol{x}$, the location of the query object q, and $\boldsymbol{o}$, the location of the object o.

The naive approach to process this query is as follows. For each target object, we consider a circle which is centered at the target point and has the radius δ. If the probability that the query object (moving robot) locates inside of the circle is greater than or equal to θ, the target object satisfies the query condition. However, we need to integrate the Gaussian distribution shown in Eq. (1) for evaluating the probability—this is quite costly because it requires numerical integration such as the Monte Carlo method.

To solve the problem, we proposed three query processing strategies to reduce the number of target objects for which numerical integration should be performed [3]. Using three strategies, we first derive three regions in which the candidate target objects are located. Then we retrieve the objects (the targets of numerical integration) which is inside of the intersection area of the three regions. In addition, the proposed algorithm can effectively use a spatial index such as an R-tree.

Let us show an experimental result. Figure 2 shows the points data obtained from the cross roads of Montgomery County of Maryland, U.S.A. We have normalized the data within 1000×1000 units. The figure shows an example query when $\delta = 50$ and $\theta = 1\%$. The ellipse shown in the figure represents the isosurface of the Gaussian distribution. Figure 3 shows the resulting points for this query.

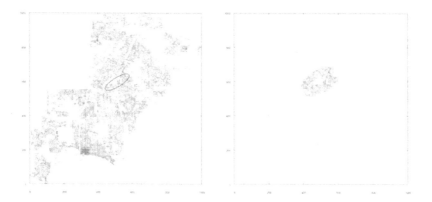

Fig. 2. Example data and query   **Fig. 3.** Example query result

3.3 Probabilistic Nearest Neighbor Queries

We also proposed the extended version of nearest neighbor queries for our context [2]. A *probabilistic nearest neighbor query (PNNQ)* is defined as follows.

Definition 3. *Given the probability density function $p_q(x)$ and the probability threshold θ $(0 < \theta < 1)$, a probabilistic nearest neighbor query $PNNQ(q, \theta)$ returns all the objects each of which satisfy the following condition: the probability that the object becomes the nearest neighbor of q is greater than or equal to θ. Let $\mathcal{O}$ be the set of data objects and let $\mathrm{Pr}_{NN}(o)$ be the probability that an object $o \in \mathcal{O}$ becomes the nearest neighbor of q:*

$$\mathrm{Pr}_{NN}(q, o) = \mathrm{Pr}(\forall o' \in \mathcal{O}, o' \neq o, \|x - o\|^2 \leq \|x - o'\|^2), \tag{3}$$

A probabilistic nearest neighbor query *is defined as follows:*

$$PNNQ(q, \theta) = \{n \mid n \in \mathcal{O}, \mathrm{Pr}_{NN}(q, n) \geq \theta\}. \tag{4}$$

We have proposed two query processing strategies both of which are based on the Voronoi diagram [15]. Figure 4 shows the result of a query. The data set is same as the former example. In the proposed method, the candidate objects (the targets of numerical integration) correspond to the intersection of the Voronoi regions which overlap with the rectangle shown in the figure and the Voronoi regions with bold borders. The shaded Voronoi regions correspond to the resulting nearest neighbor objects.

Fig. 4. Result for probabilistic nearest neighbor queries

4 Conclusions

In this paper, we described the approaches to spatial database query processing based on uncertain location information. We introduced the background and several existing studies and then described the research performed by our group. Currently, our group is working for extending the previous work. For example, we are developing query processing method when both of a query object and target objects obey Gaussian distributions and a general indexing method for processing location-based queries based on Gaussian-based distributions. We would like to report their results in the future publications.

Acknowledgments

This research is partly supported by the Grant-in-Aid for Scientific Research on Priority Areas, Japan (21013023).

References

1. Thrun, S., Burgard, W., Fox, D.: Probabilistic Robotics. MIT Press, Cambridge (2005)
2. Iijima, Y., Ishikawa, Y.: Finding probabilistic nearest neighbors for query objects with imprecise locations. In: Proc. MDM, pp. 52–61 (2009)
3. Ishikawa, Y., Iijima, Y., Yu, J.X.: Processing spatial range queries for objects with imprecise Gaussian-based location information. In: Proc. ICDE, pp. 676–687 (2009)
4. Duda, R.O., Hart, P.E., Stork, D.G.: Pattern Classification, 2nd edn. Wiley, Chichester (2000)
5. Pfoser, D., Jensen, C.S.: Capturing the uncertainty of moving-object representations. In: Güting, R.H., Papadias, D., Lochovsky, F.H. (eds.) SSD 1999. LNCS, vol. 1651, pp. 111–131. Springer, Heidelberg (1999)
6. Beskales, G., Soliman, M.A., Ilyas, I.F.: Efficient search for the top-k probable nearest neighbors in uncertain databases. In: Proc. VLDB, pp. 326–339 (2008)
7. Chen, J., Cheng, R.: Efficient evaluation of imprecise location-dependent queries. In: Proc. ICDE, pp. 586–595 (2007)
8. Tao, Y., Xiao, X., Cheng, R.: Range search on multidimensional uncertain data. ACM TODS 32(3) (2007)
9. Rigaux, P., Scholl, M., Voisard, A.: Spatial Databases with Application to GIS. Morgan Kaufmann, San Francisco (2001)
10. Cheng, R., Kalashnikov, D.V., Prabhakar, S.: Querying imprecise data in moving object environments. IEEE TKDE 16(9), 1112–1127 (2004)
11. Cheng, R., Chen, J., Mokbel, M., Chow, C.Y.: Probabilistic verifiers: Evaluating constrained nearest-neighbor queries over uncertain data. In: Proc. ICDE, pp. 973–982 (2008)
12. Kriegel, H.P., Kunath, P., Renz, M.: Probabilistic nearest-neighbor query on uncertain objects. In: Kotagiri, R., Radha Krishna, P., Mohania, M., Nantajeewarawat, E. (eds.) DASFAA 2007. LNCS, vol. 4443, pp. 337–348. Springer, Heidelberg (2007)
13. Ngai, W.K., Kao, B., Chui, C.K., Cheng, R., Chau, M., Yip, K.Y.: Efficient clustering of uncertain data. In: Proc. ICDM (2006)
14. Cheng, R., Xie, X., Yiu, M.L., Chen, J., Sun, L.: UV-diagram: A Voronoi diagram for uncertain data. In: Proc. ICDE (2010)
15. Aurenhammer, F.: Voronoi diagrams: A survey of a fundamental geometric data structure. ACM Computing Surveys 23(3), 345–405 (1991)

Skyline Sets Query and Its Extension to Spatio-temporal Databases

Yasuhiko Morimoto and Md. Anisuzzaman Siddique

Hiroshima University
1-7-1 Kagamiyama, Higashi-Hiroshima, 739-8521, Japan
morimoto@mis.hiroshima-u.ac.jp
http://www.morimo.com/morimo-ken

Abstract. Given a set of objects, a skyline query finds the objects that are not dominated by others. We consider a skyline query for sets of objects in a database in this paper. Let s be the number of objects in each set and n be the number of objects in the database. There are $_nC_s$ sets in the database. We consider an efficient algorithm for computing convex skyline of the $_nC_s$ sets, which we call "convex skyline sets". Recently, we have to aware individual's privacy. Sometimes, we have to hide individual values and are only allowed to disclose aggregated values of objects. In such situation, we cannot use conventional skyline queries. The proposed function can be a promising alternative in decision making in a privacy aware environment. In addition, if we consider sets of objects, we can extend the idea of the skyline query to spatio-temporal databases. For example, we can retrieve sets that are not dominated by another set in respect of time-interval or spatial-area. In this paper, we propose temporal and spatial skyline sets query.

Keywords: Skyline Query, Skyline Sets, Skyline Portfolio, Spatial Skyline, Temporal Skyline, Privacy.

1 Introduction

Skyline queries retrieve a set of skyline objects so that a user can choose promising objects from them and make further inquiries. Given a k-dimensional database DB, an object p is said to be in skyline of DB if there is no object q in DB such that q is better than p in all k dimension. If there exist such object q, then we say that p is dominated by q or q dominates p. Figure 1 shows a typical example of skyline. The table in the figure is a list of hotels, each of which contains two numerical attributes "distance" and "price". In the list, the best choice usually comes from the skyline, i.e., one of $\{h_1, h_3, h_4\}$ (See Figure 1 (b)). A number of efficient algorithms for computing skyline have been reported in the literature [1,2,3,4,5,6].

Recently, to preserve individuals' privacy is one of important data management issues. Sometimes, we have to hide individual values to preserve privacy and we are only allowed to disclose aggregated values of objects. In such situation, we cannot use conventional skyline queries.

S. Kikuchi, S. Sachdeva, and S. Bhalla (Eds.): DNIS 2010, LNCS 5999, pp. 317–329, 2010.

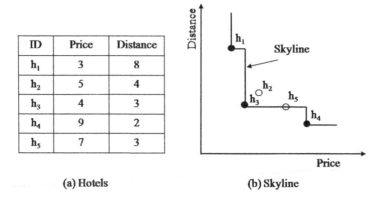

Fig. 1. Skyline Example

In this paper, we consider a skyline query for sets of objects in a database. Let s be the number of objects in each set and n be the number of objects in the database. There are $_nC_s$ sets in the database. We consider an efficient algorithm for computing convex skyline of the $_nC_s$ sets, which we call "convex skyline sets". This function does not disclose individual values. Instead, it discloses aggregated values of s objects. It will be one of the most promising alternatives for decision making in a privacy aware environment.

In addition, sometimes users are not interested in individual objects but sets of objects. In such cases, the skyline sets query can be utilized. For example, if a user has to reserve multiple hotels at a time, she/he has to find a set of hotels that she/he prefers. Then, she/he needs to consider a problem of choosing a preferable set of objects in a database.

Assume an event organizer has to reserve rooms in three different hotels around the event venue. Look at the example in Figure 1 again. The conventional skyline query, which outputs h_1, h_3, and h_4, doesn't provide sufficient information for the set selection problem. To solve the problem, we propose a function called *convex skyline sets query* that computes sets that lie in the convex hull of all object sets.

Figure 2 (a) is a list of 3-sets, in which all of the combinations of three hotels are listed. The "h_{123}" denotes a set of $\{h_1,h_2,h_3\}$. "Distance" and "Price" of "h_{123}" are the sum of the "Distance" and "Price" of respective hotels in the set. The event organizer prefers a set in the skyline of the combinations of three hotels, i.e., one of $\{h_{123}, h_{135}, h_{235}, h_{234}, h_{345}\}$ as we can see in Figure 2 (b). Our *convex skyline sets query* efficiently computes those convex skyline sets.

The problem of finding attractive sets is important especially in financial databases. For example, investors are always considering their best portfolio. Asset managers have to provide attractive financial funds by combining stocks, bonds, derivatives, currencies, commodities, and so on. Skyline sets query can provide them some important clues.

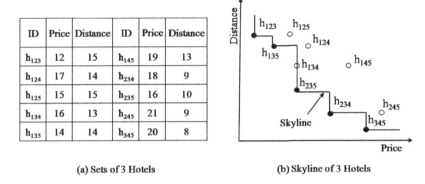

ID	Price	Distance	ID	Price	Distance
h_{123}	12	15	h_{145}	19	13
h_{124}	17	14	h_{234}	18	9
h_{125}	15	15	h_{235}	16	10
h_{134}	16	13	h_{245}	21	9
h_{135}	14	14	h_{345}	20	8

(a) Sets of 3 Hotels   (b) Skyline of 3 Hotels

Fig. 2. Skyline of 3-Set

2   Skyline Sets Problem

We consider the database DB having k attributes and n records. Let a_1, a_2, ..., a_k be the k attributes of DB. Without loss of generality, we assume that smaller value in each attribute is better.

Let s-set be an object set whose size is s. We assume s is a relatively small number such that $2 \leq s \leq 10$ though we can compute s-sets for much larger s. Let S be the set of all s-sets in DB. Note that the number of s-sets in DB is $_nC_s = \frac{n!}{(n-s)!s!}$, we denote the number by $|S|$. We assume a virtual database of S on the k dimensional space of DB. Each record of the database is an s-set whose value of each attribute (dimension) is the sum of s values of corresponding s objects. We denote $p.a_l$ as the l-th attribute value of a record p in S.

An s-set $p \in S$ is said to dominate another s-set $q \in S$, denoted as $p \preceq q$, if $p.a_r \leq q.a_r$ $(1 \leq r \leq k)$ for all k attributes and $p.a_t < q.a_t$ $(1 \leq t \leq k)$ for at least one attribute. We call such p as dominant s-set and q as dominated s-set between p and q.

An s-set $p \in S$ is said to be a skyline s-set if p is not dominated by any other s-set in S.

2.1   Convex Skyline

We can consider a record in S to be a point in k-dimensional vector space. Convex hull for the set of $|S|$ points is the minimum polyhedron containing the set. The dotted line polygon in the left of Figure 3 is an example of convex hull in two dimensional space.

In the figure, $a1_{min}$ and $a2_{min}$ are the point that has the minimum value of attribute $a1$ and $a2$, respectively. Notice that such points must be in the convex hull. We call the line between $a1_{min}$ and $a2_{min}$ the *initial facet*. Among all points in the convex hull, points that lie outside of the initial facet are skyline objects and we call such points *convex skyline* objects.

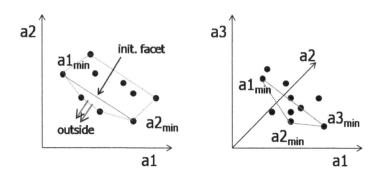

Fig. 3. Convex Hull and Convex Skyline

The triangle surrounded by $a1_{min}$, $a2_{min}$, and $a3_{min}$ in the right of Figure 3 is an example of the initial facet in three dimensional space. Convex skyline objects are points in the convex hull outside the initial facet.

In k-dimensional space, we compute such initial hyperplane surrounded by k points as the initial facet. Then, we compute convex skyline objects that lie in the convex hull outside the initial facet.

The definition of convex skyline sets problem can be simplified as follows: *Given a natural number s, find all s-sets that lies in both the convex hull and the skyline of S.*

Conventional skyline queries do not solve this problem since some non-skyline object (dominated object) can be a member of a convex skyline set. For example, as we can see in Figure 2 (b), after computing the convex skyline of 3 hotels we get five hotel sets h_{123}, h_{135}, h_{235}, h_{234}, and h_{345}. Here, both of h_2 and h_5 are the members of a convex skyline 3-sets but they are not in skyline of individual hotels.

3   Algorithm for Convex Skyline Sets

If we compute all of the s-sets from the original database and make a database containing $|S|$ records, the problem can be solved by conventional skyline query algorithms. However, $|S|$ is unacceptably large when the original database size is large. Therefore, we consider an algorithm for finding convex skyline sets without computing $|S|$ s-sets.

3.1   Touching Oracle

Each s-set in S can be represented as a k-dimensional point $\mathbf{x} = (x_1, x_2, ..., x_k)$ where x_i $(1 \leq i \leq k)$ is the sum of the i-th attribute's value of the s objects in DB.

Touching oracle function is a method to compute a point on the convex hull without generating S. It computes the tangent point of the convex hull of S and a $(k-1)$-dimensional hyperplane directly from DB.

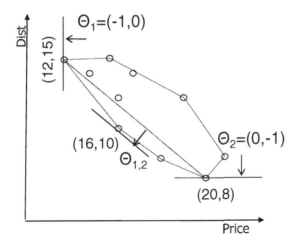

Fig. 4. Touching Oracle in 2D Space

Consider the examples of Figure 1 and Figure 2 again. There are 10 points in S if $s = 3$. Since there are two attributes in the database, those 10 points are in two-dimensional space as in Figure 2 (b). The dotted polygon in Figure 4 is the convex hull of the 10 points. In the example, there are five records in the original databases DB as in Figure 1 (a). Each of the five records is also represented as a two-dimensional point, which we call an atomic point, which is denoted as $\mathbf{a}$.

Assume there is a $(k-1)$-dimensional hyperplane which is a line if $k = 2$, whose normal vector is $\Theta_1 = (-1, 0)$ in the two-dimensional space. In order to find the tangent point with the 1-dimensional hyperplane (line) and the convex hull without precomputing all points in S, we compute $(\Theta_1, \mathbf{a})$, i.e., inner products of the normal vector and each atomic point as in the second column of Table 1. We choose the top three inner products, i.e., $\{h_1, h_2, h_3\}$. Those top three inner products composes the tangent point $(12, 15)$, which is the 3-set, h_{123}. Similarly, for a line with $\Theta_2 = (0, -1)$, we can find $\{h_3, h_4, h_5\}$ is the top three in the inner products of $(\Theta_2, \mathbf{a})$. It composes the tangent point $(20, 8)$, which is the tangent point of the convex hull and the 1-dimensional hyperplane (line) whose normal vector is Θ_2.

In k-dimensional case, we choose a $(k - 1)$-dimensional hyperplane composed by k points on the convex hull and compute normal vector of the hyperplane. If there is a point of the convex hull outside the hyperplane, we can touch the point with a $(k - 1)$-dimensional hyperplane that has the same normal vector. By using the normal vector, we can find the tangent point with the tangent $(k - 1)$-dimensional hyperplane by using the touching oracle function, which chooses top s inner products from n atomic points in DB. Since s is negligible small constant, we can compute the tangent point by scanning n atomic points only once, which is $O(n)$.

Table 1. Inner Product with Tangent Lines

a	$(\Theta_1, \mathbf{a})$	$(\Theta_2, \mathbf{a})$	$(\Theta_{1,2}, \mathbf{a})$
h_1	-3	-8	-85
h_2	-5	-4	-67
h_3	-4	-3	-52
h_4	-9	-2	-79
h_5	-7	-3	-73

3.2   Convex Hull Search

Next, we discuss how to use the touching oracle function to compute all convex skyline s-sets. First of all, we compute initial k tangent points that can be computed by touching oracle with initial k vectors $\Theta_x = (\theta_1, \theta_2, ..., \theta_k)$ where $\theta_i = -1$ if $i = x$, otherwise $\theta_i = 0$ for each $x = 1, ..., k$. Note that those k initial tangent points are on the horizon of the initial facet $((k-1)$-dimensional hyperplane). Convex skyline s-sets are points lie outside of the initial facet and are in the convex hull.

Next, we compute the normal vector of the initial facet. In the example above, we have initial two tangent points: we have $p_1 = (12, 15)$ with the normal vector $\Theta_1 = (-1, 0)$ and we have $p_2 = (20, 8)$ with the normal vector $\Theta_2 = (0, -1)$. Using the facet containing the two initial points, we can compute the normal vector of the facet as $\Theta_{1,2} = (-(15 - 8), (12 - 20)) = (-7, -8)$, which directs outside of the facet. Using this normal vector, we can find new tangent point h_{235}, which is $(16, 10)$. The new tangent point expands the initial facet into two facets, which are the facet surrounded by $p_1 = (12, 15)$ and $(16, 10)$ and the facet surrounded by $(16, 10)$ and $p_2 = (20, 8)$.

We recursively compute tangent points for each of the expanded facet. If we find new point outside the facet, we expand the facet further. We continually adopt the recursive operation while we can find new tangent point outside the facet. Finally, we can find all convex skyline s-sets. We can apply this recursive operation for higher k-dimensional space [7]. In the k-dimensional case, new tangent point, which is found by the touching oracle, divides the initial facet into k facets. In high dimensional case, the normal vector of each facet can be computed as follows:

Three Dimensional Case: Assume we have a facet surrounded by three points $P1 = (p1_1, p1_2, p1_3)$, $P2 = (p2_1, p2_2, p2_3)$ and $P3 = (p3_1, p3_2, p3_3)$. We assume that P1, P2 and P3 are clockwise order when we look the facet from outside of the convex hull. Now, we can compute two edge vectors by using the three points as follows. Suppose the edges vectors are $V1 = (v1_1, v1_2, v1_3) = (p2_1, p2_2, p2_3) - (p1_1, p1_2, p1_3)$ and $V2 = (v2_1, v2_2, v2_3) = (p3_1, p3_2, p3_3) - (p1_1, p1_2, p1_3)$. The outside normal vector of this facet is computed as the expansion of the following symbolic determinant.

$$V1 \otimes V2 = \begin{vmatrix} e_1 & e_2 & e_3 \\ v1_1 & v1_2 & v1_3 \\ v2_1 & v2_2 & v2_3 \end{vmatrix}$$

In the formula, e_1, e_2, and e_3 are the elementary vectors $(1,0,0)$, $(0,1,0)$ and $(0,0,1)$ respectively. Using this normal vector, we can divide this facet into three facets if we can find a new tangent point outside of the facet by the touching oracle function. If P is found outside of the facet, then the three new facets are $(P1, P, P3)$, $(P1, P2, P)$ and $(P, P2, P3)$. The normal vectors of these three facets are $(P - P1) \otimes (P3 - P1)$, $(P2 - P1) \otimes (P - P1)$ and $(P2 - P) \otimes (P3 - P)$ if points in each facet are clockwise order when we look the facet from outside of convex hull.

Four Dimensional Case: We can use the idea into higher dimensional case analogically.

Assume that we have a facet surrounded by four points $P1 = (p1_1, p1_2, p1_3, p1_4)$, $P2 = (p2_1, p2_2, p2_3, p2_4)$, $P3 = (p3_1, p3_2, p3_3, p3_4)$ and $P4 = (p4_1, p4_2, p4_3, p4_4)$. Using similar operations of 3D case, we can compute three vectors $V1 = (v1_1, v1_2, v1_3, v1_4) = P2 - P1$, $V2 = (v2_1, v2_2, v2_3, v2_4) = P3 - P1$ and $V3 = (v3_1, v3_2, v3_3, v3_4) = P4 - P1$. Then, the normal vector that directs outside can be computed as the expansion of the following determinant.

$$V1 \otimes V2 \otimes V3 = \begin{vmatrix} e_1 & e_2 & e_3 & e_4 \\ v1_1 & v1_2 & v1_3 & v1_4 \\ v2_1 & v2_2 & v2_3 & v2_4 \\ v3_1 & v3_2 & v3_3 & v3_4 \end{vmatrix}$$

In the determinant, the value of e_1, e_2, e_3 and e_4 are $(1,0,0,0)$, $(0,1,0,0)$, $(0,0,1,0)$ and $(0,0,0,1)$ respectively. If P is found outside of the facet, then the four new facets are $(P1, P2, P3, P)$, $(P1, P2, P, P4)$, $(P1, P, P3, P4)$ and $(P, P2, P3, P4)$. The normal vectors of these four facets are $(P2 - P1) \otimes (P3 - P1) \otimes (P - P1)$, $(P2 - P1) \otimes (P - P1) \otimes (P4 - P1)$, $(P - P1) \otimes (P3 - P1) \otimes (P4 - P1)$ and $(P2 - P) \otimes (P3 - P) \otimes (P4 - P)$.

k-Dimensional Case: We can expand above operations for k-dimensional case. Assume we have a facet surrounded by k points, $P1 = (p1_1, p1_2, ..., p1_k)$, $P2 = (p2_1, p2_2, ..., p2_k)$, $\cdots$, $Pk = (pk_1, pk_2, ..., pk_k)$. We can calculate $(k - 1)$ vectors like $V1$, $V2$, $\cdots$, $V(k - 1)$. Then, the normal vector of the facet that directs outside can be computed as the expansion of the following determinant.

$$V1 \otimes \cdots \otimes V(k - 1) = \begin{vmatrix} e_1 & \cdots & e_k \\ v1_1 & \cdots & v1_k \\ \cdots & \cdots & \cdots \\ v(k - 1)_1 & \cdots & v(k - 1)_k \end{vmatrix}$$

If P is found outside of the facet, then the k new facets are $(P, P2, \cdots, Pk - 1, Pk)$, $\cdots$, $(P1, P2, \cdots, Pk - 1, P)$. The normal vectors of these k facets are

$((P2 - P) \otimes \cdots \otimes (Pk - 1 - P) \otimes (Pk - P)), \cdots, ((P2 - P1) \otimes \cdots \otimes (Pk - 1 - P1) \otimes (P - P1)).$

4   Spatio-temporal Expansion of Skyline Sets Query

We can expand the idea of skyline query into spatio-temporal data for the sky-line sets problem. Table 2 is an example of such spatio-temporal database. The "time" column shows an attribute that contains a time stamp information. The "lat." and "lon." are latitude and longitude of each object's location, respectively.

Table 2. Spatio-Temporal Database

obj.	time	lat.	lon.	att_1	att_2	...
o_1	2	35.742	135.221	3	8	
o_2	5	38.421	134.822	5	4	
o_3	6	39.012	138.500	4	3	
o_4	3	35.985	138.159	9	2	
o_5	9	36.058	133.318	7	3	

4.1   Temporal Skyline Set

Assume we are considering $_5C_3 = |S|$ 3-sets in the database containing 5 objects as in Table 2. Following table is the projected aggregated list of 3-sets.

3-set	Twidth	att_1	att_2	...	3-set	Twidth	att_1	att_2	...
o_{123}	4	12	15		o_{145}	7	19	13	
o_{124}	3	17	14		o_{234}	3	18	9	
o_{125}	7	15	15		o_{235}	4	16	10	
o_{134}	4	16	13		o_{245}	6	21	9	
o_{135}	7	14	14		o_{345}	6	20	8	

In the list, "o_{123}" denotes a set of $\{o_1, o_2, o_3\}$. "Twidth" is width of 3 time stamps of each 3-set. And, "att_1" and "att_2" is the sum of "att_1" and "att_2" of each 3-set, respectively.

Temporal skyline 3-set query outputs convex 3-sets of the list for the database in Table 2 without computing the list of $|S|$ 3-sets.

Temporal Touching Oracle

Assume we are considering three dimensional space of the "Twidth", "att_1", and "att_2". In the three dimensional space, we specify a vector Θ and consider a two dimensional plane whose normal vector is Θ. *Temporal touching oracle* function computes the tangent point the two dimensional plane and the convex hull of all of the $|S|$ 3-sets without computing the $|S|$ 3-sets.

Let $\Theta = (\theta_1, \theta_2, \theta_3)$ be the given normal vector. Same as the touching oracle in Section 3.1, the tangent 3-set for Θ consists of three atomic points whose inner product with Θ is in the top three.

Let the θ_1 be the coefficient for the "Twidth" value of the inner product, i.e.,

$$\theta_1 * Twidth + \theta_2 * att_1 + \theta_3 * att_2.$$

We precompute partial inner product $g = \sum_{i=1}^{2}(\theta_{i+1} * att_i)$ for each atomic point and sort them by the "time" value. If $\Theta = (-1, -1, -1)$, the sorted atomic points are as follows:

obj.	o_1	o_4	o_2	o_3	o_5
time	2	3	5	6	9
$g = \sum_i(\theta_{i+1} * att_i)$	-11	-11	-9	-7	-11

In order to compute the optimal s-set that maximizes the inner product with Θ, we use a dynamic programming that maintains the best interval between the 1st and the i-th and the best interval that ends with the i-th atomic point in the sorted point sequence. Figure 5 is the outline of the dynamic programming algorithm. In the algorithm, $best_i$ is the best s-set from the 1st to the i-th and $best$-end-$with_i$ is the best s-set that ends with the i-th point.

1   initialize $best_i$ and $best$-end-$with_i$ to be from the 1st to the s-th.
2   for $i = s + 1 \ldots n$
3   update $best$-end-$with_i$
4   if $best$-end-$with_i$ is better than $best_i$ then replace $best_i$

Fig. 5. Temporal Touching Oracle Algorithm

Figure 6 shows how to update $best$-end-$with_i$ in the **Step 3** of the dynamic programming. Assume that p'_i, p''_i, and p'''_i are the top-3 points that maximize the inner product of time interval that ends in the i-th time stamp. Among the three points, we assume that p'_i is the earliest stamp point and p'''_i is the latest stamp point. If time interval from p'_i to the i-th is w'_i, then the inner product value is $\theta_1 * w'_i + g'_i + g''_i + g'''_i$, where g'_i, g''_i, and g'''_i are the partial inner product of p'_i, p''_i, and p'''_i, respectively. Note that p'''_i may be different from p_i.

The $best$-end-$with_{i+1}$ must be the top three among p'_i, p''_i, p'''_i and p_{i+1}. If p'_i is in the top three, then the inner product value is $\theta_1 * (w'_i + w)$ and the sum of partial inner product of the top three atomic points. Otherwise, the inner product is $\theta_1 * (w''_i + w) + g''_i + g'''_i + g_{i+1}$. Since s is small constant, this update procedure can be done in constant time. Therefore, time complexity of the temporal touching oracle function is $O(n)$.

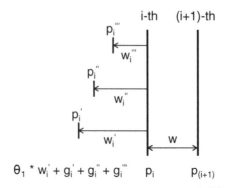

Fig. 6. Update of *best-end-with*$_i$ in DP

By using this temporal touching oracle function recursively as in Section 3.2, we can compute all the convex temporal skyline sets.

4.2   Spatial Skyline Set

By using spatial information such as "lat." and "lon." in Table 2, we can compute spatial skyline sets. Following table is the projected aggregated list of 3-sets for spatial skyline sets.

3-set	area	att_1	att_2	...	3-set	area	att_1	att_2	...
o_{123}	13.53	12	15		o_{145}	1.53	19	13	
o_{124}	8.94	17	14		o_{234}	11.13	18	9	
o_{125}	5.10	15	15		o_{235}	15.31	16	10	
o_{134}	11.78	16	13		o_{245}	11.79	21	9	
o_{135}	16.95	14	14		o_{345}	15.69	20	8	

In the list, "area" is the area of the minimum bounding rectangular of 3 locations of each 3-set. Spatial skyline 3-set query outputs convex 3-sets of the list.

5   Experiments

We conduct a series of experiments to evaluate the performance of skyline sets query by using different types of datasets. The proposed algorithm was implemented using Java J2SE V6.0. We conduct experiments on a PC with an Intel(R) Core2 Duo, 2 GHz CPU and 3 GB main memory, running on Microsoft Windows XP operating systems. Each experiment is repeated five times and the average is taken.

As benchmark databases, we use the synthetic databases proposed by Borzsonyi et al. [1]. Objects in the synthetic databases have three value distributions: "correlated", "anti-correlated", and "independent".

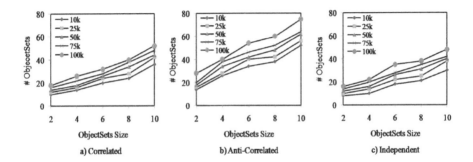

Fig. 7. Number of Retrieved Convex Skyline Sets

Figure 7 shows the number of retrieved sets of 2D case on different distributions and different database size. Five different synthetic datasets with cardinality 10k, 25k, 50k, 75k, and 100k are used in this experiment. We vary the set size "s" between 2 to 10. The result shows that the number of convex skyline sets maintains a positive correlation with the size, i.e., when "s" increases, the number of returned sets also increase as well.

We evaluate the response time of the convex skyline sets algorithm on the three different distributions. Figure 8 shows the results of 2D, 3D, 4D, and 5D cases for synthetic datasets with 100k. We observe that our algorithm becomes gradually slow if "s" increases. As in Figure 7, if "s" becomes large the number of retrieved skyline sets also increases. Considering this fact, we can conclude that our algorithm can perform well even if "s" becomes large.

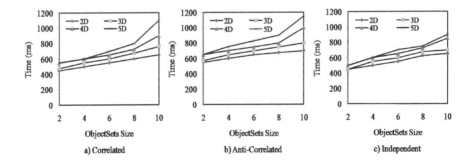

Fig. 8. Time varying Set Size

Next, we evaluate the effect of database size. We used synthetic datasets with cardinality 10k, 25k, 50k, 75k, and 100k. In this experiment, we fix "s" to 10. Figure 9 shows the results. We observe that the response time increases if the database size increases. We also observe that it gradually increases if the dimension, "k", increases. Similar to the previous experiments, the total *elapsed time* increases if database size and dimension increase.

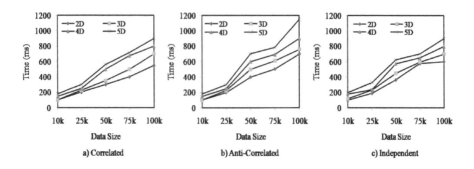

Fig. 9. Time varying Database Size

6 Conclusion

We consider a convex skyline query for sets of objects in a database in this paper. Especially in privacy aware environments, we have to hide individual values and are only allowed to disclose aggregated values of objects. In such situations, the proposed function can be a promising alternative in decision-making. The proposed function is also important for set optimization problems such as portfolio analysis in financial databases. The proposed algorithm can compute all convex skyline s-sets without making all sets.

Moreover, by using sets of objects, we can expand the idea of skyline query to spatio-temporal databases. Temporal skyline sets help us to find optimal set of objects within a small time interval. Such temporal skyline sets can be a powerful tool for many application domains such as financial databases. Spatial skyline helps us to find optimal set of objects within a small area, which can be used for location analysis such as area marketing.

Currently, our spatial skyline sets query is not so efficient and the time complexity is almost equivalent to compute all combinations of s-sets like the aggregated list in Section 4.2. We are now working to improve the performance. We are also considering skyline regions for much expressive shape than a rectangular for example rectilinear regions and x-monotone regions as we considered for two dimensional association rules [8].

References

1. Borzsonyi, S., Kossmann, D., Stocker, K.: The skyline operator. In: Proceedings of ICDE, pp. 421–430 (2001)
2. Kossmann, D., Ramsak, F., Rost, S.: Shooting stars in the sky: An online algorithm for skyline queries. In: Proceedings of VLDB Conference, pp. 275–286 (2002)
3. Papadias, D., Tao, Y., Fu, G., Seeger, B.: An optimal and progressive algorithm for skyline queries. In: Proceedings of ACM SIGMOD Conference, pp. 467–478 (2003)
4. Tan, K.-L., Eng, P.-K., Ooi, B.C.: Efficient Progressive Skyline Computation. In: Proceedings of VLDB Conference, pp. 301–310 (2001)

5. Li, C., Ooi, B.C., Tung, A.-K.H., Wang, S.: DADA: A Data Cube for Dominant Relationship Analysis. In: Proceedings of ACM SIGMOD Conference, pp. 659–670 (2006)
6. Pei, J., Jin, W., Ester, M., Tao, Y.: Catching the Best Views of Skyline: A Semantic Approach Based on Decisive Subspaces. In: Proceedings of VLDB Conference, pp. 253–264 (2005)
7. Morimoto, Y., Fukuda, T., Matsuzawa, H., Yoda, K., Tokuyama, T.: Algorithms for Mining Association Rules for Binary Segmentations of Huge Categorical Databases. In: Proceedings of VLDB Conference, pp. 380–391 (1998)
8. Fukuda, T., Morimoto, Y., Morishita, S., Tokuyama, T.: Data mining with optimized two-dimensional association rules. ACM Trans. on Database Systems 26(2), 179–213 (2001)

Implementation of Geographic Search Methods for Lunar Objects

Junya Terazono

The University of Aizu
Tsuruga, Ikki-Machi, Aizu-Wakamatsu, Fukushima 965-8580, Japan
Terazono@u-aizu.ac.jp

Abstract. The search system to find surface features on the moon and other planets are essential for assisting research activities by scientists. Many research studies are developing improved system to search the lunar nomenclature. In this report, we present the recent progress on search system using three different approaches based on two major scheme of database search policy, QBE (Query-By-Example) and QBO (Query-By-Object). The different approaches for the implementations use the same data.

Keywords: The moon, query interface, Web GIS, QBE, QBO.

1 Introduction

The names of the feature are the fundamental elements. These identify the position of surface on the moon and the planets. Even more, the location name information is commonly used in research. It can determine the feature the researchers are referring to.

Considering the commonly available systems, the main function of search in such systems is location-based map. In these systems, map is used for identifying the location by clicking any points on it. These systems set little importance on the search from features based on object names, which are considered to be basic information for identifying the locations on the moon.

We have studied technique to improve access to map objects using lunar nomenclature search system using database technology [1][2]. In this system, users can search lunar feature object such as craters, mare (sea ") and mons (mountains) and show their detailed information. Additionally, our search system has combination function which can make geographical computation with two or more objects. Users can search objects which are "near" the target objects.

However, the system has some limitations and shortcomings. The largest disadvantage upon the system is that it has no display of the current search situation. Due to lack of the capability, users can lose the current point of search during the operation. As the search becomes complex with combining two or more objects, the lack of the search status display will become serious problem.

On the other hand, the new framework of database search, called QBO (Query-By-Object) [3] is emerging. The QBO handles each search result as an object, and they can be combined (and calculated) such as addition ("and") and joining ("or"). These

S. Kikuchi, S. Sachdeva, and S. Bhalla (Eds.): DNIS 2010, LNCS 5999, pp. 330–335, 2010.

systems are different from the conventional approach called QBE (Query-By-Example) [4].

As a nature of QBO concept, QBO is particularly useful when the search is done as a combination of small searches. This means the QBO is useful for complex search and search from the portable devices which cannot handle many search options.

2 Implementation Strategy

To be able to test the new implementation, we configured three different approaches to evaluate the benefit and shortcomings of the QBO and QBE concept. Also, as the enhancement of the current search system is necessary, we prepared another search system as the enhancement version of current search.

Therefore, in our project, we are now implementing the following three systems which have different nature.

- QBO based system. This system has native implementation of QBO.
- QBE based search. This system is straight implementation based on QBE concept.
- The extension of current system (hybrid approach of QBO and QBE). This system is based on current lunar nomenclature search system, with several feature extension.

3 System Overview

The three systems share the same database table in the backend. The data used in these implementations are based on the published data on USGS (US Geological Survey) Astrogeology Research Program website [5] which provides lunar and planetary feature name information approved by IAU (International Astronomical Union).

In this site, all data on lunar and planetary features can be downloaded with CSV forms with respect to each celestial body. In this list, moon has 18 different features (such as craters, mare and mons). Every feature list was downloaded and loaded into the database table.

PostgreSQL, the open-source relational database system, is used as the backend database. Java Server Pages (JSP) are used to create front-end web pages, in conjunction with rapid application development tools such as S2JDBC [6], JUnit [7] and SAStruts [8].

4 The System

The detail of three search system are as follows:

4.1 QBO-Based System

This system, called "Moon Seeker", starts from the simple form with several option to choose the feature, the beginning letter, and location range (currently not implemented), as shown in Figure 1.

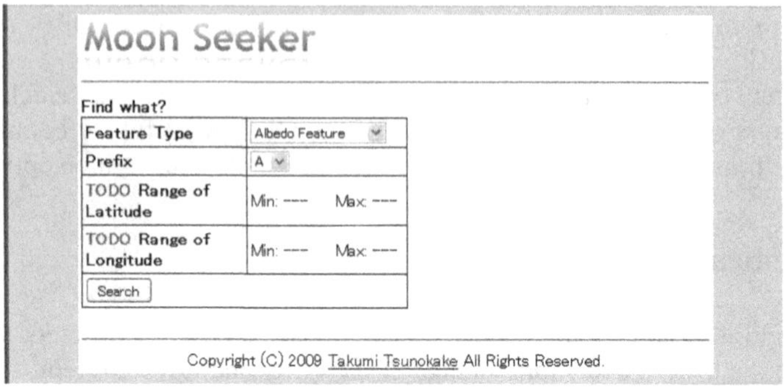

Fig. 1. The top page of the QBO-based search system "Moon Seeker"

Upon pressing "search" botton, the search result is displayed. Then we can add the next search from this screen as shown in Figure 2.

Fig. 2. The search result and button which can start the new search

By starting the new search, the same page appeared in Fig. 1. is dislayed. The second and subsequent search results can be combined and we can narrow down the search target. After the narrowing down the search targets, we can show the detail of the result by clicking the link attached in each target

4.2 QBE-Based System

In this system, the search starts from the selection of number of feature to be combined. Fig. 3. shows the starting web page. In each pull-down menu, the number is displayed that shows the number of features to be combined.

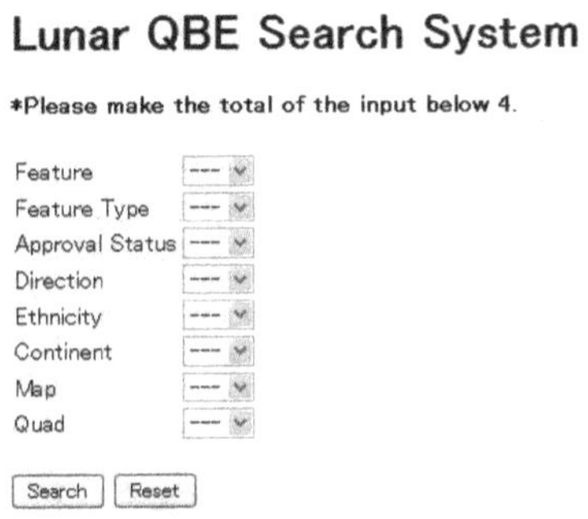

Lunar QBE Search System

*Please make the total of the input below 4.

Feature
Feature Type
Approval Status
Direction
Ethnicity
Continent
Map
Quad

[Search] [Reset]

Fig. 3. The start page of QBE-based search system. These pull-down menu selects the number to be combined. Up to four objects can be combined.

Next, the detailed search screen appears upon the request dispatched in previous screen as shown in Fig. 4. In this case, the user selected "Feature", "Feature Type" and "Ethnicity" as search objects. As the search option are very detailed, users can make very detailed search as they like.

Search

[Search] [Select Table]

Fig. 4. Detailed search screen based on QBE concept. In this case, users are attempting search for the crater which has an alphabet "a" in the name with the ethnicity is "American".

By dispatching the Search, the result is displayed. The combination search is now implementing.

4.3 Hybrid Approach

The hybrid approach is the natural extension of the existing search system with some enhancements. Currently, we successfully implemented the linkage capability with lunar GIS infrastructure, called "WISE-CAPS" [9] in The University of Aizu. Fig. 5. shows the display with the detailed information of the result and the map (image) of the location.

In the example of Fig. 5., craters which has the ethnicity of "Japanese" were selected and listed in the left column. The middle column is the detailed information of the crater "Asada", and the right image is the cropped map of "Asada" crater on the

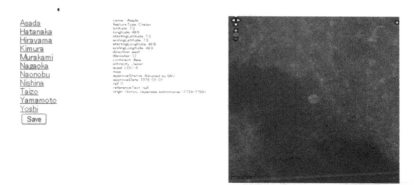

Fig. 5. An example of final result display of the hybrid search system with the image map

moon using map rendering capability of the WISE-CAPS. The system enables us to display image map from search result which are useful not only in research field but also public purpose.

5 Discussions

As we mentioned in last section, these three systems implement the natural method of each approach, except the hybrid one. The hybrid system is straight extension of existing system with several new functions.

However, we are not sure which system will be accepted by users, mainly by planetary scientists. Also, the usability will differ in the environment and users. For example, the QBO-based approach will be favored in mobile environment which has some limitations in displays and input devices. On the other hand, classical QBE-based approach may be natural and more accepted for professional users who usually use their personal computers to access to the network.

Hybrid approach seems to be promised answer as it joins merit of QBE and QBO based system. However, the assessment should be done by main users, planetary researchers, from total points including the system accessibility and design.

Considering these points, we cannot assert which system is the most acceptable for the search system. Moreover, as web page design enhancements and more choice of user input is necessary, the usability evaluation in parallel to the development of these three systems are necessary.

6 Summary and Conclusions

We have successfully built the improved search system of lunar features with three different approaches, QBE-based, QBO-based and Hybrid approach. These systems make maximum use of merits of these systems, and shows the advantage of each search method.

These three systems are currently under intensive implementation and development. There are many bugs and incomplete and unimplemented features. We are discussing on the difficulty point for the implementation.

Also, we will conduct hearing from researchers of lunar and planetary science about refinement request and additional function which they need from their point of view. The integration with the WISE-CAPS environment, seamless operation between Web-GIS and search systems, is also the pressing theme. We will continue the action for evolution of the search system.

References

1. Terazono, J., Izumita, T., Asada, N., Demura, H., Hirata, N.: Exploring Structural and Dimensional Similarities Within — Lunar Nomenclature System Using Query Interfaces. In: Bhalla, S. (ed.) DNIS 2007. LNCS, vol. 4777, pp. 48–53. Springer, Heidelberg (2007)
2. Terazono, J., Bhalla, S., Izumita, T., Asada, N., Demura, H., Hirata, N.: Construction of Lunar Nomenclature Search System. In: The 26th International Symposium on Space Technology and Science, 2008-k-67
3. Rahman, S.A., Bhalla, S., Hashimoto, T.: Query-By-Object interface for Dynamic Access and Information Requirement Elicitation. In: Proc. Fourth International Conference on Mobile Business (2005)
4. Rahman, S.A., Bhalla, S.: Spatial QBE interface for Web GIS. In: Proc. 2005 Fifth International Conference on Computer and Information Technology (2005)
5. Gazetter of Planetary Nomenclature, USGS Astrogeology Research Program, http://planetarynames.wr.usgs.gov/
6. S2JDBC, http://s2container.seasar.org/2.4/en/s2jdbc.html
7. JUnit, http://www.junit.org/
8. Super Aglie Struts (SAStruts) (in Japanese), http://sastruts.seasar.org/
9. Terazono, J., Nakamura, R., Kodama, S., Yamamoto, N., Demura, H., Hirata, N., Ogawa, Y., Sobue, S., Okumura, H.: Lunar and Planetary Informatics System Based on Geographical Information. In: ISPRS Working Group IV/7 meeting (2009)

Author Index